I0056614

Fruit Production: Horticultural Practices

Fruit Production: Horticultural Practices

Edited by Pearl Hunter

SYRAWOOD
PUBLISHING HOUSE

New York

Published by Syrawood Publishing House,
750 Third Avenue, 9th Floor,
New York, NY 10017, USA
www.syrawoodpublishinghouse.com

Fruit Production: Horticultural Practices
Edited by Pearl Hunter

© 2019 Syrawood Publishing House

International Standard Book Number: 978-1-68286-710-5 (Hardback)

This book contains information obtained from authentic and highly regarded sources. Copyright for all individual chapters remain with the respective authors as indicated. All chapters are published with permission under the Creative Commons Attribution License or equivalent. A wide variety of references are listed. Permission and sources are indicated; for detailed attributions, please refer to the permissions page and list of contributors. Reasonable efforts have been made to publish reliable data and information, but the authors, editors and publisher cannot assume any responsibility for the validity of all materials or the consequences of their use.

Trademark Notice: Registered trademark of products or corporate names are used only for explanation and identification without intent to infringe.

Cataloging-in-Publication Data

Fruit production : horticultural practices / edited by Pearl Hunter.
    p. cm.
Includes bibliographical references and index.
ISBN 978-1-68286-710-5
1. Fruit-culture. 2. Horticulture. I. Hunter, Pearl.
SB354.8 .F78 2019
634--dc23

TABLE OF CONTENTS

Permissions

List of Contributors

Index

PREFACE

This book has been an outcome of determined endeavour from a group of educationists in the field. The primary objective was to involve a broad spectrum of professionals from diverse cultural background involved in the field for developing new researches. The book not only targets students but also scholars pursuing higher research for further enhancement of the theoretical and practical applications of the subject.

Horticulture is a scientific field concerned with the production of vegetables, flowers and fruits. Some of its significant aspects include plant conservation, garden design, arboriculture, soil management, etc. Fruits are commonly cultivated under the branch of horticulture. Pomology is a horticultural practice which focuses on the development and cultivation of all types of fruit trees. It also delves into their physiological study. This book unfolds the innovative aspects of horticulture, which will be crucial for the progress of this field. It elucidates new techniques of fruit production and their application in a comprehensive manner. The extensive content of this book provides the readers with a thorough understanding of the subject.

It was an honour to edit such a profound book and also a challenging task to compile and examine all the relevant data for accuracy and originality. I wish to acknowledge the efforts of the contributors for submitting such brilliant and diverse chapters in the field and for endlessly working for the completion of the book. Last, but not the least; I thank my family for being a constant source of support in all my research endeavours.

Editor

SOME POMOLOGICAL AND CHEMICAL PROPERTIES OF LOCAL PEAR VARIETIES IN UŞAK, TURKEY

Volkan OKATAN[1], Mehmet POLAT[2], Sezai ERCİŞLİ[3], Mehmet Atilla AŞKIN[2]

[1]Usak University, Sivasli Vocational School, Uşak, Turkey
[2]Suleyman Demirel University, Agricultural Faculty, Department of Horticulture, Isparta, Turkey
[3]Ataturk University, Agricultural Faculty, Department of Horticulture, Erzurum, Turkey

Corresponding author email: okatan.volkan@gmail.com

Abstract

This study described some desirable pomological and chemical traits of 10 local pear varieties from Uşak in 2016. In these local pear varieties, some pomological and chemical properties were determined such as fruit weight, fruit height, fruit widths, fruits firmness, total soluble solids, titratable acidity, DPPH (2,2-diphenyl-1-picryl-hydrazyl-hydrate) free-radical values and ascorbic acid content. The values for fruit weights ranged from 93.12 ± 3.26 g to 287.93 ± 8.36 g, fruit height ranged from 6.12 ± 1.37 cm to 11.62 ± 1.15 cm, fruit widths ranged from 4.38 ± 0.82 cm to 8.52 ± 0.63 cm, fruits firmness ranged from 4.26 ± 0.18 lb to 11.19 ± 0.32 lb total soluble solid contents ranged from % 8.62 ± 0.32 to % 17.20 ± 0.83 and, titrable acid contents ranged from % 0.43 to % 2.63. DPPH values were determined and varied from 12.37 ± 0.73 to 26.74 ± 1.68; ascorbic acid content situated between 108.53 ± 3.12 and 263.37 ± 4.71 mg. We conclude that some local pear varieties may be recommended for the next breeding studies.

Key words: local, pear, dpph, pomological, chemical.

INTRODUCTION

Due to the favorable climate and soil conditions that our country has, it has the chance to grow a large number of species and varieties in terms of fruit.

Today, Turkey is one of the important fruit producer countries of the world in terms of fruit variety and number of varieties and production amount (Güleryüz, 1977; Özbek, 1977).

The number of pear varieties in the world is over 5000 and this number is 600 in Turkey. However, among them, the number of those are high in quality and productivity, and cultivated commercially (Buyukyilmaz, 1993).

Pear is the most mild climate fruit species after apple in the world. Pear fruit can be consumed both fresh and dry. It is also used in pear liquor, vinegar, fruit juice, fruit salad, jam, jelly, dessert, cake, pastry and canned food industry (Özçağıran et al., 2005).

The aim of this research was to determine the some pomological and chemical properties of local pear varieties grown in Uşak region in Middle Eagean of Turkey.

MATERIALS AND METHODS

The climate of the Uşak province shows a transition characteristic between the Aegean and Central Anatolian regions. More continental climate prevails. The summers are warm, the winters are long and hard.

The annual rains is between 430 mm and 700 mm. Most of the rains fall in winter. Summer rainfall is rather low.

The research district is situated between 38°40' North latitude and , 29°23' East longitude in Middle Eagean of Turkey. Uşak region is nearly 906 meters above sea level (Wikipedia participants, 2017).

The data of microclimates of the research district are presented Table 1.

Fruit material

Local pear genotypes were selected according to fruit sizes. Ten local pear cultivars were harvested in different region of Uşak, Middle Eagean, Turkey, in Semptember-October 2016. The harvested fruits were then transported to the laboratory for analysis.

Table 1. Average monthly air temperature (°C) and total rainfall (kg/m^2) in the Uşak – 1916 to 2016.

USAK	Jan	Feb	Mar	Apr	May	Jun	Jul	Aug	Sep	Oct	Nov	Dec
Maximum (°C)	6.8	8.1	12.0	17.0	22.0	27.0	30.0	31.0	26.0	20.0	14.0	8.7
Minimum (°C)	-1.3	-1.0	1.3	5.2	9.2.0	13.0	15.0	16.0	12.0	7.9	3.8	0.6
Rainfall (kg/m^2)	75.0	66.0	59.0	50.0	48.0	28.0	15.0	9.6	16.0	40.0	59.0	82.0

(Source: www.mgm.gov.tr)

Methods

The matured fruits were selected based on morphological characteristics for fruit analyses. Desirable morphological characteristics such as fruit weight (g), fruit height (mm), fruit weight (mm), fruit firmness (lb), total soluble solid content (SSC,%) and acidity (%) were determined according to Cemeroglu (1992).

Determination of vitamin C

After pureeing and filtering, the fruit juices samples were obtained. The juices were used for vitamin C analysis.

The samples were homogenized by centrifuge and 400 µL oxalic acid (0.4 %) and 4.5 ml 2,6 - diclorofenolindofenol solution were added upon supernatant.

The data were read spectrophotometrically at the wavelength of 520 nm against the blank.

Determination of radical – scavenging activity:

In the 1,1-diphenyl-2-picrylhydrazyl (DPPH) assay, antioxidants were capable to reduce the stable radical DPPH• to the yellow coloured diphenylpicrylhydrazine (DPPH-H). The test is based on the reduction of an alcoholic solution of DPPH in the presence of a hydrogen dona-ting antioxidant due to the formation of the non-radical form DPPH-H (Gulcin, 2007). The DPPH radical-scavenging activity was estima-ted after Blois (1958). Briefly, 0.1 mL of each sample extract was mixed with 0.9 mL of 0.04 mg/mL methanolic solution of DPPH. The mixtures were left for 20 min at room tempe-rature and its absorbance then measured at 517 nm against a blank. All measurements were carried out in triplicate. The percentage of DPPH scavenging activity was calculated using the following equation:

$$\% \ DPPH = [(Ac - As)/Ac] \times 100$$

where Ac was the absorbance of the negative control (contained extraction solvent instead of the sample), and As was the absorbance of the samples.

RESULTS AND DISCUSSUION

The fruit weight (g), fruit height (mm), fruit width (mm) and fruit firmness (lb) measure-ment in fruits of 10 local pear genotypes were examined in this study (Table 2).

According to the measurements, among pear genotypes were found to be; for fruit weights ranged from 93,12 ± 3,26 g to 287,93 ± 8,36 g, fruit heights ranged from 6,12 ± 1,37 cm to 11,62 ± 1,15 cm, fruit widths ranged from 4,38 ± 0,82 cm to 8,52 ± 0,63 cm, fruits firmness ranged from 4,26 ± 0.18 lb to 11,19 ± 0,32 lb average.

In the study on pear varieties in Yalova, there were identified fruits in diameters ranging from 42.61 to 83.54, fruit heights ranging from 72.83 to 108.25, fruit weight ranging from 57.26 to 410.75 g, SSC between 11.17 and 14.06 and TA 0.23 to 0.40 g / 100 ml (Akçay et al., 2009). In pear fruit which is very rich in nutritional value, the amount of SSC is 14.63-19.5 % and the titratable acidity value is 0.154-0.462 % (Özbek, 1978; Şen et al., 1992).

Table 2. Pomological characteristics of pear genotypes

Genotypes	Fruit weight (g)	Fruit height (mm)	Fruit width (mm)	Fruit firmness (lb)
64 USAK 01	117.26 ± 5.13	9.52 ± 0.83	5.24 ± 0.68	5.19 ± 0.14
64 USAK 02	287.93 ± 8.36	11.62 ± 1.15	8.52 ± 0.63	7.26 ± 0.56
64 USAK 03	185.83 ± 6.19	6.12 ± 1.37	6.81 ± 0.92	6.84 ± 0.27
64 USAK 04	114.85 ± 9.12	7.26 ± 1.08	4.97 ± 0.47	4.26 ± 0.18
64 USAK 05	247.51 ± 12.75	8.46 ± 0.72	7.16 ± 0.54	9.10 ± 0.59
64 USAK 06	93.12 ± 3.26	10.36 ± 0.78	4.38 ± 0.82	8.74 ± 0.47
64 USAK 07	176.28 ± 15.16	11.35 ± 1.03	8.26 ± 1.04	6.57 ± 0.26
64 USAK 08	139.68 ± 7.28	6.28 ± 0.47	5.26 ± 0.41	11.19 ± 0.32
64 USAK 09	162.44 ± 6.17	8.51 ± 1.25	6.29 ± 0.49	10. 23 ± 0.55
64 USAK 10	192.74 ± 9.84	9.84 ± 0.42	8.06 ± 0.82	9.61 ± 0.92

SSC, TA, DPPH and ascorbic asid measuretments in genotypes are presented in Table 3. According to the measurements, among pear genotypes were found to be; for total soluble solid contents ranged from % 8.62 ± 0.32 to % 17.20 ± 0.83 and, titrable acid contents ranged from % 0.43 ± 0.05 to % 2.63 ± 0.12. DPPH values were determined between from 12.37 ± 0.73 to 26.74 ± 1.68 and ascorbic acid content were found between 108.24 ± 3.12 and 263.36 ± 4.71 mg average.

Table 3. Chemical characteristics of pear genotypes.

Genotypes	SSC %	TA %	DPPH	Ascorbic Acid (mg)
64 USAK 01	12.84 ± 0.23	2.63 ± 0.06	21.82 ± 1.26	147.18 ± 2.15
64 USAK 02	16.15 ± 0.56	1.79 ± 0.05	12.37 ± 0.73	162.93 ± 5.24
64 USAK 03	9.62 ± 0.08	1.52 ± 0.09	17.19 ± 1.38	108.53 ± 3.12
64 USAK 04	8.62 ± 0.32	0.82 ± 0.01	16.25 ± 1.24	125.28 ± 4.81
64 USAK 05	11.69 ± 1.24	0.91 ± 0.02	18.22 ± 0.86	263.37 ± 4.71
64 USAK 06	15.27 ± 0.85	0.43 ± 0.15	24.15 ± 1.76	214.26 ± 7.59
64 USAK 07	13.60 ± 0.16	1.35 ± 0.09	26.74 ± 1.68	185.21 ± 4.19
64 USAK 08	17.20 ± 0.83	2.14 ± 0.27	18.21 ± 1.25	192.17 ± 9.14
64 USAK 09	16.21 ± 1.54	2.37 ± 0.11	17.91 ± 1.17	251.66 ± 4.10
64 USAK 10	15.28 ± 0.84	2.18 ± 0.29	13.57 ± 1.19	182.24 ± 6.43

According to the average of the two years in the examined varieties; the total soluble solids ranged from 10.6% (Etrusc) to 15.75% (Hosui). In studies similar to this, the total soluble solid ratio was 14.60% - 19.90% (Güleryüz, 1972). In a study on pear varieties in JAPAN, They were identified as high DPPH radical scavenging activity 19.76% (Ieguchi et al., 2015). Szete et al. (2002) reported an average for ascorbic acid of 60 mg in a study of pear varieties in China.

CONCLUSIONS

As a result, fruit growing culture is based on very ancient histories in Uşak province. For centuries, many civilizations and a large number of local fruit genotypes have hosted in this region, so there has an important selection potential.

REFERENCES

Akçay, M. E., Büyükyılmaz, M., & Burak, M. (2009). Marmara Bölgesi için ümitvar armut çeşitleri-IV. Bahçe, 38(1), 1-10.

Blois M. S., 1958. Antioxidant determinations by the use of a stable free radical, Nature, 181, 1199 - 1200.

Büyükyılmaz, M. Armut çeşit kataloğu. T.C. Tarım ve Köyişleri Bakanlığı Yayın No: 360/19. Ankara 47 s. 1993.

Cemeroglu B: Meyve ve Sebze İşleme Endüstrisinde Temel Analiz Metodları. Ankara: Bıltav Yayınları; 1992:381.

Güleryüz, M., 1972. Erzincan'da Yetiştirilen Bazı Önemli Elma ve Armut Çeşitlerinin Pomolojileri ile Döllenme Biyolojileri Üzerinde Araştırmalar, Atatürk Üniversitesi Fen Bilimleri Enstitüsü, Doktora Tezi, Erzurum, 216 s.

Güleryüz, M., 1977. Erzincan'da Yetiştirilen Bazı Önemli Elma ve Armut Çeşitlerinin Pomolojileri ve Döllenme Biyolojileri. Atatürk Üniversitesi Ziraat Fakültesi, Yayın No:229, Erzurum.

Gulcin I., 2007. Comparison of in vitro antioxidant and antiradical activities of L-tyrosine and L-Dopa, Amino acids, 32: 431-438.

Ieguchi, T., Takaoka, M., Nomura, K., Uematsu, C., & Katayama, H. (2015). Pear (Pyrus L.) Genetic Resources from Northern Japan: Evaluation of Antioxidant capacity. Acta Horticulturae.

Özbek, S., 1977. Genel Meyvecilik. Çukurova Üniversitesi Yayınları No: 111, Adana.

Özbek, S., 1978. Özel Meyvecilik. Çukurova Üniv. Ziraat Fak. Yayın No:128, Adana. 486 s

Özçağıran, R., Ünal, A., Özeker, E., ve İsfendiyaroğlu, M., 2005. Ilıman İklim Meyve Türleri. Cilt-3, Ege Üniversitesi Yayınları, İzmir.

Szeto, Y. T., Tomlinson, B., & Benzie, I. F. (2002). Total antioxidant and ascorbic acid content of fresh fruits and vegetables: implications for dietary planning and food preservation. British journal of nutrition, 87(01), 55-59.

Şen, S. M., Cangi, R., Bostan, S. Z., Balta, F., Karadeniz, T., 1992. Van ve çevresinde yetiştirilen seçilmiş bazı Mellaki ve Ankara armut çeşitlerinin fenolojik, morfolojik ve pomolojik özellikleri üzerinde araştırmalar. Yüzüncü Yıl Üniversitesi Ziraat Fakültesi Dergisi, 2(2):29-40.

Wikipedia participants (2017). Servant. Wikipedia, The Free Encyclopedia. Access date 20:00, April 4, 2017 url: //en.wikipedia.org/wiki/U%C5%9Fak

THE DETERMINATION OF POMOLOGICAL AND TOTAL OIL PROPERTIES OF SOME OLIVE CULTIVARS GROWN IN ISPARTA, TURKEY

**Adnan Nurhan YILDIRIM[1], Fatma YILDIRIM[1], Gülcan ÖZKAN[2],
Bekir ŞAN[1], Mehmet POLAT[1], Hatice AŞIK[2], Tuba DILMAÇÜNAL[1]**

[1]Suleyman Demirel University, Faculty of Agriculture, Horticultural Science,
32260, Isparta, Turkey
[2] Suleyman Demirel University, Faculty of Engineering, Food Engineering Department,
32260, Isparta, Turkey

Corresponding author email: adnanyildirim@sdu.edu.tr

Abstract

The aim of this study was to determine the physical characteristics of three olive cultivars' fruits at 3 different harvest time (skin green with pink spots, pulp white-skin black, skin black and pulp purple) growing in Mediterrenean region of Turkey in Isparta/Sütçüler at the same garden and growing conditions. Ayvalık and Memecik olive cultivars are grown in large areas of Turkey. The third cultivar Topakaşı, is a local cultivar and is cultivated limitedly in the research area. Thus, in the study, the differences between the varieties adapted to the region's ecology and the varieties brought from different regions were investigated. According to mean values, the highest individual fruit weight was found in Memecik (4.99 g) followed by Topakaşı (3.49 g) and Ayvalık (3.48 g). Ayvalık had the lowest (0.65 g) kernel weight followed by Memecik (0.76 g) and Topakaşı (0.86 g). In terms of fruit / kernel ratio, the Memecik cultivar has the best result (84.77%). The highest amount of dry matter was found in Topakaşı (53.37 g/100 g), followed by Ayvalık (39.00 g/100 g) and Memecik (38.53 g/100 g). The total amount of oil was highest in Ayvalık (57.46 g/100 g), followed by the Memecik (54.19 g/100 g) and Topakaşı (53.84 %).

Key words: olive cultivars, physical characteristics, Ayvalık, Memecik, Topakaşı.

INTRODUCTION

The olive (*Olea europaea*) is a native to the coastal areas of the eastern Mediterranean Basin and it is estimated that the cultivation of olive trees began more than 7000 years ago (Ercişli et al., 2012). Olive production in the Mediterranean basin accounts for more than 95 % of world's olive production. Located on the northeastern coast of the Mediterranean Sea, Turkey is a major olive-producing country. Olives originated from the coast of Eastern Mediterranean Sea and, to date, more than 1250 cultivars have been recognized worldwide for olive production. Most of these cultivars are present in countries located in the Mediterranean basin. The presence of 87 local olive cultivars has been documented in Turkey (İpek et al., 2012).

The increasing health consciousness and more cosmopolitan society explains the rising consumption of olive oil around the world and hence the rapid growth of the olive industry.

The beneficial health properties of olive oil have been known for centuries, particularly in the Mediterranean region. Olives and olive oil are an inherent part of Mediterranean culture and diet, and hence the decreased incidence of cardiovascular disease in this area (being one of the lowest in the Western Hemisphere) has been attributed to their consumption (Ryan and Robards, 1998). The positive effects of olive oil on health are linked to the presence of monounsaturated fatty acids (oleic acid) and a high antioxidant source, as well as high vitamin (A, D, E, K) content (Oktar et. al., 1983; Ryan and Robards, 1998; Salvador et. al., 2003). It also contains leucine, aspartic acid and glutamic acid, among other essential amino acids. Olive oil is the only vegetable oil that can be consumed without being refined (raw) and has its own odor, color and texture. These properties of olive oil are determined by chemical constituents such as fatty acids, phenolic substances, tocopherols, carotenoids and chlorophyll (Servili and Montedoro, 2002;

Ayton et. al., 2006; Turaa et. al., 2007). The chemical composition of olive oil is highly influenced by genotype, geographical region and its ecological conditions, cultural processes, harvest time and oil extraction methods (Mousa et. al., 1996; Boskou, 2000; Ayton et. al., 2006; Selvili et al., 2007; Al-Maaitah, 2009; Keçeli, 2013). In this study, it was aimed to determine the physical characteristics and oil yield of three olive cultivars at 3 different harvest periods (Pink spots on green ground, pink-violet, purple-black) grown in the same orchard and maintenance conditions in Sütçüler/Isparta located in Mediterranean Region of Turkey. The two of the olive varieties investigated in this research (Ayvalık and Memecik) are the most grown varieties and are grown in large geographical areas in Turkey. The third (Topakaşı) is a local cultivar and grown only in the research area. Thus, in the study, the differences between the cultivars adapted to the region ecology and brought from different regions are also revealed.

MATERIALS AND METHODS

The study was carried out in Sütçüler / Isparta (37°29'40"N   30°58'54"E) located in Mediterranean Region of Turkey. Ayvalık, Memecik and Topakaşı, grown in the commercial orchard conditions where the same cultural practices were applied (irrigation, fertilization, pruning, etc.), were used as plant material. The trees are 10 year old and the planting spacing is 5x4 m. The altitude of the orchard is 250 m. Fruit samples were taken at 3 different stages of maturity according to the coloring of fruit peel and fruit flesh. These are; (1) Pink spots on green ground (Maturity index: 2-3), (2) pink-violet (maturity index: 4-5), and (3) purple-black (maturity index: 6-7). The fruits were harvested by hand. Samples were brought to the laboratory immediately after harvest on ice. Fruit weight (g), fruit width (mm), fruit length (mm), shape index, seed weight (g), seed length (mm), seed width (mm) and seed/fruit flesh ratio (%) were measured. Dry matter and total oil ratios were determined in fruit samples as well. The measurements were made at each harvest period with 50 fruit samples in each triplicates.

Dry matter. The flesh of olive fruit samples were dried at 105°C in a vacuum oven until the weight reached to a constant weight. The amount of dry matter was calculated as %.

Total oil. 2 g of dried and milled fruit flesh sample was extracted with 200 ml of hexane for 4 hours in a Soxhlet apparatus and then evaporated (Guinda et al., 2003). The total crude oil was calculated as % dry sample.

RESULTS AND DISCUSSIONS

Cultivar and harvest period interactions were found statistically significant in terms of fruit weight (Table 1). As the maturity progressed, statistically significant increase was found in the fruit weight of Memecik. The highest fruit weight was determined in the second period of fruit harvest in Ayvalık.

Table 1. Some fruit characteristics of cultivars

Cultivars	Harvest Period I	Harvest Period II	Harvest Period III	**Mean**
	Fruit Weight (g)			
Ayvalık	3.49bAB	3.66bA	3.30cB	3.48
Memecik	4.69aB	5.09aA	5.19aA	4.99
Topakaşı	3.40b	3.39b	3.67b	3.49
Mean	3.86	4.05	4.05	**Lsd:0.3202**
	Fruit Height (mm)			
Ayvalık	19.17b	19.47b	18.57c	19.07
Memecik	25.45aB	27.19aA	26.59aAB	26.41
Topakaşı	19.94b	19.85b	20.33b	20.04
Mean	21.52	22.17	21.83	**Lsd:1.513**
	Fruit Width (mm)			
Ayvalık	14.78b	14.89b	14.59b	14.76
Memecik	16.20aB	17.45aA	17.55aA	17.07
Topakaşı	14.11b	13.89b	14.40b	14.13
Mean	15.03	15.41	15.52	**Lsd:1.125**

Each value is expressed as mean ±standard deviation. Means followed by different capital letters (years) in the row are significantly different (p<0.05). Means followed by different small letters in the columns (cultivars) are significantly different (p<0.05).

The highest fruit weight for the Topakaşı was determined in the third period. However, the differences between the harvest periods were not significant. Memecik had the biggest fruits (4.99 g) in all of the three harvest periods. The fruit sizes of Ayvalık and Topakaşı were similar. There was a significant difference in the fruit length and fruit width between the harvest periods only for Memecik. While the highest fruit length (27.19 mm) was detected in the second harvest period, the highest fruit width was found in the third harvest period (17.55 mm).

A significant difference was found between the cultivars in terms of seed weight. The highest average seed weight (0.86 g) was determined in the Topakaşı.

Table 2. Some seed characteristics of cultivars

Cultivars	Harvest Period I	Harvest Period II	Harvest Period III	Mean
		Seed Weight (g)		
Ayvalık	0.73	0.61	0.61	0.65b
Memecik	0.73	0.78	0.76	0.76ab
Topakaşı	0.89	0.84	0.86	0.86a
Mean	0.78	0.74	0.74	**Lsd:0.1663**
		Seed Height (mm)		
Ayvalık	13.03c	12.33c	12.17c	12.51
Memecik	18.00aB	18.00aB	19.87aA	18.62
Topakaşı	14.80bB	14.13bB	17.07bA	15.33
Mean	15.28	14.82	16.37	**Lsd:1.605**
		Seed Width (mm)		
Ayvalık	6.56ab	6.33ab	6.33b	**6.41**
Memecik	5.86bB	6.00bB	8.27aA	**6.71**
Topakaşı	6.83aB	6.83aB	8.73aA	**7.47**
Mean	6.42	6.39	7.78	**Lsd:0.7267**
		Fruit flesh/seed ratio (%)		
Ayvalık	79.08	83.33	81.52	81.32
Memecik	84.43	84.68	85.36	84.77
Topakaşı	73.82	75.22	76.57	75.36
Mean	**79.11**	**81.08**	**81.15**	

Each value is expressed as mean ±standard deviation. Means followed by different capital letters (years) in the row are significantly different (p<0.05). Means followed by different small letters in the columns (cultivars) are significantly different (p<0.05).

There was no significant relationship between seed weight and harvesting periods. While there was a significant increase in the third harvest period in the seed length and width parameters of Memecik and Topakaşı, a insignificant decrease was determined in Ayvalık..

While the longest seed size was determined in Memecik (18.62 mm), the largest seed width (7.47 mm) was found in Topakaşı. As the harvest progressed, the fruit flesh ratio increased.

The highest fruit flesh ratio was found in Memecik (84.77%) and lowest was in Topakaşı (75.36%) (Table 2). Significant differences were found between the cultivars in terms of leaf characteristics. The highest leaf area was determined in Topakaşı (Table 3).

Table 3. Leaf characteristics of cultivars

Cultivars	Leaf Length (mm)	Leaf Width (mm)	Leaf Area (mm^2)
Ayvalık	55.72b	15.02	470.80b
Memecik	55.21b	14.57	619.60ab
Topakaşı	66.50a	16.85	841.90a
Mean	59.14	15.48	644.10
LSD	6.083	2.207	313.5

The differences among the averages indicated with different letters in each column are statistically significant at the level of 5 %

Cultivar and harvest period interactions were found significant in terms of dry matter and total oil (Table 4).

The results of the study showed that the dry matter accumulation varies at different harvest periods according to the cultivars.

The amount of dry matter increased with increasing maturity in Memecik and the highest value was found in the third harvest period when the fruits were the most mature.

A fluctuation was found in Ayvalık and the highest amount of dry matter was determined at the first and third harvest periods and the values were close to each other.

According to the average values, Topakaşı had the highest (over 50%) dry matter content. The results of the research showed that Topakaşı had the ability to accumulate high dry matter in the early harvest period. The reason for this is thought to be the result of more photosynthesis due to the higher leaf area of Topakaşı. It is observed that Ayvalık also completed the accumulation of dry matter in the early period. As expected, the total amount of oil increased in all of the three cultivars as the maturity progressed (Table 4).

Table 4. Some chemical characteristics of cultivars

Cultivars	Harvest Period I	Harvest Period II	Harvest Period III	Mean
	Dry Matter (g/100 g)			
Ayvalık	40.93bA	36.89bB	39.20cA	**39.00**
Memecik	34.62cC	37.45bB	43.51bA	**38.53**
Topakaşı	54.85aA	52.56aB	52.72aAB	**53.37**
Mean	**43.47**	**42.30**	**45.14**	**Lsd:2.183**
	Total Oil (g/100 g dry matter)			
Ayvalık	51.10aC	57.92aB	63.35aA	57.46
Memecik	46.50bC	53.80bB	62.26aA	54.19
Topakaşı	51.09aB	53.61bAB	56.81bA	53.84
Mean	49.57	55.11	60.81	**Lsd:3.582**

Each value is expressed as mean ±standard deviation. Means followed by different capital letters (years) in the row are significantly different (p<0.05). Means followed by different small letters in the columns (cultivars) are significantly different (p<0.05).

The highest amount of oil was obtained at the stage of full ripeness (third harvest period, black purple fruit). The highest average oil ratio (57.46%) was found in Ayvalık. Memecik and Topakaşı had oil contents close to each other (Table 4).

Although the total oil content was highest as percentage in Ayvalık, the obtained dry matter content as above 50% on average in Topakaşı led to the conclusion that this cultivar had the higher oil yield (average 217 g / kg dry fruit flesh) than Ayvalık. In addition, the study results indicated that Topakaşı cultivar, which has completed its dry matter accumulation in the early period, can be harvested at early harvest period (second harvest period: pink-purple) without loss of excess oil yield. On the other hand, it has been concluded that harvesting of Ayvalık and Memecik at the full ripe stage (third period) should be more appropriate in terms of oil yield.

As in other fruit species, especially in table olives, the physical properties of the fruit can vary depending on the cultivar, maturation status and environmental factors. It is possible to see the effects of these factors on olive varieties in previous studies. Likewise, Nas and Gökalp (1990) found the average fruit length, fruit width, fruit/flesh ratio, total dry matter content and total oil content between 17.33-20.62 mm, 12.57-16.09 mm, 61.20%-74.38%, 38.3%-73.0%, 6.0%-24.6%, respectively in a research conducted on some fruit characteristics of some table olive cultivars. Erbay et al. (2010) reported that the fruit width of green olives varied between 13.4 and 16.9 mm. Gümüşoğlu et al. (2006) found that fruit lengths of Domat and Gemlik olive varieties varied between 22.78-27.96 mm and 16.90-23.34 mm, respectively in a research on fruit characteristics of Domat and Gemlik. Kaya and Mutlu (2010) reported that the fruit width, fruit length, fruit/flesh ratio and total oil content were varied between 16.0-19.0 mm, 22.0-24.0 mm, 70%-80%, 6.10%-26.60%, respectively in a research conducted on olives grown in İznik. Özdemir et al. (2011) reported that the fruit/flesh ratio and total oil content were varied between 3.15%-4.87% and 17.53%-32.05%, respectively in a research that aimed to determine the physicochemical changes of olive fruits collected at the different ripening stages. Aşık and Özkan (2011) investigated the fruit characteristics of Memecik olive cultivar and found that the average fruit length, fruit width, seed weight, fruit weight and total oil ratio were 2.55 mm, 1.88 mm, 0.95 g, 5.98 g, 44.74%, respectively.

CONCLUSIONS

The aim of this study was to determine the physical characteristics and total oil of three olive cultivars' fruits at 3 different harvest maturity (skin green with pink spots, pulp white-skin black, skin black and pulp purple) which were grown in Mediterrenean region of Turkey in Isparta/Sütçüler at the same garden and growing conditions.

According to mean values, the highest fruit weight was found in Memecik (4.99 g) followed by Topakaşı (3.49 g) and Ayvalık (3.48 g).

Ayvalık had the lowest (0.65 g) seed weight followed by Memecik (0.76 g) and Topakaşı (0.86 g).
In terms of fruit / seed ratio, Memecik has the best result (84.77%) according to mean values. The highest amount of dry matter was found in Topakaşı (53.37 g/100 g), followed by Ayvalık (39.00 g/100 g) and Memecik (38.53 g/100 g). The total amount of oil was highest in Ayvalık (57.46 g/100 g), followed by Memecik (54.19 g/100 g) and Topakaşı (53.84 %).

ACKNOWLEDGEMENTS

The study was supported by the Research Project Coordination Unit under the project number 2601-M-10, at Suleyman Demirel University.

REFERENCES

Al-Maaitah M.I., Al-Absi K.M., Al-Rawashdeh A., 2009. Oil Quality and Quantity of three olive cultivars as influenced by harvesting date in the Middle and Southern Parts of Jordan. International Journal of Agriculture &Biology, 11 (3): 266-272.

Aşık H.U., Özkan G., 2011. Physical, chemical and antioxidant properties of olive oil extracted from Memecik Cultivar . Academic Food Journal, 9 (2): 13-18.

Ayton J., Mailer R.J., Haigh A., Tronson D., Conlan D., 2006. Quality and oxidative stability of Australian olive oil according to harvest date and irrigation. J. Food Lipids, 14: 138-156.

Boskou D., 2000. Olive oil. In mediterranean diets, World. Rev. Nutr. Diet. 56–77.

Erbay B., Üçgül İ., Küçüksayan S., Küçüköner E., 2010. Physical, sensorial, color and rehydration properties of dried green olive slices. Journal of Natural and Applied Sciences, 14 (3): 246-250.

Ercisli, S., Bencic, D., Ipek, A., Barut, E., Liber, Z., 2012. Genetic relationships among olive (Olea europaea L.) cultivars native to Croatia and Turkey. Journal of Applied Botany and Food Quality 85: 144 – 149.

Guinda A., Dobarganes M.C., Ruiz-Mendez M.V., Mancha M., 2003. Chemical and physical properties of a sunflower oil with high levels of oleic and palmitic acids. European Journal of Lipid Science and Technology, 105 (3-4): 130-137.

Gümüşoğlu G., İnce A., Güzel E., 2006. Domat ve Gemlik zeytin çeşitlerinde bazı fiziksel özelliklerinin olgunlaşma periyodu süresince değişimi. Tarım Makinaları Bilimi Dergisi, 2(3): 239-244.

Ipek, A., Barut, E., Gulen, H., Ipek, M., 2012. Assessment of inter- and intra-cultivar variations in olive using SSR markers. Sci. Agric., 69(5): 327-335

Kaya Ü., Mutlu T.K., 2010. İznik'te yetiştirilen gemlik zeytininin ve yağının bazı fiziksel, kimyasal ve antioksidan özelliklerinin belirlenmesi. Ç.Ü. Fen Bilimleri Enstitüsü Dergisi, 22(1):199-206.

Keceli T.M., 2013. Influence of time of harvest on 'adana topagi', 'gemlik' olives, olive oil properties and oxidative stability. Journal of Food and Nutrition Research, 1(4): 52-58.

Mousa Y.M., Gerasopoulos D., Metzidakis I., Kiritsaki A., 1996. Effect of altitude on fruit and oil quality characteristics of 'Mastoides' olives. Journal of the Science of Food and Agriculture, 71 (3): 345-350.

Nas S., Gökalp H.Y., 1990. Yusufeli yöresinde üretilen sofralık siyah zeytinlerin bazı fiziksel, kimyasal ve duyusal özellikleri. Gıda, 15 (3): 155-160.

Oktar A., Çolakoğlu A., Işikli T., Acar H.,1983. Zeytinyağı ve teknolojisi. Zeytincilik Araştırma Enstitüsü Yayınları, Bornova-İzmir, 75.

Özdemir Y., Özkan M., Kurultay Ş., 2011. Olgunlaşmayla Gemlik zeytininde oluşan fizikokimyasal değişimler. Bahçe, 40(2): 21-28.

Ryan D., Robards K., 1998. Phenolic compounds in olives. Analyst, 123: 31-44.

Salvador M.D., Aranda F., Gómez-Alonso S., Fregapane G., 2003. Influence of extraction system, production year and area on cornicabra virgin olive oil: a olive oil quality. Eur. J. Lipid Sci. Technol., 104: 602–613.

Servili M., Esposto S., Lodolini E., Selvaggini R., Taticchi A., Urbani S., Montedoro G., Serravalle M., Gucci R., 2007. Irrigation effects on quality, phenolic composition, and selected volatiles of virgin olive oils Cv. Leccino. J. Agric. Food Chem., 55: 6609-6618.

Servili M., Montedoro G., 2002. Contribution of phenolic compounds to virgin olive oil quality. Eur. J. Lipid Sci. Technol., 104: 602–613.

Turaa D., Gigliottib C., Pedoa S., Faillaa O., Bassia D., Serraioccoc A., 2007. Influence of cultivar and site of cultivation on levels of lipophilic and hydrophilic antioxidants in virgin olive oils (Olea europea L.) and correlations with oxidative stability. Scientia Horticulturae, 112 (1): 108-119.

Vossen, P.M., 2007. Current Opportunities in the California Olive Oil Industry. Proceedings of the 2007 Plant and Soil Conference of the California Chapter of the American Society of Agronomy, 157-167.

DETERMINATION OF SOME CHEMICAL PROPERTIES OF 'SWEET ANN' AND 'KABARLA' STRAWBERRY CULTIVARS IN HIGHLAND CLIMATE

Mehmet POLAT[1], Abdullah KANKAYA[2], Mehmet Atilla AŞKIN[1]

[1]Süleyman Demirel University, Faculty of Agriculture, Isparta, Turkey
[2]Elma Tarım LTD Isparta, Turkey

Corresponding author email: mehmetpolat@sdu.edu.tr

Abstract

This study was conducted to determine of some chemical properties of 'Sweet Ann' and 'Kabarla' strawberry cultivars in highland climate. This research was carried out in Isparta region with highland climate conditions. Both varieties of strawberry cultivars used in the research are neutral day plants. Due to the rich polyphenol content of the berries, the positive effects on human health have begun to be explored in recent years. Especially high anthocyanin contents are important. It is well known that the strawberries are rich in polyphenol compounds. In this research, two different production methods were applied, open field and cover cultivation. The total phenolic, total anthocyanin and ascorbic acid content were determined. Total phenolic of 'Sweet Ann' strawberry cultivar varied between 620.44mg GAE 100 g^{-1} FW to 786.64 mg GAE 100 g^{-1} FW. Total anthocyanin of 'Sweet Ann' strawberry cultivar and ascorbic acid ranged between 58.15 µg/g to 98.88 µg/g and 76.42 mg/100 g to 94.39 mg/100 g respectively. Total phenolic of 'Kabarla' strawberry cultivar varied between 448.01 mg GAE 100 g^{-1} FW to 1050.48 mg GAE 100 g^{-1} FW. Also total anthocyanin of 'Kabarla' strawberry cultivar ranged between 47.16 µg/g to 74.44 µg/g. Ascorbic acid content was determined by 125.09 mg/100 g to 134.81 mg/100 g in 'Kabarla' strawberry cultivar. Highest value of total anthocyanin and ascorbic acid were obtained from open field cultivation for both varieties.

Key words: anthocyanin, ascorbic acid, phenols.

INTRODUCTION

Having a very important place in human nutrition, strawberry is a rich source of phenolic, anthocyanin and C vitamin. Thanks to the antioxidant substances it contains, this fruit is also beneficial for health. It has been reported by many researchers that the strawberries have the high antioxidant activity. (Cordenunsi et al., 2002; Wang et al., 1996, Cordenunsi et al., 2005).It is important to consume fresh in order to make more use of high antioxidant activity. Because the strawberry is a product that cannot be stored for a long time after being harvested, it will decay immediately. In order to make more use of phenolic substances, vitamin C and anthocyanins found in strawberries, it is necessary to give importance to fresh consumption. Because of the short shelf life, fresh consumption is very important (Cordenunsi et al., 2005). Phenolic substances, vitaminC and anthocyanins, which are very important in terms of health, are affected by

environmental factors, harvest maturity, storage conditions as well as by genetic characteristics (Pradas et al., 2015). Hence, this study was conducted to see the effects of two different production methods (open field cultivation and under cover cultivation) on the level of total phenolic, total anthocyanin and ascorbic acid content.

MATERIALS AND METHODS

In this study, "Kabarla" and "Sweet Ann" strawberry varieties were used as plant material. Both are neutral day varieties.

"Kabarla" variety has bears large, hard, sweet fruits with bright red color. It's fruits bearing starts slightly later than the other day variety and it keeps bearing for a long period of time. It bears fruit throughout the summer in highland regions (Anonymous, 2017).

"Sweet Ann" cultivar fruit bears in uplands and passageway regions throughout the summer. It has large, hard, oval and conical shaped bright red colored fruits (Anonymous, 2017). The variety is characterized by vigorous plants which produce high yields of large to very large. It has sweet fruit with an excellent flavor. And well-shaped, long and conical fruits (Bagdasarian, 2012)

In this research, two different production methods were applied that, open field cultivation and under cover cultivation. Fruits were harvested at the same maturity stage in both varieties. Total phenolic, vitamin-c and anthocyanin were analyzed in fruit samples.

Total phenolic were determined by using the Folin–Ciocalteau reagent according to the method of Singleton and Rossi (1965). Results were explicated as mg GAE 100 g^{-1} FW.

Total anthocyanin was determined by pH differential spectroscopic method (Cheng and Breen, 1991).

Vitamin C was determined spectrophoto-metrically at 525 nm according to the procedure of Hodges et al. (2001).

The trial was run in triplicate and statistical analysis was done by using the Minitab 17 software package version (Minitab 17 Statistical Software 2010). Differences between means were analyzed by ANOVA test and Tukey test was applied (P < 0.05).

RESULTS AND DISCUSSIONS

The total phenolic, anthocyanin and ascorbic acid contents were determined in the "Kabarla" and "Sweet Ann" cultivars grown by two different production methods (Table 1).As you can see from Table 1, the differences between the methods of growing in both varieties in terms of vitamin C are statistically significant. Vitamin C content in "Sweet Ann" cultivars varied from 76.42 to 94.39 mg/100g. Vitamin C content of the "Kabarla" variety was determined between 125.09 mg/100g and 134.81 mg/100g. The "Kabarla" variety has higher vitamin C content in both growing methods than the "Sweet Ann" cultivar. On the other hand, vitamin C content in under cover cultivation is higher than open field cultivation for both varieties.

The amounts of vitamin C in strawberry fruits were reported by Asami et al. (2003), Van de Velde et al. (2013) and Tonutare et al. (2009) as 27.1-32.6 mg/100g, 39.9-44.5 mg/100g and 44-60 mg/100g respectively. Our results are well above these values. And Polat et al. (2016)

reported as 94.10-118.87 mg/100g in strawberry fruits. The findings we obtained are consistent with these values.

The total amount of anthocyanin in the "Sweet Ann" variety was found to be 98.88 µg/g in open field and 58.15 µg/g in under cover. In the "Kabarla" varieties, the total anthocyanin content ranged from 47.16 (under cover) to 74.44 µg/g (open field) (Table 1).

In terms of total anthocyanin, the "Sweet Ann" variety has higher values than the "Kabarla" variety. However, total anthocyanin content of open-grown "Kabarla" fruits is higher than "Sweeten" fruits grown under cover (Table 1).

Table 1. Some chemical characteristics of "Sweet Ann" and "Kabarla" cultivars

Cultivar	Growing Methods	Vitamin-C (mg/100g)	Total Anthocyanin (µg/g)	Total Phenolic (mg/100 g FW)
Sweet Ann	Open field	76.42±0.03* b	98.88±0.05* a	786.64±0.52* a
	Under cover	94.39±0.07a	58.15±0.23b	620.44±0.13b
Mean		85.41 B	78.52 A	703.54 B
Kabarla	Open field	125.09±0.03 b	74.44±0.05a	1050.48±0.02 a
	Under cover	134.81±0.02 a	47.16±0.04b	448.01±0.52b
Mean		129,95 a	60,80 B	749,25 A

*The differences between the numbers shown in the same column with different letters are statistically significant (P<0.01)

Polat et al. (2016) are reported that the total amount of anthocyanin as 37.41-105.58 µg/g in fresh strawberry fruits. The data we obtain in our study is in consistent with these values. However, our data are lower than those reported by some other researchers. For example, in a study conducted by Tonutare et al. (2014), the total anthocyanin content was determined as 27.79-60.05 mg/100g in strawberries. In similar researches, Rekika et al. (2005), Zheng et al. (2007) and Wang and Lin (2000) reported that the total amounts of anthocyanin in strawberries are 190.5-841.26 µg/g, 20.07 mg/100g and 38.9 g/100g, respectively. Gill et al. (1997) noticed the total amount of anthocyanin as 113.7-153.5 µg/g. We think that as mentioned by Voca et al. (2014) , the reason of the our findings lower than some of the values reported in the literature, harvesting at different maturity stages is affect the total amount of anthocyanin. As can be seen in Table 1, in terms of the total phenolic contents, open-grown "Kabarla" fruits showed higher values than the "Sweet Ann" variety. The highest total phenolic value was found in open-grown "Kabarla" fruit (1050.48

mg GAE 100g-1 FW). However, both "Sweet Ann" fruits grown in open field (786.64 mg GAE 100g-1 FW) and under cover (620.44 mg GAE 100g-1 FW) have higher total phenolic content than "Kabarla" fruits grown under cover (448.01 mg GAE 100g-1 FW).

The total amounts of phenolic of strawberry cultivars have reported as 308 to 353 mg/100 g FW by Cordenunsi et al. (2005). Our datas are higher than these values in terms of total phenolic. Polat et al. (2016) reports that, the total phenolic content between 474.97 mg GAE 100g-1 FW and 896.85 mg GAE 100g-1 FW. The findings we obtained in our study are consistent with these values.

CONCLUSIONS

Strawberry fruits have an important place in daily diet nowadays with its rich phenolic substance, anthocyanin and vitamin C content. Some chemical contents of "Kabarla" and "Sweet Ann" varieties have been determined in our research carried out in highland climate conditions. In our research, we tried to reveal the effects of open field cultivation and under cover cultivation on this rich content.

According to the results obtained, we can recommend under cover cultivation to producers and researchers in terms of vitamin C for both varieties. For higher total anthocyanin and total phenolic level, we are recommended the open field cultivation to each two cultivars in highland climate condition.

REFERENCES

Asami D.K., Hong Y.J., Barrett, D.M., Mitchell A.E., 2003. Comparison of the Total Phenolic and Ascorbic Acid Content of Freeze-Dried and Air-Dried Marionberry, Strawberry, and Corn Grown Using Conventional, Organic, and Sustainable Agricultural Practices. J. Agric. Food Chem., 51, 1237–1241

Bagdasarian, J. H., 2012. United States Plant. Patent. Patent No: US PP22,472 P3

Cheng G.W., Breen B.J., 1991. Activity of phenylalanyl ammonialyase (PAL) and concentrations of anthocyanins and phenolics in developping strawberry fruit. J. Am. Soc. Hort. Sci. 116, 865–868.

Cordenunsi B.R., Nascimento J.R.O., Genovese, M. I., Lajolo,F. M., 2002. Influence of cultivar on quality parameters and chemical composition of strawberry fruits grown in Brazil. Journal of Agricultural and Food Chemistry, 50, 2581–2586.

Cordenunsi B. R., Genovese, M. I., Nascimento, J. R. O., Mariko, N., Hassimotto, A., Jose´ dos Santos, R. and

Lajolo, F. M., 2005. Effects of temperature on the chemical composition and antioxidant activity of three strawberry cultivars. Food Chemistry 91 (2005) 113–121

Gil M. I., Holcroft D.M., Kader A.A., 1997. Changes in Strawberry Anthocyanins and Other Polyphenols in Response to Carbon Dioxide Treatments. J. Agric. Food Chem. (45), 1662–1667

Hodges D.M., Wismer W. V., Forney C.F., 2001. Antioxidant responses in harvested leaves of two cultivars of spinach differing in senescence rates. Journal of the American Society for Horticultural Science, 126, pp. 611–617

Minitab 17 Statistical Software, 2010. State College, PA: Minitab, Inc

Polat, M., Okatan, V., Güçlü, S. F., Çolak, A. M., Korkmaz, N. And Aşkın, M. A., 2016. Effects of GA3 treatments on some chemical properties of "amiga" and "festival" strawberry cultivars. Lucrări Ştiinţifice – vol. 59(2)/2016, seria Agronomie

Pradas, I., Jesús M.J., Víctor O., Manuela M.R.J., 2015. 'Fuentepina' and 'Amiga', two new strawberry cultivars: Evaluation of genotype, ripening and seasonal effects on quality characteristics and health-promoting compounds. Journal of Berry Research, vol. 5, no. 3, pp. 157-171

Rekika,D., Khanizadeh, S., Deschênes, M., Levasseur, A. and Charles, M. T., 2005. Antioxidant Capacity and Phenolic Content of Selected Strawberry Genotypes. HortScience 40(6):1777-1781

Singleton V.L., Rossi J.J.A., 1965. Colorimentry of total phenolics with phosphomolybdic-phosphotungstic acid reagents. American Journal of Enology and Viticulture, 16 (1965), pp. 144–158

Tõnutare T., Moor, U., Mölder, K., Põldma, P., 2009. Fruit composition of organically and conventionally cultivated strawberry 'Polka'. Agronomy Research 7(Special issue II), 755–760

Tonutare T., Moor, U., Szajdak, L., 2014. Strawberry anthocyanin determination by ph differential spectroscopic method – how to get true results. Acta Sci. Pol., Hortorum Cultus 13(3), 35-47

Wang H., Cao G., Prior R.L., 1996. Total antioxidant capacity of fruits. Journal of Agricultural and Food Chemistry, 44, 701–705.

Wang S.Y., Lin H.S., 2000. Antioxidant Activity in Fruits and Leaves of Blackberry, Raspberry, and Strawberry Varies with Cultivar and Developmental Stage. J. Agric. Food Chem. (48), 140–146

Van De Velde, F., Tarola A.M., Güemes D., Pirovani M.E., 2013. Bioactive Compounds and Antioxidant Capacity of Camarosa and Selva Strawberries (Fragaria x ananassa Duch.) Franco. Foods, 2, 120-131

Voća S., Žlabur J.S., Dobričević N., Jakobek L., Šeruga M., Galić A., Pliestić,S., 2014. Variation in the Bio-active Compound Content at Three Ripening Stages of Strawberry Fruit. Molecules, (19), 10370-10385

Zheng Y., Wang S.Y., Wang C.Y., Zheng W., 2007. Changes in strawberry phenolics, anthocyanins, and antioxidant capacity in response to high oxygen treatments. LWT (40) 49–57.

THE DETERMINATION OF OIL PROPERTIES OF SOME OLIVE CULTIVARS GROWN IN SÜTÇÜLER, ISPARTA REGION

Adnan Nurhan YILDIRIM[1], Fatma YILDIRIM[1], Gülcan ÖZKAN[2],
Bekir ŞAN[1], Mehmet POLAT[2], Hatice AŞIK[2], Tuba DİLMAÇÜNAL[1]

[1]Suleyman Demirel University, Faculty of Agriculture, Horticultural Science,
32260, Isparta, Turkey
[2]Suleyman Demirel University, Faculty of Engineering, Food Engineering Department,
32260, Isparta, Turkey
Corresponding author email: tubadilmacunal@sdu.edu.tr

Abstract

The aim of this study was to assess the fatty acid compositions, tocopherol contents and some biochemical properties of 'Ayvalık', 'Memecik', and 'Topakaşı' olive cultivars' fruits grown in Mediterrenean region of Turkey, Sütçüler/Isparta. According to mean values the highest value of oleic acid (73.88 %), which is the most dominant acid in olive fruit, was found in Memecik. The highest value for alpha tocopherol content was obtained from Topakaşı (213.63 %) whereas the highest values for beta, gamma and delta tocopherol contenst were obtained from Memecik (2.46, 4.19 and 0.31 % respectively). Memecik had the highest values for chlorophyll (0.47), carotenoid (0.31) and pheophytin a (2.29) contents according to mean values. According to the knowledge obtained from the research, the fatty acid composition and the quality characteristics of the olive oil are mainly depended on the growing conditions, harvest period and the oil extraction methods. In the study, it is concluded that 2nd harvest period for Memecik, 2nd and 3rd harvest periods for Ayvalık and Topakaşı would be more suitable under the Isparta, Sütçüler growing conditions for high-quality olive oil. This research is the first detailed research on olive in the research area and it is considered that it will be the basis to future scientific studies.

Key words: *Ayvalık, biochemical properties, fatty acid compositions, tocopherol contents.*

INTRODUCTION

The olive (*Olea europaea* L.) is known the oldest cultivated tree in the world (Özbek, 1975). *Olea europaea* is one of the most important and widespread fruit trees in the Mediterranean basin (Conde et al., 2008), especially in Spain, Italy, Greece and Turkey (Dıraman and Dibeklioğlu, 2009).

Olive oil is an essential constituent of Mediterranean diet and is obtained from the fruit of several cultivars. Each one of these cultivars exhibits specific physical and biochemical characteristics, providing oils with different compositions and performances (Matos et al., 2007). Olive oil is a good source of several bioactive compounds such as mono/poly-unsaturated fatty acids, phenols, phytosterols, carotenoids and tocopherols. Due to their antioxidant properties, these bioactive compounds have reducing effect on the risk of chronic degenerative diseases such as coronary heart disease, cancer, obesity, immune and inflammatory responses (Dag et al., 2015).

Turkey is the sixth largest producer of olive oil in the World (4.6 %) (Sevim et al., 2013) and exported 8% on the average between 2004 and 2010. While the production was 130.000 tonnes in 2008, it increased to 160.000 tonnes in the 2010 (Alkan et al., 2012). The economically important Turkish olive cultivars include Ayvalık, Memecik, Gemlik, Erkence, Nizip Yağlık, and Uslu (Dıraman and Dibeklioğlu, 2009). Olive cultivar 'Memecik' has more than 45% of orchard area in Turkey. Memecik is used both as table olive and for extraction of oil. Olive oil is classified as extra virgin, virgin, olive oil. This classification is carried out according to some quality characteristics (acidity, peroxide value, K_{232}, K_{270}, DK values, alkyl esters) (Caporaso et al., 2015).

The quality of olive oil is affected by many factors such as olive variety, climate and soil characteristics of the geographical region, maturity level, cultural practices and extraction methods (İlyasoğlu and Özçelik, 2011). Olives for oil production are generally harvested in November and December in Turkey. However, olives are sometimes harvested in early period (October) for production high quality olive oil. So, high quality olive oil was obtained in terms of phenolics and other quality parameters (Yıldırım et al., 2016).

The aim of this study was to determine the variation in the fatty acids, tocopherols and biochemical properties of oils obtained from olive cultivars 'Ayvalık', 'Memecik' and 'Topakaşı' harvested at 3 different maturity stages: (1) early harvest period-1 (green skin with pink spots in less than half of the fruit— Beginning of spotting), (2) early harvest period-2 (pink or purple skin in more than half of the fruit—End of spotting), and (3) optimum harvest period (black skin, less than half of pulp to be purple).

MATERIALS AND METHODS

The study was conducted in the Mediterranean region of Turkey, Sütçüler / Isparta (37° 29'40 "N 30° 58'54" E). In this study, olive cultivars 'Ayvalık', 'Memecik' and 'Topakaşı' grafted on seedling rootstocks were used as materials. Olive trees planted at spacing distance of 5 x 4 m were 10 years old. The altitude of the orchard was 250 m. While Ayvalık and Memecik cultivars are grown in a wide area in Turkey, Topakaşı is only grown in a narrow area of Sütçüler-Isparta, Anamur and Tarsus regions of Turkey. Fruit samples were harvested at 3 different stages of maturity according to the coloring of fruit peel and fruit flesh. These are; (1) Pink spots on green ground (maturity index: 2-3, (2) pink-violet (maturity index: 4-5), and (3) purple-black (maturity index: 6-7).

The obtained natural extra virgin olive oil was placed in dark bottle glasses and kept at -80 °C until analysis. The fatty acid composition of olive oil was determined according to Marquard (1987). 1 ml Na- Methylate (0.5 g Na-methylate + 80 ml methanol + 20 ml iso-octane) solution was added on to 1 ml oil sample and esterification was carried out. 0.25 ml of iso-octane was added before injection and the tube was well-shaken. Then 0.5 ml sample was drawn from the upper phase, which became clear, and by means of a microinjection and injected into the GC apparatus (Pelkin Elmer Autosystem XL). GC condidtions: FID

dedector; Cp WAX 52 CB 50 m x 0.32 mm. 1.2 μm column; injector and dedector temperatures: 250°C; carrier gas: He; flow rate: 15 mL/min. The oven temperature was held at 60 °C for 4 min and increased to 175 with increasing 4 °C per min. After holding at this temperature for 27 min, the temperature was increased to 215 °C. After holding at 215 °C for 5 min, the temperature was gradually increased to 240 °C. Peaks were identified by taking in to account the relative retention times of standards (Sigma-Aldrich Chemicals 189-19) and results were expressed as percentages of peak areas.

The method of Lampi et al. (1999) was used for tocopherol analysis. Tocopherols (α, γ, δ, and β) were determined by HPLC with a RF-10AXL fluorescence detector (Ex 295-Em 330 nm). 20 μl of the oil was injected into the HPLC device equipped with Luna Silica column (250 × 4.6 mm, 5 μm particle size). Mobile phase: heptane: THF (95:5) (v/v); mobile phase flow rate, 1.2 ml/min; Peak identification was performed according to standards (Cabliochem, Germany). The quantity of tocopherols was calculated according to peak area and expressed as mg kg^{-1} oil.

Analysis of carotenoids and chlorophyll were carried out using spectrophotometer (T70+UV/VIS Spectrometer, PG Instruments Ltd-England). The method defined by Minguez-Mosquera et al. (1991) was used for the determination of carotenoids. 7.5 grams of olive oil was dissolved in cyclohexane and was completed to 25 ml. Samples for total carotenoid were read at 470 nm wavelength and the results were calculated as mg carotenoid/kg fat. The oil samples prepared in cyclohexane were measured at 630 nm, 670 nm and 710 nm wavelengths and the amounts of chlorophyll were calculated as mg of pheophytin a/kg oil using the formula given below (Pokorny et al., 1995).

The amount of chlorophyll (equivalent of mg of pheophytin a / kg fat) = 345.3x [A670-(A630 + A710)] / L

A refers to absorbance, L means the ray path (cell thickness, mm) in the equation.

The method defined by Anonymous (2001) was used for absorbance values. 50 mg of oil sample was weighed and 5% cyclohexane was added to prepare a 1% solution of oil in cyclohexane. The specific absorbance values in UV radiation was determined using K_{232} and K_{270} nm spectrophotometer. The obtained data were analyzed according to One-Way Anova variance analysis method, and significant differences between the averages were determined according to Duncan's test.

RESULTS AND DISCUSSIONS

1. Fatty acid composition
In this study, oleic (71.80-74.19%), palmitic (11.80-13.72%), linoleic (5.92-8.96%) and stearic (2.05- 2.78%) acid were the major fatty acid in the olive oils. Other fatty acids detected in olive oil were tricosanoic, palmitoleic, linolenic, arachidic, gamalinolenic, heptadecaenoic, eicosatrienoic, behenic, heptadecanoic and myristic acid (Table 1). For all fatty acids except for behenic acid, the cultivar x harvest period interaction was found significant. The oleic and linoleic acid contents of the cultivars were differed according to the fruit maturity periods. The content of oleic acid, which is the main acid in olive cultivars, increased significantly with progress of the maturity in Ayvalık, whereas it decreased in Memecik. The highest oleic acid content was found in Ayvalık as 74.19%, followed by Memecik 73.44% and Topakaşı 72.69% at commercial harvest period (third harvest period). Gurdeniz et al. (2008) reported that the oleic acid contents of Ayvalık and Memecik as 69.58 and 66.32%, respectively at commercial harvest period. Dıraman et al. (2009) found that the amount of oleic acid content of Ayvalık changed between 61.44% and 74.68% at different regions and years. Köseoğlu et al. (2016) investigated the oleic acid component of Memecik at 3 stages of ripening according to skin pigmentation as green, purple, and black and found as 72.37%, 71.23% and 68.92%, respectively.

These different results in literature revealed that the growing conditions and harvest time have important effects on the fatty acid compositions and quality attributes of an olive cultivar.

While the content of linoleic acid decreased in Ayvalık, it increased in Memecik with the progress of maturity. According to average values, the highest content of oleic acid among the varieties was determined as 73.88% in Memecik. The cultivars of Ayvalık and Topakaşı had similar contents of oleic acid. While oleic acid was stable in Topakaşı, linoleic acid increased significantly in the third harvest period. In third maturity stage, oleic acid increased while linoleic acid decreased according to average values.

The contents of palmitic acid decreased at significant levels in Ayvalık and Topakaşı with increased maturity. However, no significant difference was obtained in Memecik. The highest average content of palmitic acid was determined as 13.36% in Ayvalık. Similar to our findings, Dıraman et al. (2009) reported that the variation of palmitic and linoleic acids, which are other major fatty acids for Ayvalık, were as 8.94% - 17.77% and 4.68% - 15.14% respectively in different regions and years. In addition, Uğurlu and Özkan (2011) found that the content of palmitic acid in Memecik was 11.38%.

The content of stearic acid in Ayvalık increased significantly in the third harvest period. The palmitic, linoleic and stearic acid contents were found as 12.65%, 5.92% and 2.78%, respectively at the commercial harvest period (third harvest period) (Table 1).

The contents of tricosanoic acid differed according to the harvest periods of the cultivars. Tricosanoic acid increased in Ayvalık with increasing maturity, while it fluctuated in Memecik.

A similar situation was also observed in Topakaşı. The results of the study showed an increase in palmitoleic acid content increasing maturity. The highest values were determined during the third harvesting period in all cultivars. The highest palmitoleic acid content was determined in Ayvalık with an average of 1.12%, followed by Topakaşı (0.83%) and Memecik (0.79%). Similar to our result, Uğurlu and Özkan (2011) found that palmitoleic acid content was 0.50% in Memecik. The content of linolenic acid increased in all cultivars with incresing maturity. The highest average linolenic acid content was found in Memecik as 0.65%, followed by Ayvalık as 0.54%. Tanılgan et al. (2007) found the linolenic acid content as 0.2% in Ayvalık.

Table 1. Composition of fatty acids (%) obtained from Ayvalık, Memecik and Topakaşı olive cultivars in three different harvesting periods.

Olive cultivars	First harvest	Second harvest	Third harvest	Mean Values
Oleic acid (C18:1)				
Ayvalık	71.80bB	71.93bB	74.19aA	72.64
Memecik	74.72aA	73.48aB	73.44bB	73.88
Topakaşı	72.29b	72.88a	72.69c	72.62
Mean	72.93	72.76	73.44	Lsd:0.7012
Palmitic acid (C16:0)				
Ayvalık	13.72aA	13.70aA	12.65B	13.36
Memecik	11.80b	12.01c	12.28	12.03
Topakaşı	13.59aA	12.96bB	12.44C	12.99
Mean	13.03	12.89	12.46	Lsd:0.4984
Linoleic acid (C18:2)				
Ayvalık	8.96aA	8.01bB	5.92cC	7.63
Memecik	7.41cC	8.75aA	8.26bB	8.14
Topakaşı	8.27bB	8.24bB	8.76aA	8.42
Mean	8.21	8.33	7.65	Lsd:0.4416
Stearic acid (C18:0)				
Ayvalık	2.46aB	2.38aB	2.78aA	2.54
Memecik	2.18bA	2.05bB	2.09cAB	2.11
Topakaşı	2.28b	2.33a	2.39b	2.33
Mean	2.31	2.25	2.42	Lsd:0.1168
Tricosanoic acid (C23:0)				
Ayvalık	0.77cC	0.94B	1.13aA	0.94
Memecik	1.28aA	0.93C	1.14aB	1.11
Topakaşı	0.90b	0.94	0.90b	0.91
Mean	0.98	0.93	1.05	Lsd:0.08924
Palmitoleic acid (C16:1)				
Ayvalık	0.85aB	1.21aA	1.32aA	1.12
Memecik	0.71bB	0.81bAB	0.85bA	0.79
Topakaşı	0.77ab	0.83b	0.89b	0.83
Mean	0.78	0.95	1.02	Lsd:0.1282
Linolenic acid (C18:3)				
Ayvalık	0.46bB	0.49bB	0.67aA	0.54
Memecik	0.54aB	0.68aA	0.72aA	0.65
Topakaşı	0.47bAB	0.44bB	0.51bA	0.47
Mean	0.49	0.54	0.63	Lsd:0.06176
Arachidic acid (C20:0)				
Ayvalık	0.33bC	0.39B	0.46A	0.39
Memecik	0.44a	0.41	0.45	0.43
Topakaşı	0.46a	0.43	0.44	0.44
Mean	0.41	0.41	0.45	Lsd:0.04317
Gamma-linolenic acid (C18:3)				
Ayvalık	0.22bC	0.30bB	0.38aA	0.30
Memecik	0.38a	0.37a	0.41a	0.38
Topakaşı	0.36a	0.33b	0.35b	0.34
Mean	0.32	0.33	0.38	Lsd:0.03939
Heptadecanoic acid (C17:0)				
Ayvalık	0.18B	0.21aA	0.19bAB	0.19
Memecik	0.17A	0.15bA	0.09cB	0.13
Topakaşı	0.18B	0.23aA	0.23aA	0.21
Mean	0.18	0.19	0.17	Lsd:0.02659
Eicosatrienoic acid (C20:1)				
Ayvalık	0.10cB	0.18aA	0.08bC	0.12
Memecik	0.24aA	0.18aB	0.06cC	0.16
Topakaşı	0.18bA	0.13bB	0.18aA	0.16
Mean	0.17	0.16	0.11	Lsd:0.01174
Behenic acid (C22:0)				
Ayvalık	0.12	0.15	0.13	0.13
Memecik	0.12	0.13	0.14	0.13
Topakaşı	0.13	0.14	0.13	0.13
Mean	0.12B	0.14A	0.13AB	Lsd:0.01509
Heptadecenoic acid (C17:1)				
Ayvalık	0.04bB	0.13aA	0.12aA	0.09
Memecik	0.04bB	0.07bA	0.07bA	0.06
Topakaşı	0.14aA	0.13aA	0.10aB	0.12
Mean	0.07	0.11	0.09	Lsd:0.02176
Myristic acid (C14:0)				
Ayvalık	0.010cB	0.020A	0.010bB	0.013
Memecik	0.017bB	0.020A	0.020aA	0.019
Topakaşı	0.020a	0.020	0.020a	0.020
Mean	0.016	0.020	0.020	Lsd:0.002998

Each value is expressed as mean ±standard deviation. Means followed by different capital letters (years) in the row are signifi cantly different (p<0.05). Means followed by different small letters in the columns (cultivars) are signifi cantly different (p<0.05).

Our findings are similar those of İlyasoğlu and Özçelik (2011) in terms of palmitic, palmitoleic, stearic, oleic, linoleic and linolenic acid components of Memecik.

The arachidic and gamma-linolenic acid contents increased only in Ayvalık at the significant level with increasing maturity. However, there was no significant difference between the harvest periods for the other two cultivars. The highest average arachidic acid was found in Topakaşı and the highest gamma-linolenic acid was found in Memecik. Dıraman and Dibeklioğlu (2014) found similar results in terms of arachidic acid contents of Ayvalık and Memecik.

In terms of heptadecanoic acid content, there was an increase in Topakaşı in the second harvest period and this increase remained constant during the third harvest period. On the other hand, Ayvalık showed an increase in terms of heptadecanoic acid in the second harvest period and but it was decreased in the third harvest period. The content of heptadecanoic acid decreased in Memecik with increasing maturity and it decreased at significant level in the third harvest period. The highest average heptadecanoic content was determined in Topakaşı as 0.21%, followed by Ayvalık as 0.19%.

Eicosatrienoic acid content increased significantly at the second harvest period in Ayvalık and Topakaşı. the content of eicosatrienoic acid showed a decreasing at significant level in Memecik with progress of maturity. The highest eicosatrienoic content was found in Memecik and Topakaşı as 0.16%. There was no significant difference between cultivars and harvest periods in terms of behenic acid content.

In terms of heptadecanoic content, significant increases observed in the second harvest period in Ayvalık and Memecik and this increase remained stable in the third harvest period. On the contrary, a significant decrease was observed in Topakaşı in the third harvest period. The highest average content of heptadecanoic acid was determined in Topakaşı as 0.12%.

While the content of myristic acid increased during the second harvest period, it decreased in third harvest period in Ayvalık. The fatty acid components of Ayvalık and Memecik obtained in this research were similar with those of Dıraman (2010) and Dıraman and Dibeklioğlu (2014).

2. Tocopherol contents

One of the most important minor chemical components in olive oils is tocopherols. Beside their health benefits, they also enhance the oxidative stability of olive oils due to their antioxidant properties (Dag et al., 2015). In terms of all tocopherol components, the cultivar x harvest period interaction was found significant.

Among the different tocopherol forms, alpha (α) tocopherol was the major one in all cultivars. The highest α-tocopherol contents were determined in the second harvest period in Ayvalık and Memecik while the highest value was found in the third harvest period in Topakaşı.

The highest average α-tocopherol content was determined in Topakaşı (213.63 mg/kg oil), followed by Ayvalık (212.97 mg/kg oil) (Table 2). Beta (β) tocopherol content increased significantly during the second harvest period in all cultivars and decreased to the lowest level in the third harvest period. The highest average content of β-tocopherol was determined in Memecik (2.46 mg/kg oil). While the content of gamma (γ) tocopherol showed an increasing in Ayvalık with increasing maturity, a decreasing was found in Topakaşı. It increased at second harvest period and then decreased at third harvest period in Memecik.

The highest average content of γ-tocopherol was determined in Memecik (4.19 mg/kg oil), followed by Ayvalık (2.31 mg/kg oil). The highest delta (δ) tocopherol content was determined in the first harvest period in Ayvalık and Topakaşı. The highest average δ-tocopherol content was found in the Memecik as 0.31 mg/kg oil (Table 2). Uğurlu and Özkan (2011) reported that the values of α, β, γ and δ-tocopherol were 205.45 mg/kg oil, 1.645 mg/kg oil, 6.065 mg/kg oil and 0.325 mg/kg oil, respectively in Memecik. These results are similar to our average results obtained in Memecik.

The highest K_{232} values were determined in the second harvest period in all cultivars and the lowest values were determined in the third harvest period. The highest average K_{232} value

was determined as 1.57 in the Ayvalık while the lowest value was 1.38 in Memecik (Table 3). The highest average K_{270} value was determined as 0.14 in Ayvalık. Similar to our results, oil quality parameters of K_{232} and K_{270} were found as 1.493 and 0.098, respectively in Memecik by Uğurlu and Özkan (2011).

Table 2. Composition of tocopherol obtained from Ayvalık, Memecik and Topakaşı olive cultivars in three different harvest periods.

Olive cultivars	First harvest	Second harvest	Third harvest	Mean Values
Alpha (α) tocopherol (mg/kg oil)				
Ayvalık	186.85cC	229.65aA	222.40bB	212.97
Memecik	205.50bB	215.25bA	200.25cC	207.00
Topakaşı	210.15aB	201.30cC	229.45aA	213.63
Mean	200.83	215.40	217.37	Lsd:3.581
Beta (β) tocopherol (mg/kg oil)				
Ayvalık	1.57cB	1.77aA	1.42bC	1.59
Memecik	2.45aB	2.76aA	2.17aC	2.46
Topakaşı	1.72bB	1.79bA	1.27cC	1.59
Mean	1.91	2.10	1.62	Lsd:0.01341
Gamma (γ) tocopherol (mg/kg oil)				
Ayvalık	1.64bC	2.44bB	2.87bA	2.31
Memecik	3.67aB	5.28aA	3.63aC	4.19
Topakaşı	1.64bA	1.56cB	1.14cC	1.45
Mean	2.32	3.09	2.54	Lsd:0.02448
Delta (δ) tocopherol (mg/kg oil)				
Ayvalık	0.13cA	0.11bB	0.10bC	0.11
Memecik	0.24aC	0.42aA	0.28aB	0.31
Topakaşı	0.17bA	0.07cC	0.09cB	0.11
Mean	0.18	0.19	0.16	Lsd:0.004240

Each value is expressed as mean ±standard deviation. Means followed by different capital letters (years) in the row are signifi cantly different (p<0.05). Means followed by different small letters in the columns (cultivars) are signifi cantly different (p<0.05).

Table 3. Some biochemical properties of cultivars

Olive cultivars	First harvest	Second harvest	Third harvest	Mean Values
Chlorophyll (mg/kg oil)				
Ayvalık	0.31cA	0.24cB	0.17cC	0.24
Memecik	0.52aA	0.43aB	0.46aB	0.47
Topakaşı	0.37bA	0.32bB	0.22bC	0.30
Mean	0.40	0.33	0.28	Lsd:0.02665
Carotenoid (mg/kg oil)				
Ayvalık	0.26cA	0.23cB	0.20bC	0.23
Memecik	0.35aA	0.28aB	0.29aB	0.31
Topakaşı	0.29bA	0.26bB	0.18cC	0.25
Mean	0.30	0.26	0.22	Lsd:0.008654
Pheophytin-a (mg/kg oil)				
Ayvalık	1.08cA	0.74cB	0.34cC	0.72
Memecik	2.69aA	2.04aC	2.13aB	2.29
Topakaşı	1.46bA	0.90bB	0.46bC	0.94
Mean	1.74	1.23	0.98	Lsd:0.01731
K_{232}				
Ayvalık	1.58aA	1.59aA	1.56aB	1.57
Memecik	1.40cA	1.41bA	1.35cB	1.38
Topakaşı	1.51bB	1.58aA	1.50bB	1.53
Mean	1.49	1.53	1.47	Lsd:0.01095
K_{270}				
Ayvalık	0.15aA	0.13aB	0.15aA	0.14
Memecik	0.12bA	0.11bAB	0.10cB	0.11
Topakaşı	0.12b	0.13a	0.12b	0.13
Mean	0.13	0.13	0.13	Lsd:0.01187

Each value is expressed as mean ±standard deviation. Means followed by different capital letters (years) in the row are signifi cantly different (p<0.05). Means followed by different small letters in the columns (cultivars) are signifi cantly different (p<0.05).

İlyasoğlu and Özçelik (2011) also found similar result to our finding in term of K_{270} value. In this research, the interaction between chlorophyll, carotenoid and pheophytin-a contents and harvest periods were found significant. Chlorophyll, carotenoid and pheophytin-a contents decreased significantly with increasing maturity. The highest average values of chlorophyll, carotenoid, and pheophytin-a were determined in Memecik (0.47, 0.31 and 2.29, respectively) (Table 3). Our findings are similar to those of Uğurlu and Özkan, (2011) which reported the average value of chlorophyll content of Memecik as 0.49 mg/kg of oil.

CONCLUSIONS

In this research, the effects of cultivar and harvest period on the fatty acid composition were significantly determined.
According to the knowledge obtained from the research, the fatty acid composition and the quality characteristics of the olive oil are mainly depended on the growing conditions, harvest period and the oil extraction methods.
In the study, it is concluded that 2nd harvest period for Memecik, 2nd and 3rd harvest periods for Ayvalık and Topakaşı would be more suitable under the Isparta/Sütçüler growing conditions for high-quality olive oil.

ACKNOWLEDGEMENTS

The study was supported by the Research Project Coordination Unit under the project number 2601-M-10, at Suleyman Demirel University.

REFERENCES

Alkan D., Tokatli F., Ozen B., 2012. Phenolic characterization and geographical classification of commercial extra virgin olive oils produced in Turkey. J. Am. Oil Chem. Soc. 89: 261–268.

Caporaso N., Savarese M., Paduano A., Guidone G., De Marco E., Sacchi R., 2015. Nutritional quality assessment of extra virgin olive oil from the Italian retail market: Do natural antioxidants satisfy EFSA health claims? Journal of Food Composition and Analysis 40: 154–162.

Anonymous, 2001. Codex Alimentarius Commission. Codex Standard 12-1981, Rev. 2.

Conde C., Delrot S., Geros H., 2008. Physiological, biochemical and molecular changes occuring during olive development and ripening. J. Plant Physiol. 165, 1545-1562.

Dag C., Demirtas I., Ozdemir I., Bekiroglu S., Ertas E., 2015. Biochemical characterization of Turkish extra virgin olive oils from six different olive varieties of ıdentical growing conditions. J. Am. Oil Chem. Soc. 92:1349–1356.

Dıraman H., Dibeklioğlu H., 2009. Characterization of Turkish virgin olive oils produced from early harvest olives. J. Am. Oil Chem. Soc. 86:663–674.

Dıraman H., Özdemir D., Hışıl Y., 2009. Characterization of early harvest virgin olive oils produced from ayvalık cultivar based on their fatty acid profiles by chemometrics. Electronic Journal of Food Technologies 4 (3): 1-11.

Dıraman H., 2010. Characterization by chemometry of the most important domestic and foreign olive cultivars from the national olive collection orchard of Turkey. Grasas y Aceites, 61 (4): 341-351.

Dıraman H., Dibeklioglu H., 2014. Using lipid profiles for the characterization of Turkish monocultivar olive oils produced by different systems. International Journal of Food Properties, 17: (5): 1013-1033.

Gurdeniz G., Ozen B., Tokatli F., 2008. Classification of Turkish olive oils with respect to cultivar, geographic origin and harvest year, using fatty acid profile and mid-IR spectroscopy. Eur. Food Res. Technol. 227: 1275–1281.

İlyasoğlu H., Özçelik, B.B, 2011. Biochemical characterization of Memecik olive oils. Food, 36 (1): 33-41.

Köseoğlu O., Sevim D., Kadiroğlu P., 2016. Quality characteristics and antioxidant properties of Turkish monovarietal olive oils regarding stages of olive ripening. Food Chemistry 212: 628–634.

Lampi A.M., Kataja L., Kamal-Eldin A., Vieno P., 1999. Antioxidant activities of α- and γ- tocopherols in the oxidation of rapeseed oil triacylglycerols. Journal of the American Oil Chemists' Society, 76 (6): 749-755.

Marquard R., 1987. Qualitatsanalytik im dienste der ölpflanzenzüchtung. Fat.Sci. Technol. 89: 95-99.

Matos L.C., Pereira J.A., Andrade P.B., Seabra R.M., Oliveira M.B.P.P., 2007. Evaluation of a numerical method to predict the polyphenols content in monovarietal olive oils. Food Chemistry 102, 976–983.

Minguez-Mosquera M.I., Rejano-Navarro L., Gandul-Rojas B., Gomez A.H.S., Garrido-Fernandez J., 1991. Color- pigment correlation in virgin olive oil. Journal of the American Oil Chemists Society, 68 (5): 332-336.

Özbek S., 1975. General fruit growing. Çukurova University Agricultural Faculty Publications, 111. Textbook: 6, p.386, Ankara.

Pokorny J., Kalinova L., Dysseler P., 1995. Determination of chlorophyll pigments in crude vegetable oils: Results of a collaborative study and the standardized method (technical report). Pure and Applied Chemistry, 67 (10): 1781–1787.

Sevim D., Köseoğlu O. Öztürk G.F., 2013. Effect of different growing area on triacylglycerol composition of cv. Gemlik olive oil in Turkey.

Journal of Agricultural Faculty of Uludag University, 27 (1): 49-54.

Tanılgan K., Özcan M.M., Ünver A., 2007. Physical and chemical characteristics of five Turkish olive (Olea europea L.) varieties and their oils. Grasas y Aceites, 58 (2): 142- 147.

Uğurlu A.H., Özkan, G., 2011. Physical, chemical and antioxidant properties of olive oil extracted from Memecik cultivar. Academıc Food Journal 9 (2): 13-18.

Yıldırım F., Yıldırım A.N., Özkan G., Şan B., Polat M., Aşık H., Karakurt Y., Ercişli S., 2016. Early harvest effects on hydrophilic phenolic components of extra virgin olive oils cvs. 'Ayvalık,' 'Memecik,' and 'Topakaşı'. Biochem. Genet., DOI 10.1007/s10528-016-9784-3.

5

PHYSICAL AND BIOCHEMICAL CHANGES IN POMEGRANATE (*PUNİCA GRANATUM* L. cv. 'HİCAZNAR') FRUITS HARVESTED AT THREE MATURITY STAGES

Berna BAYAR, Bekir ŞAN

Suleyman Demirel University, Faculty of Agriculture, Horticultural Science, 32260, Isparta, Turkey
Corresponding author email: bekirsan@sdu.edu.tr

Abstract

This study was carried out to determine the physical and biochemical changes in fruits harvested at three different maturity stages in pomegranate cultivar 'Hicaznar'. Fruits were harvested at the periods of 1) the time of the beginning of the color change in arils (August 15), 2) the time of the pink color of arils (September 6) and 3) the time of the red color of arils (October 3). In the study, fruit weight, fruit juice, colour (L*, a*, b*), titratable acidity, pH, total soluble solids, total phenolics, phenolic composition (gallic, chlorogenic, ellagic and syringic acids) and organic acids (malic and citric acids) were investigated. Significant increases in fruit weight, total soluble solids, malic acid and colour value of a* were detected with progress of maturity. On the contrary, titratable acidity, total phenolics, gallic, ellagic and sitric acids were significantly decreased with progress of maturity. The changes in chlorogenic and syringic acids were not statistically significant.

Key words: pomegranate, maturity, phenols, organic acids.

INTRODUCTION

Pomegranate belongs to the *Punica* genus of Punicaceae family, the most important species being *Punica granatum* L. Pomegranate is one of the oldest known fruit species and its cultural history dates back to around 3000 BC. It can grow up to 1000 m above sea level in tropical and subtropical climates. Pomegranate is cultivated widely in the Mediterranean Basin, South West Asia and America in the world. Pomegranate plants have many advantages such as being easily adaptable to various climatic and soil conditions, being easy to replicate and having high productivity. Pomegranates are also grown as ornamental plants or hedge plants in Turkey. Pomegranate production, which is 59 000 tonnes in 2000 in Turkey, reached 445 750 tonnes in 2015 with a big increase. Pomegranate is a very rich fruit species in terms of phenolic substances, flavonoids, tannins, fatty acids, aromatic compounds, amino acids, tocopherols, sterols and terpenoids (Ozgen et al., 2008; Wang et al., 2010).

In addition to many factors such as increasing environmental pollution, ultraviolet rays and smoking, stressful living conditions, environmental and psychological factors negatively affect human health and cause free radicals formation. It has became important to consume natural nutrients instead of using drugs to remove the negative effects of free radicals and to prevent disease formation (Hochstein and Atallah, 1988; Benzie, 2003). Fruits that are rich in terms of phenolic substances prevent these diseases by preventing these free radicals and strengthen the immune system. In addition to that, they have positive effects on health due to their antimicrobial and antioxidative effects. The most important factor affecting the nutritional content of fruits evaluated as functional food is genotype. However, it is a known fact that environmental conditions also influence the nutritional contents of fruits. Researchers have reported that the antioxidant capacities of fruits vary significantly with respect to fruit maturity levels as well as ecological (temperature, soil characteristics and night-time temperature difference) and cultural (irrigation, fertilization) conditions (Gao et al., 2012). There are many studies interested in the increase of the nutrient contents of fruits by different applications (Rossi et al., 2003, Ancos et al., 2000). It has been reported that the nutritional content of fruit species changes significantly with maturity level. For example, in a study of two different species of *Zizipus*, it was determined that the amounts of phenolic compounds in green fruits were higher than those in ripe ones in *Z. mauritiana* and *Z. nummularia* species (Choi et al., 2012; Wu et al., 2012).

The objective of this study was to evaluate the pomegranate fruit harvested at different maturity levels in terms of their physical and biochemical contents.

MATERIALS AND METHODS

Materials

In the study, the fruits harvested in three maturity levels of pomegranate cultivar Hicaznar were used. The fruits were obtained from a farmer's orchard in Serik region of Antalya, Turkey.

Methods

In the study, fruits were harvested at 3 different maturity levels. In this regard, pomegranate fruits were harvested at 1) the time of the beginning of the color change in arils (August 15), 2) the time of the pink color of arils (September 6) and 3) the time of the red color of arils (commercial harvest time, October 3).

Determination of physical properties. Fruit width, fruit height, fruit weight, peel weight and aril weight were determined in the harvested fruit. Peel color of fruits were measured using a colorimeter (Chroma Meter CR-400, Minolta) and expressed as L*, a* and b* values.

Determination of biochemical properties. In order to determine the biochemical properties, the arils separated from their peels were squeezed and the obtained fruit juice was filtered through filter paper. The pH of the fruit juice was determined by a pH meter (Hanna). Total soluble solids were measured with a hand refractometer and expressed as %. Titratable acid content was determined according to the

method described by Karaçalı (1990) and calculated as % citric acid.

The total phenolic content of fruit juices was determined using the Folin-Ciocalteu method. For this purpose, the aril juice was diluted 1:5 with etanol. 100μl of fruit juice was added and 3ml of purified water was added. Then, 200 μL of Folin-Ciocalteu (0.2N) and 100 μL of sodium carbonate (20%) were added and incubated in the dark for 2 hours. The absorbance values were then read on a spectrophotometer adjusted to a wavelength of 765 nm. The total amount of phenolic substances in pomegranate juice was calculated from the standard calibration curve. For the determination of standard calibration curve, 50, 100, 150, 200, 250, 300, 350 and 400 mg/L gallic acid solutions were prepared and their absorbances were read at 765 nm in a spectrophotometer by the same method.

Phenolic contents were analyzed according to the modified procedure of Caponio et al. (1999). The 5 ml of fruit juice was mixed with 10 ml of methanol (80%). The sample was incubated in an ultrasonic bash for 10 min and centrifuged at 4000 rpm for 10 min. The upper phase was filtered through a 0.45 μm membrane filter (Millipore) and 20μl of the sample was injected into an HPLC (Shimadzu Inc) equipped with a diode array detector (lmax = 278), Agilent Eclipse XDB-C18 column (250x4,6 mm, 5μm) operated at 30°C, a SIL-10AD vp autosampler, a LC-10AD vp pump, a CTO-10Avp column oven, and a DGU-14A degasses. The mobile phase consisted of 3% acetic acid (A) and methanol (B). The flow rate was 0.8 mL/min. The gradient program was given in Table 1. The peaks were identified by comparison with the peak of standard of gallic acid, catechine, chlorogenic acid, vanillic acid, syringic acid, ellagic acid, quercetin and kaempherol (Sigma Chemical Co) (Figure 1). The phenolics were expressed as μg per g fruit juice.

Table 1. The linear solvent gradient system used in HPLC analysis of phenolics.

Time (min)	0.1	20	28	35	50	60	62	70	73	75	80	81
A *(%)	93	72	75	70	70	67	58	50	30	20	0	93
B (%)	7	28	25	30	30	33	42	50	70	80	100	7

*Solvent A: 3% Acetic acid, Solvent B: Methanol

Table 2. Some physical properties of fruits of pomegranate cultivar 'Hicaznar' harvested at 3 different periods

Harvest date	Fruit weight (g)	Fruit height (mm)	Fruit width (mm)	Shape index	Peel weight (g)	Aril weight (g)	Aril yield (%)
August 15	260.8 c*	70.7 c	79.5 c	1.1	131.6 c	129.1 c	0.49
September 6	326.9 b	76.7 b	86.7 b	1.1	172.6 b	154.2 b	0.47
October 3	438.5 a	87.9 a	97.2 a	1.1	235.3 a	203.2 a	0.46

* Means followed by different letters in the same column are significantly different from each other ($p \leq 0.05$).

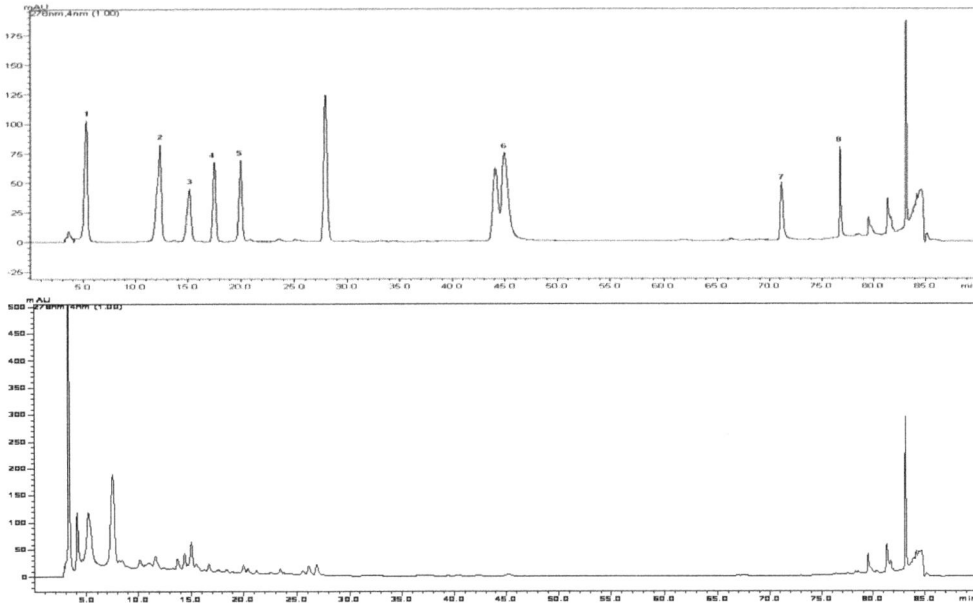

Figure 1: HPLC chromatograms of phenolic standards and extracts of pomegranate fruit juice. 1; Gallic acid 2; Catechin 3; Chlorogenic acid 4; vanilic acid 5; syringic acid, 6; ellagic acid, 7; quercetin 8; Kaempferol

Organic acid contents of pomegranate juice were analized according to the modified procedure of Alhendawi et al. (1997) and Kordis-Krapez et al. (2001). 2 ml of fruit juice was diluted with 2 ml H_3PO_4 (2%) and then 1ml of sample was diluted with 1 ml of extraction solution (0.01M KH_2PO_4, pH: 8.0). 20µl of sample was injected into an HPLC (Shimadzu Inc) equipped with SPD-10Avp UV-VIS detector (210nm), SIL-20AC prominence auto sampler, LC-20AT prominence system controller, LC-20AT prominence Pump, DGU-20A5 degasser and Prodigy ODS-2 (250x4.6mm, 5µm) column operated at $30^{\circ}C$. The mobile phase was distiled water adjusted to pH 2.25 with phosphoric acid. The flow rate was 0.8 mL/min. Peaks were identified by comparison with the peak of standard of tartaric, malic, ascorbic, citric and sucsinic acids (Sigma

Chemical Co) (Figure 2). The organic acids were expressed as µg per g fruit juice.

Data analysis: The experiment was planned according to a completely randomized design with three replications. The data were subjected to the analysis of variance using the MINTAB software (MINITAB Inc.) and the means were separated from each other by Tukey's test at the 5 % level of significance.

RESULTS AND DISCUSSION

Fruit weight, fruit height, fruit width, peel weight and aril weight values of pomegranate cultivar 'Hicaznar' fruits harvested in 3 different periods increased significantly with maturity. Fruit shape index and aril yield were not changed significantly by maturity stages in the study (Table 2). The results obtained in the study were found to be similar to those of Özsayın (2012). It has been determined that the fruit weight increased about 2 times in the last 48 days (from 15 August to 3 October).

Table 3. Changes in values of fruit peel color of pomegranate cultivar 'Hicaznar' harvested at 3 different periods

Harvest Date	L^*	a^*	b^*
August 15	55.88 b*	-1.28 c	33.18
September 6	59.47 a	13.31 b	33.09
October 3	58.04 ab	30.49 a	31.39

* Means followed by different letters in the same column are significantly different from each other ($p \leq 0.05$).

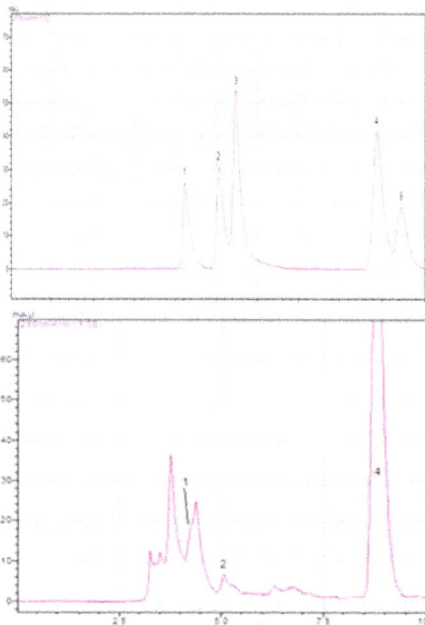

Figure 2: HPLC chromatograms of organic acid standards and extracts of pomegranate fruit juice. 1; tartaric acid 2; malic acid 3; ascorbic acid 4; sitric acid 5; succinic acid

With regard to the fruit peel color, the b* value did not change while a significant increase in a* value was determined with maturity. A relative increase was observed with maturity in terms of L* value (Table 3). It has been determined that the highest pH value (2.77) of the fruit juice was in the commercial harvesting period (October 3). Correspondingly, the content of titratable acidity has also decreased regularly with the progress of the maturity. As expected, the total soluble solid content of juices increased steadily with the progress of the maturity and this increase was found to be statistically significant. The content of total soluble solid was found to be 15.7% for the fruits harvested in the commercial harvesting period. Similar to our findings, Kulkarni and Aradhya (2005) reported that the content of total soluble solids increased, while the content of titretable acidity decreased with the progress of the maturity in pomegranate. In another study, it was reported that the total soluble solid content of pomegranate cultivar 'Hicaznar' varied between 14% and 18.2% in the commercial harvest time (Özsayın, 2012), which was similar to our results. The malic acid content of fruits significantly increased with the progress of the maturity. While malic acid content was 317.2 µg/g on August 15, this value increased to 545.2 µg/g at commercial harvest date (October 3). The citric acid content of the fruit was 25323 µg/g on August 15 and decreased by half to 12666 µg/g at the time of commercial harvest (October 3). Tartaric acid was not found in the fruits harvested on August 15 and September 6, but was found to be 466.8 µg/g on October 3 (Table 4). The contents of malic, sitric and tartaric acids of fruits observed in our study were similar to those reported by Karaca (2011). Karaca (2011) reported that sitric acid, malic acid and tartaric acid contents of mature fruit were 17360 µg/g, 500 µg/g and 590 µg/g, respectively. It has also been reported that the content of organic acids in myrtle decreases with maturity (Mulas et al., 2013).

The total phenolic contents of 'Hicaznar' fruit juice were found to be higher in the immature fruit than in the mature fruit. In the study, the total phenolic content in fruit juice was the highest (8308 ± 335 µg / g) at the beginning of coloring (15 August) and decreased significantly with maturity. Similarly, the amounts of gallic acid and ellagic acid in the fruit juice were found to be the highest (97.2 and 13.6 µg/g, respectively) in fruits harvested on August 15 and were reduced to 29.47 and 4.46 µg/g, respectively in fruits harvested on October 3 (commercial harvest date). There were no statistically significant differences in the fruits harvested at the different maturity stages in terms of chlorogenic and syringic acids (Table 5). In support of our results, Al-Maiman and Ahmad (2002) found that the total phenolic amount of pomegranate fruits decreased significantly with the progress of maturity.

Table 4. Changes in pH, total soluble solid, titretable acidity and orhanic acids of pomegranate cultivar 'Hicaznar' harvested at 3 different periods

Harvest Date	pH	Total Soluble Solid (%)	Titretable Acidity (%)	Malic acid (µg/g)	Sitric acid (µg/g)	Tartaric acid (µg/g)
August 15	2.70 ab*	10.93 b	3.07 a	317.2 b	25323 a	-
September 6	2.57 b	11.76 b	2.49 ab	354.6 b	16271 ab	-
October 3	2.77 a	15.70 a	1.95 b	545.2 a	12666 b	466.8

* Means followed by different letters in the same column are significantly different from each other ($p \leq 0.05$).

Table 5. Changes in phenolic compounds of pomegranate cultivar 'Hicaznar' harvested at 3 different periods

Harvest Date	Total phenolics (µg/g)	Gallic Acid (µg/g)	Chlorogenic Acid (µg/g)	Syringic Acid (µg/g)	Ellagic Acid (µg/g)
August 15	8308 a*	97.2 a	83.9	5.7	13.6 a
September 6	5896 b	50.8 ab	84,2	6.3	4.8 b
October 3	5696 b	29.5 b	66.3	5.0	4.5 b

* Means followed by different letters in the same column are significantly different from each other ($p \leq 0.05$).

Similarly, Siriamornpun et al. (2015) reported that immature green fruits contained more total phenolic substance than mature fruits in jujube. Moreover, it was reported that total phenolic, gallic acid and ellagic acid contents of myrtle decreased with the progress of maturity which was similar to our results (Fadda and Mulas, 2010; Babou et al., 2016). On the other hand, unlike our findings, it was reported that gallic acid content did not change with the progress of maturity in 'Hicaznar' (Özhan-Tümer, 2006). It has been found that the content of flovonols in *Morus alba* species generally decreases with maturity (Lee and Choi, 2012), but not in *Vitis vinifera* species (Doshi et al., 2006). It has been reported in many studies that the amounts of phenolic substances in fruit species may vary significantly according to genotype, ecological conditions and analysis method (Karaca, 2011; Wang et al., 2010).

CONCLUSIONS

As a result, the biochemical contents of the 'Hicaznar' fruits vary significantly with their maturity level. Malic acid content increases with maturity, while citric acid content decreases. Tartaric acid was not detected in immature fruit. It has been found that the total phenolic substance, gallic acid and ellagic acid contents decrease with maturity. Fruits should still be harvested at commercial harvest time for fresh consumption, even though the amount of some phenolics and total phenolic substance are redused with progress of maturity.

ACKNOWLEDGEMENTS

The Research Project Coordination Unit of Suleyman Demirel University financially supported the presented study under the project number 4586-YL2-16.

REFERENCES

Alhendawi R.A., Römheld V., Kirkby E.A., Marschner H., 1997. Influence of increasing bicarbonate concentrations on plant growth, organic acid accumulation in roots and iron uptake by barley, sorghum and maize. Journal of Plant Nutrition, 20 (12): 1731-1753.

Al-Maiman S.A., Ahmad D., 2002. Changes in physical and chemical properties during pomegranate (*Punica granatum* L.) fruit maturation. Food Chemistry, 76: 437-441.

Ancos B., González E.M., Cano M.P., 2000. Ellagic acid, vitamin c, and total phenolic contents and radical scavenging capacity affected by freezing and frozen storage in raspberry fruit. Journal of Agricultural and Food Chemistry, 48(10): 4565-4570.

Babou L., Hadidi L., Grosso C., Zaidi F., Valentao P., Valentao P.B., 2016. Study of phenolic composition and antioxidant activity of myrtle leaves and fruits as a function of maturation. European Food Research Technology, 242: 1447-1457.

Benzie I.F., 2003. Evolution of dietary antioxidants. Comparative Biochemistry and Physiology Part A: Molecular and Integrative Physiology, 136(1): 113-126.

Caponio F., Alloggio V., Gomes T., 1999. Phenolic compounds of virgin olive oil: influence of paste preparation techniques. Food Chemistry, 64: 203-209.

Choi S.H., Ahn J.B., Kim H.J., Im N.K., Kozukue N., Levin C.E., Friedman M., 2012. Changes in free amino acid, protein, and flavonoid content in jujube (*Ziziphus jujuba*) fruit during eight stages of growth and antioxidative and cancer cell inhibitory effects by extracts. Journal of Agriculture and Food Chemistry, 60: 10245-10255.

Doshi P., Adsule P., Banerjee K., 2006. Phenolic composition and antioxidant activity in grapevine parts and berries (*Vitis vinifera* L.) cv. Kishmish Chornyi (Sharad Seedless) during maturation. International Journal of Food Science and Technology, 41: 1-9.

Fadda A., Mulas M., 2010. Chemical changes during myrtle (*Myrtus communis* L.) fruit development and ripening. Scientia Horticulturae, 125(3): 477-485.

Gao Q.H., Wu C.S., Wang M., Xu B.N., Du L.J., 2012. Effect of drying of jujubes (*Ziziphus jujuba* Mill.) on the contents of sugars, organic acids, alpha-tocopherol, beta-carotene, and phenolic compounds. Journal of Agricultural and Food Chemistry, 60: 9642-9648.

Hochstein P., Atallah A.S., 1988. The nature of oxidants and antioxidant systems in the inhibition of mutation and cancer. Mutation Research/ Fundamental and Molecular Mechanisms of Mutagenesis, 202(2): 363-375.

Karaca E., 2011. The effect on phenolic compounds applied some of the procedures during production of pomegrane juice concentrate. Institute of Natural and Applied Science, University of Çukurova, MSc. Thesis, 144 pages, Adana, Turkey.

Karaçalı İ., 2010. Bahçe Ürünlerinin Muhafaza ve Pazarlanması (Storage and Marketing of Horticulture Products). Ege University, Faculty of Agriculture, Publication Number: 494 (Book in Turkish).

Kordis-Krapez K.M., Abram V., Kac M., Ferjancic S., 2001. Determination of organic acids in white wines by RP-HPLC. Food Technology and Biotechnology, 39(2): 93-99.

Kulkarni A.P., Aradhya S.M., 2005. Chemical changes and antioxidant activity in pomegranate arils during fruit development. Food Chemistry, 93: 319-324.

Lee W.J., Choi S.W., 2012. Quantitative changes of polyphenolic compounds in mulberry (*Morus alba* L.) leaves in relation to varieties, harvest period, and heat processing. Preventive Nutrition and Food Science, 17: 280-285

Mulas M., Fadda A., Angioni A., 2013. Effect of maturation and cold storage on the organic acid composition of myrtle fruits. Journal of the Science of Food and Agriculture, 93: 37-44

Ozgen M., Durgac C., Serce S., Kaya C., 2008. Chemical and antioxidant properties of pomegranate cultivars grown in the Mediterranean region of Turkey. Food Chemistry, 111: 703-706

Özhan-Tümer L., 2006. Changes on the amount of phenolic compound on maturation stages of some pomegranate varieties. Institute of Natural and Applied Science, University of Çukurova, MSc. Thesis, 53 pages, Adana, Turkey.

Özsayın S., 2012. Determination of nutritional status, some fruit quality parameters and antioxidant activity of pomegranate orchards (*Punica granatum* L.) in Antalya region. Institute of Natural and Applied Science, University of Akdeniz, MSc. Thesis, 134 pages, Antalya, Turkey.

Rossi M., Giussani E., Morelli R., Lo-Scalzo R., Nani R.C., Torreggiani D., 2003. Effect of fruit blanching on phenolics and radical scavenging activity of high bush blueberry juice. Food Research International, 36: 999-1005.

Siriamornpun S., Weerapreeyakul N., Barusrux S., 2015. Bioactive compounds and health implications are better for green jujube fruit than for ripe fruit. Journal of Functional Foods, 12: 246-255.

Wang R., Ding Y., Liu R., Xiang L., Du L., 2010. Pomegranate: constituents, bioactivities and pharmacokinetics. In: Chandra R. (Ed), Pomegranate. Fruit, Vegetable and Cereal Science and Biotechnology, 4 (Special Issue 2): 77-87.

Wu C.S., Gao Q.H., Guo X.D., Yu J.G., Wang M., 2012. Effect of ripening stage on physicochemical properties and antioxidant profiles of a promising table fruit pear-jujube (*Zizyphus jujuba* Mill.). Scientia Horticulturae, 148: 177-184.

6

THE EFFECT OF CERTAIN CLIMATIC PARAMETERS ON THE APRICOT TREE

Cristina MOALE[1], Adrian ASĂNICĂ[2]

[1]Research Station for Fruit Growing Constanta, 25 Pepinierei Street,
Valu lui Traian, Romania
[2]University of Agronomic Sciences and Veterinary Medicine of Bucharest,
59 Marasti Blvd., District 1, Romania
Corresponding author email: moalecristina@yahoo.com

Abstract

The pedo-climatic conditions in the South-Eastern part of Dobrogea are favourable to the culture of the apricot tree; this species that loves warm weather has always found good conditions for growing and yielding in the South-Eastern part of Romania, and especially in Dobrogea. In this period, 6 Romanian and foreign apricot tree cultivars were studied at RSFG Constanta: 'Harcot', 'Auraş', 'Goldrich', 'Dacia', 'Fortuna' and 'Hungarian C.M.B'. Branch samples were harvested from these 6 cultivars with different ripening periods three days after the frost and they were analysed. This paper presents the manner in which certain apricot tree cultivars reacted to frost in the winter of 2012, 2013 and 2014, as well as the effect of the hail on July 11th, 2014 on the apricot production. The greatest losses caused by the frost were registered in the winter of 2012 as far as the fructiferous buds are concerned: 94% at 'Fortuna', 83% at 'Auraş', 82% at 'Goldrich' and 'Dacia', 77% at 'Harcot' and 65% at 'Hungarian C.M.B.'. The losses caused by the hail on July 11th, 2014 affected the production of the 'Dacia' cultivar by 40% and of the 'Hungarian C.M.B.' by 30%. The climatic changes that have been registered throughout the past 10 years have negatively influenced the culture of the apricot tree and the effects have been classified according to the cultivar and its biology, as well as to the topographic placement of the allotments. The studies that have been carried out, together with the obtained results demonstrate the importance of choosing the cultivar assortment taking into account the favourability of the area, as well as the importance of installing anti-hail nets when setting up fruit-growing plantations.

Key words: climate change, Prunus armeniaca L., resistance to temperature variations.

INTRODUCTION

The apricot tree is one of the most valuable fruit-growing species, due to the fact that is precocious, its production is quite significant and the fruits are very appreciated for their organoleptic qualities; are demanded by the market, both for fresh and processing, being capitalised at convenient prices.

The apricot tree has higher requirements concerning the heat, the temperature being the limitative factor in terms of extending the specie. The trees burst after a period of 7-10 days with temperatures above the biological threshold (6.5 °C), while flowering and fruit setting occur if the temperature is at least 10-12 °C (Hoza, 2000).

The studies which were carried out over a period of several years by researchers in the agro-meteorological field consider Dobrogea, as well as the entire south-eastern part of

Romania, to belong to the most favourability area for the apricot tree culture (Roman, 1992). Previous researches have revealed that the impact of climatic changes upon fruit-growing species can already be felt. For instance, by the end of the 90's, the flowering of the trees in Germany occurred several days earlier (Chmielewschi et al., 2004 and 2005). The vegetative season in Europe became longer by 10 days in the past 10 years (Chmielewschi and Rotzer, 2002). Due to the early flowering of the trees, in certain regions of Europe there was noticed an increase in the risk of damage caused by late frosts (Anconelli et al., 2004; Sunley et al., 2006; Legave and Clauzel, 2006; Legave et al., 2008; Chitu et al., 2004 and 2008) or by the disorders in the pollination and fruit setting processes (Zavalloni et al., 2006).

The Black Sea Coast is situated in the area with the largest average annual sums of day length on the country's territory, sums which surpasses 2250-2300 hours (Păltineanu et al., 2000).

Action to adapt to climate changes through an appropriate management of structure, rotation and technology of fruit crops require knowledge of regional and local characteristics of present and future climate and of assessment the associated risks. In the latest 50 years, according to the studies of the National Meteorology Administration (Bojariu, et al., 2015), the monthly average of air temperature exclusively presents growth trends, statistically significant over the whole Romania, during spring and summer. There are also rising trends of air temperature in the winter, for the central and north-eastern part of the country, but the percentage of stations showing significant trends is lower (Birsan and Dumitrescu, 2014). A gradual downward trend was also noted in the intensity of cold stress generated by minimum air temperatures below -15°C in the winter months, from 26 units of the "cold" in the 1961 to 1970 decade, to a range between 12 and 21 of "cold" units in the last four decades (1971-2010).

Chitu et al. (2015) found that between 1985-2014, the highest growth rate of both from all the months had been recorded in November, the trend (+1.3 °C in ten years for maximum and 0.94 °C for minimum) being statistically insured. It was also noted that if in the west and center part of the country, winters were becoming milder (temperature increase with more than 1.5 °C per decade, even if the trend is not statistically assured). In the eastern and especially in the southern part of Romania in the last 30 years, lower temperatures by 0.1 to 0.8 °C were recorded.

Changes in the year 2007, the whole Europe and implicit Romania will be confronted in future with a process of global warming, characterized by increasing of temperatures with -0.5 - 1.5 °C for the period 2020 – 2029 and with -2 – 5 °C for the period 2029 – 2099. In the period 2090-2099 Romania will confront with pronounced drought during the time of summer. Researches from many countries, in the frame of climatic research methodology have the approached aspects regarding climatic changes effects on growth and development of some fruit tree species (Chmielewski and

Rotzer et al., 2002; Olensen 2002; Sunley et al.2006, Chitu et al., 2010; Sumedrea et al., 2009). Climatic changes occurred also in Romania, they have determined meteorological phenomena, which are manifesting with augmented amplitude and intense frequency (severe drought, intense flooding, tornados, hail).

Throughout the entire world, the research concerning the apricot tree has among its main objectives the relationship between the climatic conditions and the apricot tree culture. In our country, this relationship has been studied by numerous authors: Burloi (1957), Bordeianu et al. (1961), Cojean (1961), Mănescu et al. (1975), Topor (1987, 2002, 2009), Stancu et al. (1989), Roman et al. (1992) and so on. The results obtained by all these studies corresponded to a certain period of time and to a certain assortment of cultivars which represented the material. As concerns the new apricot tree cultivars, the obtained results are correlated with the evolution of the main climatic elements recorded in the past ten years.

This paper deals with the manner in which frost and hail influenced the fruit production of certain apricot tree cultivars from Dobrogea in the years 2012, 2013 and 2014.

That is why each area promotes a specific assortment of cultivars, although some of the latter might be common for all the areas (Topor, 1995).

MATERIALS AND METHODS

The biological material utilised consisted of 6 apricot tree cultivars and represented 3 ripening groups:
- extra early: 'Fortuna', 'Auraş';
- early: 'Harcot', 'Goldrich';
- medium: 'Dacia', 'Hungarian C.M.B.'

Upon comparing the ripening period of the fruit we can observe that the 'Fortuna' and 'Auraş' ripened in the period between the 16th and the 30th of June, the 'Harcot' and 'Goldrich' cultivars ripened in the period between the 20th of June and the 10th of July, whereas the period for the 'Dacia' and 'Hungarian C.M.B.' cultivars ripened in the period between the 7th and the 20th of July.

The studied cultivars are new and were created at RSFG Constanta: 'Fortuna' and 'Auraş', as well as the apricot tree cultivars introduced in the zonal and national assortment: 'Harcot', 'Goldrich', 'Dacia' and 'Hungarian C.M.B.'

The trees were planted at a distance of 5 m between rows and 4 m between trees within the row (500 trees/ha).

The canopy shape is a Veronese vase and the trees were planted in 2003.

The applied culture technology is the one specific to the apricot tree: pruning, phytosanitary treatments, soil works, irrigation, harvesting, conditioning and capitalisation of the fruit.

Due to climatic changes over the past few years, the resistance of apricot trees seems to have become very different from one year to another. However, there are other factors involved as well, such as the topographic position of the orchard lot in which the apricot trees are planted (in the case of the studied cultivars the land was the same – a plateau), the alternation between minimum and maximum temperatures during winter, which renders the trees less resistant and last but not least, the severity of climatic accidents. The numbers of studied buds in the three years are shown in Figure 1.

The observations and determinations were carried out in the plots where there are some of the promoted apricot tree cultivars. Branch samples from the 6 cultivars were collected and analysed. The degree of differentiation of the flowering buds was relatively good. As concerns the soil where the plantation is placed, it is a calcareous chernozem (CZKa), with a loamy texture and a low alkaline pH (8.2) on its entire profile.

In addition, the overall climatic conditions were favourable to the growth and fructification of the trees, with exception of the years 2012-2014, when a very strong frost was registered in both January and February, leading to the loss of some of the floriferous buds, while the hail on July 11th, 2014 affected the production of the 'Dacia' and 'Hungarian C.M.B.' cultivars. With regard to these cultivars we observed the main fructification phenophases: the beginning of the blossoming, upon the appearance of the pink button; the beginning of the flowering, upon the appearance of the first open flowers; the ending of the flowering, when most of the flowers have lost their petals.

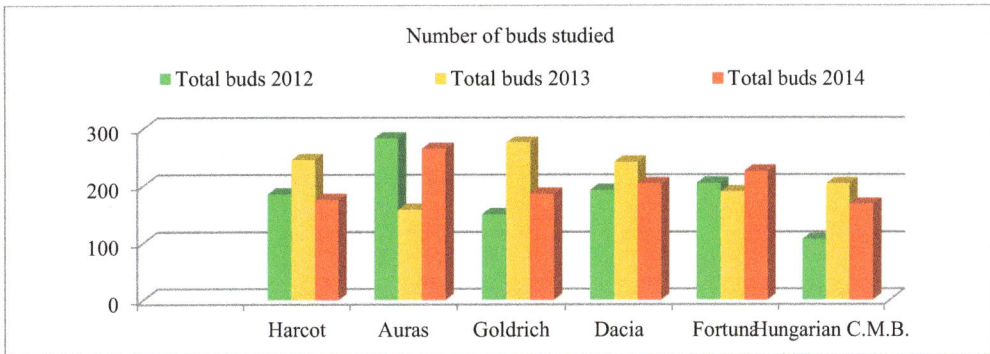

Figure 1. The number of studied buds per cultivar in the years 2012, 2013, 2014.

The duration of the flowering phenophase at a certain cultivar can vary according to the action of the maximum temperatures during the day and the intensity of the wind, correlated with the degree of differentiation of the trees (i.e. the amount of flowers per tree). The intensity of the flowering was graded on scale from 0 to 5, 0 being used when the cultivars displays no flowers at all, while 5 is used when the cultivar displays a plethora of flowers. The hardening of the core was determined by means of piercing it with a needle at regular intervals, usually 2 days. The process was carried out progressively, in the same day for all the observed cultivars. The harvesting maturity is largely influenced by a series of climatic and agro-technical factors, such as: temperature, drought, quantity of fruit per tree, shape of the head, density of the trees etc. The observations and determinations were carried out 3-5 days after the climatic accidents recorded in 2012, 2013 and 2014, respectively and the production was assessed after the hail occurrence on July 11th, 2014. The hail, with a dimension of approximately 5-20 mm, seriously damaged the fruit production of some of the apricot tree cultivars, more exactly those who had not been harvested until July 11th, 2014. The climatic data were recorded with the aid of an automatic meteorological station (the WatchDog type) and were processed as daily averages. We noticed that the resistance of apricot tree cultivars differs from one year to the next because of the climatic changes that have occurred during the past few years and it depends on the gravity of climatic accidents. The minimum and maximum temperatures during winter alternate and together with the gravity of climatic accidents lead to the weakening of the trees.

The behaviour of apricot tree cultivars towards the attack of the pathogen agents - 1) *Stigmina carpophila* (Lév.) M.B. Ellis, 2) *Cytospora cincta* Sacc andi 3) *Monilinia laxa* (Aderhol et Ruhl.) - was studied under conditions of natural infections, according to the test created by Crossa Raynaud (1968).

The evaluation technique consisted in writing down the frequency of the attacked organs and the intensity with which the symptoms manifested themselves and these two aspects were utilised in assessing the behaviour of the cultivars.

The field observations were centred on the calculation of the pathogens' frequency (F %) and intensity (I) on different tree organs such as: leaves, flowers, shoots, branches and fruits. For the intensity of the diseases marks were granted on a scale from 0 to 4.

Depending on the frequency and intensity of the disease, the studied cultivars and hybrids were categorised into 4 classes and 8 groups of resistance according to the following scale (Table 1).

F.A.= cultivars without attack (F%= 0 and I= 0)
T= tolerant cultivars (F%= 0.1-5% and I= $0 ^{\pm} +$)
S.A.= weakly attacked cultivars (F%= 5.1% - 10% and I= +)
M.A.= moderately resistant cultivars (F%= 10.1% - 25% and I= +)
S=sensitive cultivars (F%=25.1–50% and I=+ ^{2} 4)
F.S.= highly sensitive cultivars (F%= 50.1% - 100%, I= + ^{4} 4)

Table 1. Cultivar categorisation into classes and groups of resistance

Resistance class	Resistance group	Frequency (F%)	Intensity (I%)
1= tolerant (T)	1	0	0
2= medium resistance (MR)	2	0.1-11.0	+
	3	11.1-25.0	+
3= sensitive (S)	4	25.1-34.0	$+^{\pm} 2$
	5	34.1-50.0	$+^{1} 2$
4= very sensitive (VS)	6	50.1-59.0	$+^{2} 3$
	7	59.1-75.0	$+^{3} 3$
	8	75.1-100	$+^{4} 4$

RESULTS AND DISCUSSION

The triggering of the main fructification phenophases in the years 2012-2014 occurred between rather wide limits, according to the characteristics of the cultivar and the climatic characteristics of the studied years.

In the period 2012-2014 the blossoming of the floriferous buds of the apricot trees occurred between the following limits: between 08.03 and 14.03 for the 'Fortuna' cultivar, between 11.03 and 27.03 at the 'Auraș' cultivar, between 15.03 and 27.03 at the 'Harcot' cultivar, between 24.03 and 29.03 at the 'Goldrich' cultivar, between 13.03 and 28.03 at the 'Dacia' cultivar, between 24.03 and 29.03 at the 'Hungarian C.M.B.' cultivar. Calendaristically the blossoming at the apricot tree occurred between 08.03 and 29.03 (21 days) in the studied years 2012-2014 (Table 2).

Table 2. The main stagers of fructification and apricot in the 2012-2014 period

No.	Cultivar	Year	The swelling of the flowering buds	The flowering			Inten-sity	The hardening of the stone	Harvesting maturity
				Beginning	Ending	Duration (days)			
1	Fortuna	2012	08.03	18.03	05.04	17	1	04.06	14.06
		2013	14.03	16.03	01.04	15	1	10.06	27.06
		2014	12.03	21.03	02.04	11	1	07.06	25.06
		Limits	08.03-14.03	16.03-21.03	01.04-06.04	11-17	1	04.06-10.06	14.06-25.06
2	Auraş	2012	11.03	22.03	06.04	14	2	04.06	12.06
		2013	27.03	28.03	10.04	12	3	08.06	17.06
		2014	22.03	17.03	03.04	16	4	10.06	29.06
		Limits	11.03-27.03	17.03-28.03	03.04-10.04	12-16	2-4	04.06-10.06	12.06-29.06
3	Harcot	2012	15.03	25.03	10.04	15	2	02.06	18.07
		2013	29.03	28.03	16.04	18	2	08.06	16.07
		2014	26.03	20.03	03.04	13	4	08.06	27.07
		Limits	15.03-27.03	20.03-28.03	03.04-16.04	13-18	2-4	02.06-08.06	16.07-27.07
4	Goldrich	2012	26.03	04.04	17.04	13	2	06.06	13.07
		2013	29.03	09.04	23.04	14	2	10.06	18.07
		2014	24.03	20.04	28.04	8	2	08.06	16.07
		Limits	24.03-29.03	04.04-20.04	10.04-28.04	8-14	2	06.06-10.06	13.07-18.07
5	Dacia	2012	22.03	05.04	20.04	15	3	08.06	03.07
		2013	28.03	11.04	19.04	8	3	10.06	02.07
		2014	13.03	20.04	30.04	10	4	07.06	12.07
		Limits	13.03-28.03	05.04-20.04	19.04-30.04	8-15	2-4	07.06-10.06	02.07-12.07
6	Hungarian C.M.B.	2012	24.03	05.04	18.04	13	4	07.06	17.07
		2013	29.03	09.04	16.04	7	3	09.06	15.07
		2014	25.03	18.04	27.04	9	4	10.06	19.07
		Limits	24.03-29.03	05.04-18.04	16.04-27.04	7-13	3-4	07.06-10.06	15.07-19.07

The beginning of the flowering. For all the studied cultivars the beginning of the flowering in the period 2012-2014 was recorded; however, the cultivars entered this phenophases at different times, albeit not necessarily significant (a few days from one cultivar to the next), so that mutual pollination was fully ensured. The limits for this phenophase were 16.03 and 20.04.

The ending of the flowering. In the studied period 2012-2014 the ending of the flowering occurred between 01.04 and 06.04 for the 'Fortuna' cultivar, between 03.04 and 10.04 for the 'Auraş' cultivar, between 03.04 and 16.04 for the 'Harcot' cultivar, between 10.04 and 28.04 for the 'Goldrich' cultivar, between 19.04 and 30.04 for the 'Dacia' cultivar, between 16.04 and 27.04 for the 'Hungarian C.M.B.' cultivar. The dates were recorded as the days when the flowers lost their last petals.

The duration of the flowering at the peach tree (average for the three studied years) expressed in number of days varied between 7 days (the 'Hungarian C.M.B.' cultivar in 2013) and 18 days (the 'Harcot' cultivar in 2013).

The intensity of the flowering. In 2012, 2013 and 2014 the following cultivars displayed a weak intensity of the flowering: 'Fortuna' - 1 (2012, 2013 and 2014), 'Auraş' - 2 (2012), 'Harcot' - 2 (2012, 2013) and 'Goldrich' - 2 (2012, 2013 and 2014).

The hardening of the core. This phenophase occurred in the first half of the month of June (between 2th and 10th) in the years 2012, 2013 and 2014.

The harvesting maturity. Each ripening period has large variation limits from one year to another, depending on how the climatic factors determine the type of vegetation in a specific year: early, late or extra late. The harvesting maturity of the fruit had as variation limits the 14th of June and the 27th of July.

As we can notice in figure 2a, January 2012 was the coldest month, during which 9 days recorded daily average temperatures ranging from -10.2 °C and -17.6 °C. These values, together with those that were extremely varied in February (7 days with daily average temperatures of -10.4 and -16.4 °C) and eight consecutive days of hoarfrost, the ice on the branches caused the loss of 65% - 94% of the floriferous buds at the studied cultivars.

Figure 2b. reveals the fact that the coldest month in the period September 2012 - April 2013 was January 2013, when the recorded values were -13.7 °C (January 10th, 2013). These values did not significantly influence the loss of floriferous buds at the apricot tree cultivars (local observations).

In the period October 2013 - March 2014 (Figure 2c.) the lowest temperature was recorded in January: -17.6 °C (January 30th, 2014); another day when the recorded temperature was low (-9.4 °C) was February 5th, 2014. The low temperatures recorded during this period affected the 'Goldrich' cultivar (67%) and the 'Fortuna' cultivar (90%).

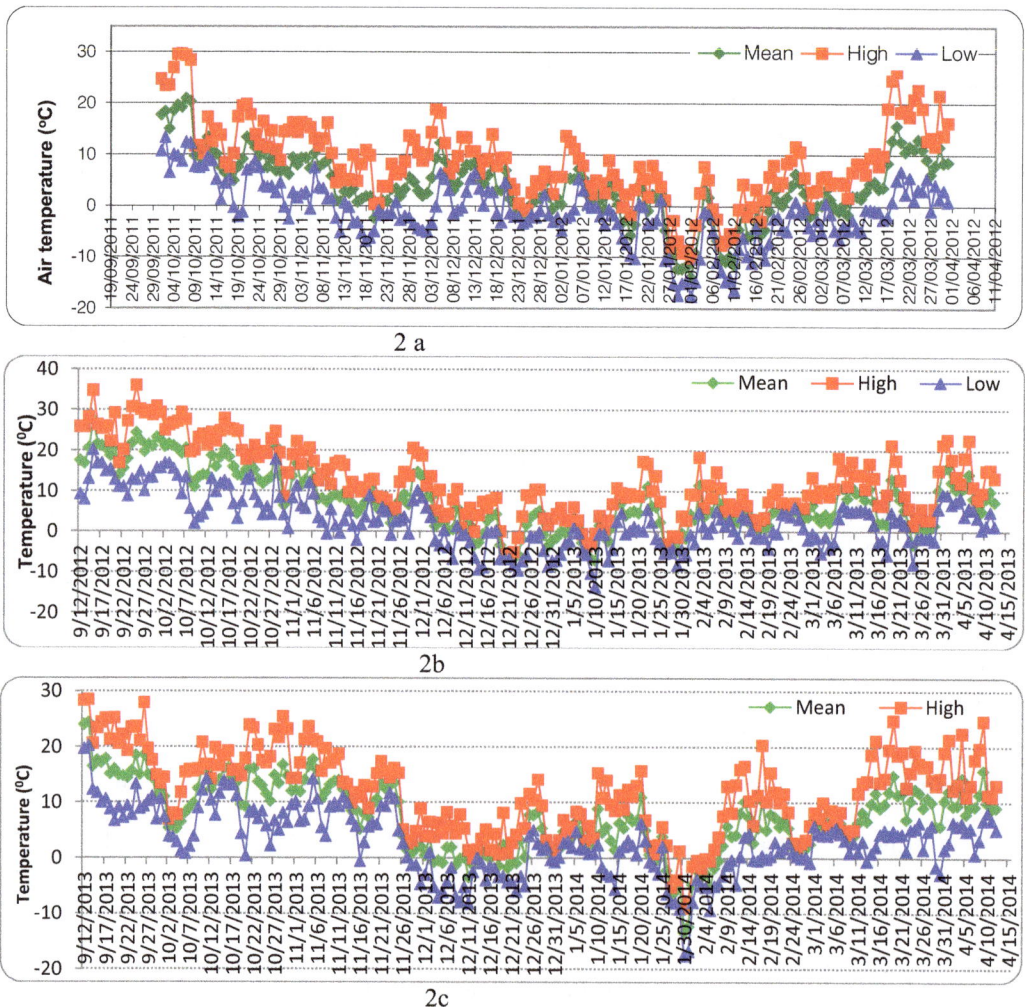

2 a

2b

2c

Figure 2. Air temperature (°C) in the cold period October 2011 – March 2012 (a), October 2012 – March 2013 (b) and October 2013 – March 2014 (c) at Valu lui Traian, Constanța

The observations were carried out with the aim of assessing the losses of floriferous buds because of temperature variations during winter and the low temperatures during the day.

Thus, for the 'Harcot' cultivar the losses recorded for 2012 were of approximately 77%, 58% for 2013 and 42% for 2014, there being difference from one cultivar to another. The winter frost caused losses for the 'Goldrich' cultivar of 82% in 2012, 69% in 2013 and 67% in 2014. For the 'Dacia' cultivar, the losses were of 82% in 2012, 63% in 2013 and 28% in 2014. The 'Auraş' cultivar recorded losses of 83% in 2012, 65% in 2013 and 24% in 2014. For the 'Fortuna' cultivar, the losses were of 94% in 2012, 92% in 2013 and 90% in 2014. The 'Hungarian C.M.B.' cultivar recorded losses of 65% in 2012, 54% in 2013 and 15% in 2014 (Figure 3).

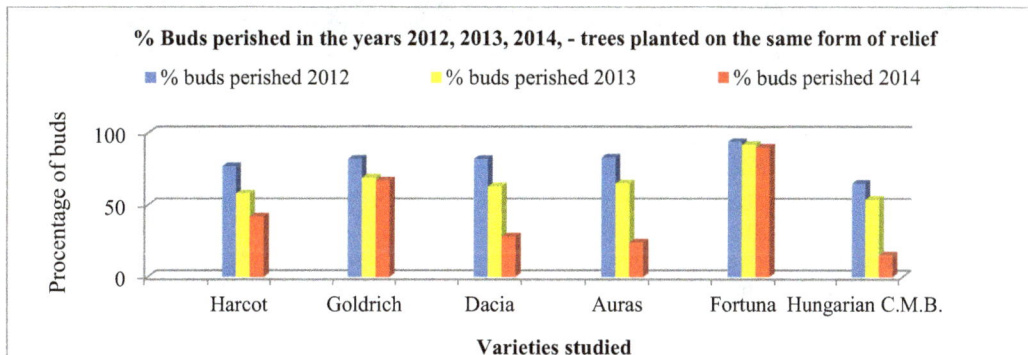

Figure 3. Procentage of apricot tree flowering buds perished due to frosts during the winter of 2012, 2013 and 2014 at Valu lui Traian, Constanta

We must bear in mind the fact that the losses caused by the winter frost of 2012, together with those caused by hoarfrosts and late frosts were very severe, taking also into account the surface of the Station's orchards cultivated with this cultivar. These losses were also caused by the warm period before the frost – in the first three weeks of January 2012 the average temperature of the air was positive, of approximately 5 °C.

A good resistance to frost during the winter of the three studied years was displayed by the apricot cultivar, with the following percentages: 'Hungarian C.M.B.'- 45%, (Figure 4).

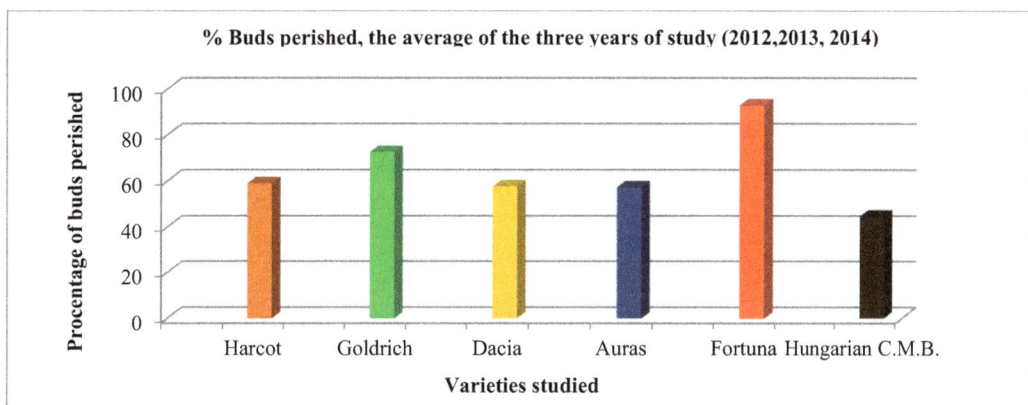

Figure 4. Procentage of apricot tree flowering buds perished because of frosts (average over the three years), Valu lui Traian

In these conditions, the 'Auraş', 'Dacia' and 'Harcot' cultivars were more than 50% damaged. The climatic accidents recorded in January and February 2012 (sudden temperatures of -16.4 °C, minimum temperature during the day) and 8 days of hoarfrost caused the damaging of the production for the early cultivars 'Fortuna' and 'Goldrich'.

At R.S.F.G. Constanta, in the second week of June 2014, more exactly on July 11th, the amount of precipitations was accompanied for 10 minutes by hail, which affected 40% of the fruit production for the 'Dacia' cultivar (the fruit were just beginning to ripe) and 30% for the 'Hungarian C.M.B.' cultivar (Figure 5). The hail bruised the fruit, the shoots and the stems, thus creating a good environment for future infections and diseases. The bruises on the fruit, despite some of them becoming scars, diminished the commercial aspect and the quality of the production.

Figure 5. The 'Dacia' cultivar affected by the hail on July 11th, 2014 (full maturity)

Table 3, presents the relative sensitivity of 6 cultivars from the demonstrative plot created within the laboratory responsible with improving the apricot tree concerning the attack of the pathogens *Stigmina carpophila*, *Cytospora cincta* and *Monilinia laxa*, under natural conditions of infection. The analysis of the data in this table highlights a variation in the apricot tree cultivars' behaviour towards a pathogen or another.

Table 3. The behaviour of apricot tree cultivars towards the attack of the main pathogens in the period 2012-2014

No.	Cultivar	Year	Intensity of attack (note)			
			Stigmina carpophila	*Cytospora cincta*	*Monilinia laxa*	*Monilinia fructigena*
a. Cultivars planted in 2003						
1.	Harcot	2012	FA	T	T	T
		2013	FA	T	T	T
		2014	FA	T	T	T
2.	Auraş	2012	T	FA	FA	FA
		2013	T	FA	FA	FA
		2014	T	FA	FA	FA
3.	Goldrich	2012	FA	S	T	T
		2013	FA	SA	T	T
		2014	FA	SA	T	T
4.	Dacia	2012	FA	S	SA	SA
		2013	FA	SA	SA	SA
		2014	FA	SA	S	S
5.	Fortuna	2012	SA	FA	FA	FA
		2013	SA	FA	FA	FA
		2014	SA	FA	FA	FA
6.	Hungarian C.M.B.	2012	FA	T	SA	SA
		2013	FA	T	SA	SA
		2014	FA	T	S	S

According to the intensity (I) of the attack the studied cultivars were classified as follows (Table 3): *Stigmina carpophila* in 6 resistance classes: **cultivars without attack (F.A.)** – 4 cultivars belong to this class, where the intensity (I) of the attack was 0: 'Harcot', 'Goldrich', 'Dacia' and 'Hungarian C.M.B.', during all three studied years (2012, 2013 and 2014). The 'Auraş' cultivar was not attacked by *Stigmina carpophila*; **tolerant cultivars (T)** – 'Auraş' in all three studied years; **weakly**

attacked (S.A) - 'Fortuna' (in 2012, 2013 and 2014.

The *Stigmina carpophila* fungus survives from one year to the next in buds, in cracks in the bark and in wounds, which facilitates the occurrence of infections in early spring, at temperatures above 8 °C and an atmospheric humidity of over 80%. Temperatures over 30°C stop the evolution of the disease.

By correlating the frequency of the attack (F%) of *Cytospora cincta* Sacc with its intensity (I), the studied cultivars were categorised as follows: **cultivars without attack (F.A.)** - both the frequency (F%) and the Intensity (I) were 0; there are 2 cultivars in this class: 'Auraş', and 'Fortuna', in all three studied years; **tolerant cultivars (T)** – 'Harcot' and 'Hungarian C.M.B.'. in the studied years 2012, 2013 and 2014; **weakly attacked (S.A)** – 'Goldrich' and 'Dacia' were weakly attacked in 2013 and 2014; **sensitive (S)** – 'Goldrich' and 'Dacia'. None of the studied cultivars could be categorised into the classes **moderately resistant (M.A.)** and **highly sensitive (F.S.)**. The disease caused by the *Cytospora cincta* Sacc fungus manifested itself through the sudden drying of the branches. The affected trees seemed to be suffering in full summer the leaves wilted, becoming yellowish-green and the drying of the branches occurred from the tip towards the base.

As far as the attack of the *Monilinia laxa* and *Monilinia fructigena* fungi is concerned (Aderh et Ruhl) Honey, the disease manifested itself on all aerial organs during spring, through the wilting of the flowers and the drying of the vegetative buds and shoots.

The strong attack of the *Monilinia laxa* fungus occurs when the maximum phase of pathogenicity or the fungus is corroborated with the maximum phase of receptivity of the trees (the flowering phase), directly conditioned by temperature and humidity. This explains why the intensity of the attack on the same cultivar varies from one year to another, remaining, though, within the limits of the minimal, medium or maximal attack levels. Certain factors favour the attack of the *Monilinia*, among which: the biological reserve of the pathogen from one year to the next, given the fact that the fungus winte rs as a mycelium within the branches, at the base of

the flowering buds, or in mummified fruit that are still on the branches; the atmospheric humidity and the air temperature during the flowering; the genetic resistance of the cultivars.

By correlating the frequency of the attack (F %) with its intensity (I), the studied cultivars were categorised as follows: **cultivars without attack (F.A.)** – 2 cultivars, for which both the frequency (F%) and the intensity (I) were 0: 'Auraş' and 'Fortuna', in all three studied years; **tolerant cultivars (T)** - 2 cultivars: 'Harcot' and 'Goldrich', in 2012, 2013 and 2014; **weakly attacked (S.A)**: 'Dacia' and 'Hungarian C.M.B.' in 2012 and 2013; **sensitive (S)** – 'Dacia' and 'Hungarian C.M.B.' in 2014. None of the studied cultivars could be categorised into the classes **moderately resistant (M.A.)** and **highly sensitive (F.S.)**.

Most studied cultivars manifested an increased resistance towards the attack of the pathogen agent *Monilinia laxa*, with the exception of 'Dacia' and 'Hungarian C.M.B.' cultivars, which were sensitive in 2014. These cultivars were affected by the hail that occurred on July 11th, 2014, which facilitated the development of moniliasis especially at cultivars that were in full harvesting maturity.

CONCLUSIONS

The greatest yield losses were recorded for the 'Fortuna' cultivar in 2012 – 94% in 2013 - 92% and 90% in 2014.

The lower losses during the three studied years were recorded by 'Hungarian C.M.B'.

The hail from July 11th, 2014, which lasted for only 10 minutes, affected the 'Dacia' cultivar (40%) and the 'Hungarian C.M.B.' cultivar (30%) and therefore we recommend the orchards to be covered with anti-hail nets.

As far as the attack of the *Stigmina carpophila* (Lév.) M.B. Ellis fungus is concerned, the following cultivars: 'Harcot', 'Goldrich', 'Dacia' and 'Hungarian C.M.B.' – proved to be tolerant in all three studied years (2013-2015).

Concerning the attacks of the *Cytospora cincta* fungus, both the sensitivity and the resistance towards the pathogen are exclusively connected to the soil.

The sensitive cultivars were 'Goldrich' and 'Dacia' in 2012, while the resistant ones, without attack were 'Auraş' and 'Fortuna' (2013, 2014, 2015).

Most studied cultivars manifested an increased resistance towards the attack of the pathogen agent *Monilinia laxa*, with the exception of 'Dacia' and 'Hungarian C.M.B.' cultivars, which were sensitive in 2014. These cultivars were affected by the hail that occurred on July 11th, 2014, which facilitated the development of moniliasis especially at cultivars that were in full harvesting maturity.

REFERENCES

Anconelli S. Antolini G. Facini O. Giorgiadis T.Merletto V. Nardino M. Palara U. Pasquali A.Pratizzoli W. Reggitori G.Rossi F. Sellini A. Linoni F, 2004: Previsione e difesa dalle gelate tardive – Risultati finali del progetto DISGELO. CRPV Diegaro di Cesena (FO). Natiziario tecnico N.70. ISSN 1125-7342. 64. pp.

Bordeanu, T., Tranavski, I., Radu, I.F. 1961. Study on Winter Rest and Biological Threshold of Apricot Flower Buds. Studies and research of Biology. Plant Biology Series no. 4 (XIII).

Burloi Niculina. 1957. Behavior apricot buds in the Bucharest. Communications of the Romanian Academy no. 9, vol. VII, Bucharest

Birsan, M.V., Dumitrescu, A., 2014. ROCADA: Romanian daily gridded climatic dataset (1961-2013) V1.0. National Meteorological Administration, Bucharest Romania, doi: 10.1594/PANGAEA.833627.

Bojariu, R., Bîrsan, V.M., Cică, R., Velea, L., Burcea, S., Dumitrescu, A., Dascălu, S.I., Gothard, M., Dobrinescu, A., Cărbunaru, F., Marin, L., 2015. Climatical changes - from physical bases to risks and adaptation. Printech Publishing House, Bucharest, ISBN 978-606-23-0363-1, 200 pp.

Chitu E., M. Butac, S. Ancu and V.Chitu 2004. Effects of low temperatures in 2004 on the buds viability of some fruit species grown in Maracineni area. Annals of the University of Craiova. Vol. IX (XLV), ISSN 1435-1275: 115-122.

Chiţu E., D. Sumedrea, Cr. Pătineanu, 2008. Phenological and climatic simulation of late frost damage in plum orchard under the conditions of climate changes foreseen for Românca. Acta Horticulturae (ISHS) 803:139-146.

Chitu, E., Giosanu D., Mateescu E., 2015. The Variability of Seasonal and Annual Extreme Temperature Trends of the Latest Three Decades in Romania. Agriculture and Agricultural Science Procedia. Volume 6, 2015, doi:10.1016/j.aaspro.2015.08.113, pages 429-437.

Chiţu, E., Elena Mateescu, Andreea Petcu, Ioan Surdu, Dorin Sumedrea, Tănăsescu Nicolae, Cristian Păltineanu, Viorica Chiţu, Paulina Mladin, Mihail

Coman, Mădălina Butac, Victor Gubandru, 2010. Methods of estimating climatic favorability for tree culture in Romania. The Publishing House INVEL Multimedia, CNCSIS accredited, ISBN 978-973-1886-52-7.

Chmielewski F.M., Muller A., Kuchler W., 2005. Possible impacts of climate change on natural vegetation in Saxony (Germany). Int. J. Biometeorol, 50:96-104.

Chmielewski F.M., Rotzer T 2002. Annual and spatial variability of the begenning of growing season in Europe in relation to air temperature changes. Clim. Res. 19(1), 257-264.

Chmielewski F.M., Muller A., Bruns E., 2004. Climate changes and trends in phenology of fruit trees and field crop in Germany, 1961-2000, Agricultural and Forest Meteorology 121 (1-2), 69-78.

Cojean Natalia. 1961. Rezistenţa la îngheţ a mugurilor floriferi de cais în legătură cu etapele de creştere şi de dezvoltare. Studii şi cercetări de Biologie. Seria Biologie vegetală nr.1.

Crossa-Raynaud, P.H. (1969). Evaluating Rezistance to *Monilinia laxa* (Aderh&Ruhl) Honey of varieties and hybrids of apricots and almonds using mean growth rate of cankers on young branches as criterion of susceptibility. J. Amer. Soc. Hort. Sci. 94, 282-284.

Hoza D., 2000. Pomologie. Ed.Prahova, Ploieşti, pp: 286, ISBN 973-99268-3-5.

Legave, J.M. and Clazel G., 2006. Long-term evolution of flowering time in apricot cultivars grown in southern France: wich future imtacts of global warming? Acta Horticulturae, 714: 47-50.

Legave, J.M., Farrera, I., Almeras, T. and Calleja, M, 2008. Selecting models of apple flowering time and understading how global warming has had an impact on this trait. Journal of Horticultural Science & Biotechnology, 83:76-84.

Olensen, J.O., Bindi, M., 2002. Consequences of climate change for European agricultural productivity, land use and policy. European Journal of Agronomy, 16, 239–262.

Mănescu Creola, Baciu Elena And Cosmin Silvia, 1975. Controlul biologic în pomicultură. Ed. Ceres, Bucureşti.

Paltineanu, Cr., Mihailescu, I.F. & Seceleanu, I. (2000A). Dobrogea, condiţiile pedoclimatice, consumul si necesarul apei de irigatie ale principalele culturi agricole. Editura Ex Ponto, Constanta, 258pp.

Roman Ana Maria, Cusursuz Beatrice, Cociu V., Elena Topor, S.A. 1992. Studiul agrometeorologic al mecanismelor exogene ce controlează înflorirea unor specii prunoidee din Romania. Eucarpia, Angers, Franta.

Stancu, T., Balan Viorica, Ivascu Antonia, Cociu, V.,1989. Resistance to frost and wintering of some apricots varieties with different geographical origin, under Romanian plain conditions. Acta Horticulturae, nr. 239.

Sunley, R.J., Atkinson, C.J. and Jones, H.G., 2006. Chill unit models and recent changes in the occurrence of winter chill and soring frost in the United Kingdom. Jurnal of Horticulturae. Science & Biotechnology, 81: 949-958.

Sumedrea D., Tănăsescu N., Chițu E., Moiceanu D., Marin Fl., Cr., 2009. Present and perspectives in Romanian fruit growing technologies under actual global climatic changes. Scientific Papers of the Research Institute for Fruit Growing Pitesti, Vol. XXV, ISSN 1584-2231, Editura INVEL Multimedia, București: 51-86.

Topor Elena, 1995. Cercetări privind îmbunătățirea sortimentului de cais în Dobrogea. Teză de doctorat ASAS București.

Topor Elena, 1987. Modificările continutului in hidrati de carbon în funcție de perioada de vegetație a 11 soiuri de cais. Vol. Omagial, S.C.D.P. Constanta.

Topor Elena, 2002. Cercetari privind relația dintre condițiile climatice si fructificarea caisului. Lucr. St. Vol.1 . Seria Horticultura, Ed. Ion Ionescu de la Brad Iasi.

Topor Elena, 2009. Cercetari privind influența schimbarilor climatice asupra culturii caisului în Dobrogea. Lucrările Simpozionului Mediul si Agricultura în regiunile aride –prima editie. ISBN 978-973-7681-68-3.(223)

Zavalloni, C., Andersen, J.A., Flore, J.A., Black, J.R. and Beedy, T.L., 2006. The pileus project: climate impacts on sour cherry production in the great lakes region in past and projected future time frames. Acta Horticulturae, 707: 101-108.

7

EXAMINATION OF THE POMOLOGICAL CHARACTERISTICS AND THE PRESENCE OF HEAVY METALS IN THE PEACH CULTIVAR "CRESTHAVEN" FROM REPUBLIC OF MACEDONIA

Viktorija STAMATOVSKA[1], Ljubica KARAKASOVA[2], Gjore NAKOV[3],
Tatjana KALEVSKA[1], Marija MENKINOSKA[1], Tatjana BLAZEVSKA[1]

[1]"Saint Clement of Ohrid" University of Bitola, Faculty of Technology and Technical Sciences,
Dimitar Vlahov bb, 1400 Veles, Republic of Macedonia
[2]"Ss. Cyril and Methodius" University, Faculty for Agricultural Sciences and Food,
Aleksandar Makedonski bb Blvd., 1000 Skopje, Republic of Macedonia
[3]"Angel Kanchev" University of Ruse, Department of Biotechnology and Food Technologies,
Branch Razgrad, 47 Aprilsko vastanie Blvd., Razgrad 7200, Bulgaria
Corresponding author email: vikistam2@gmail.com

Abstract

The pomological characteristics are very important for each type of fruit. The different types of fruits are being classified according to their shape, size, dimensions and the other features. The purpose of this paper is to examine the pomological characteristics of the peach fruits from "Cresthaven" cultivar, from Rosoman, Republic of Macedonia. The fruits are being collected in full technological maturity. The following characteristics have been determined: height, width and thickness of the fruit (using a caliper with an accuracy of 0.01 mm), weight of the fruit and weight of the mesocarp and pit (using an analytical balance with an accuracy of 0.001 g), and the yield is also mathematically calculated. Also, during the research have been conducted an analysis for the presence of heavy metals (Pb, As and Cd) in the fruits (using atomic absorption spectrometer). The research was repeated three times, in a period of three years (2011, 2012, 2013). In the three-year period of examinations, from a statistical point of view, between the calculated values of the examined parameters of the peach fruits there are no statistically significant differences, with the exception of the pit weight and the yield, for which statistically significant differences have been determined. The contents of Pb, As and Cd in the fruits in three-year testing period are in accordance with the applicable rules and prescribed norms for food safety in the Republic of Macedonia.

Key words: peach, pomological characteristics, heavy metals.

INTRODUCTION

The peach is one of the most widespread fruits in the world because of its specific pleasant taste, juiciness and nutritional values. The peach, along with the cherries, sour cherries, plums and apricots, belongs to the group of stone fruits (Kantoci, 2008; Jašić, 2007). Peach originates from China, and today is being cultivated in warm temperate regions of Europe, Asia, North America, parts of Africa and Australia. Peach fruits can be consumed fresh and processed in different ways (Mratinić, 2000).

According to the systematic classification, the peach fruit belongs to the family *Rosaceae*, subfamily *Prunoidea*e, genus *Prunus*, with a greater number of species (Bulatović-Danilović, 2007). Known varieties: 'Springtime', 'Suncrest', 'Springgold', 'Sprincrest', 'Redtop', 'Redhaven', 'Cresthaven' and other (Kantoci, 2008).

The fruit of the peach is big, usually with a round shape, juicy, with a pleasant sweet-sour taste and specific aroma. From the total weight of the fruit, which ranges from 80 to 250 g (sometimes more), the part that can be consumed amounts 93 to 98% (Mratinić, 2000). It is known that the peach fruit contains carbohydrates, organic acids, pigments, phenolic compounds, antioxidants and traces of proteins and lipids. It is a rich source of potassium, iron, fiber, vitamin A, vitamin C and other vitamins (Crisosto and Valero, 2008; Hajilou et al., 2013).

The natural conditions for the cultivation of peaches in Macedonia are very favorable. With the construction of more factories for fruit processing, peach became much demanded raw material. The increased interest in this fruit affected the increasing number of peach trees. Thus, from 327.000 in 1970 their number increased in 2010 up to 505.000, and the output was 10.200 tons. The cultivation of peaches is most widespread in the Tikves region, then in the regions of Skopje, Strumica, Radovis, Gevgelija and Kumanovo (Stojmilov and Apostolovska-Toševska, 2016). Different varieties have been cultivated, and one of the most known is the 'Cresthaven' peach cultivar from the region of Rosoman.

The features of the fruits of peach cultivar 'Cresthaven' have been presented by many authors (Mratinić, 2000; Bulatović-Danilović, 2007; Nenadović-Mratinić et al., 2003). This variety has been obtained by complex hybridization. Peach ripens from mid to late August. The fruit is round in shape, firm and belongs to the group of large fruits from 250 g. It features with an average weight of 184 g, length of 6.75 cm, width of 7.41 cm and thickness of 7.56 cm. The main color of the exocarp which is yellow is complemented with red color, which covers most of the surface of the fruit (50-60%). The mesocrap is yellow, juicy, firm, tasty and of high quality. The pit can be easily removed from the fruit's flash. Peaches can handle transportation. Fruit fresh include: 13.12% soluble dry matter, total sugars 10.43% and 0.60% total acids.

The quality of the fruits expressed by their size, appearance and taste, is very important prerequisite for their sale. A review of the fruits should include mechanical, sensorial, chemical and microbiological control, as well as remarks for the variety.

The first remark refers to the type and variety, degree of maturity, the average mass, the fruit size, the percentage of seed in the fruit, the stalk, the petals, husk, etc. Pomological characteristics of the agricultural products are parameters for determining the appropriate standards for assessment, transportation, processing and packaging (Karakašova and Babanovska-Milenkovska, 2012).

According to Crisosto et al. (2004) cited by Milošević and Milošević (2011), the size is a quantitative inherited factor for determination of the yield of fruit, the quality and the acceptability of the consumers.

The fruit is part of the daily diet, so it is very important to know the possible presence of heavy metals in it. The data indicate that heavy metals are pollutants of the fruit and such fruit consumed by humans may pose health risks (Sobukola et al., 2010; Elbagermi et al., 2012; Matei et al., 2013; Chandorkar and Deota, 2013). Lead, cadmium and arsenic are among the most widespread toxic elements present in the food and the environment.They have a long half-life after absorption in humans and animals, which can cause unpleasant effects such as damage of internal organs, nervous system, kidneys, liver and lungs (Ghazanfarirad et al., 2014).

Considering the previously mentioned, we consider that the data obtained in this examination will be of interest for the manufacturers, as well for the consumers, too. The obtained data will define the pomological characteristics of the peach fruits of the cultivar 'Cresthaven', available in the market, and at the same time will bring awareness for the presence of heavy metals in them, which is very important from a health point of view.

MATERIALS AND METHODS

The peach fruits of the cultivar 'Cresthaven', cultivated in Rosoman, Republic of Macedonia, have been used as a testing material. The fruits are collected in full technological maturity. For analysis have been taken 50 healthy fruits without major and average weight of the fruit, mesocarp and pit, using analytical balance with an accuracy of 0,001 g, average fruit sizes (height, width and thickness), using a caliper with an accuracy of 0,01 mm, have been determined. The fruits have been examined in the Laboratory for fruit and vegetables processing at the Faculty of Agricultural Sciences and Food in Skopje, Republic of Macedonia. The yield is mathematically calculated. The mass ratio between the useful part and the part that is not used, expressed as a percentage represents the yield (Karakašova, 2011). The contents of As, Pb and Cd have been determined at the Institute of Food at the Faculty of Veterinary Medicine - Skopje using atomic absorption spectrometry according SOP 392. The research was repeated three times, in a period of three years.

The results of the examination are presented, analyzed and statistically processed using the computer program Microsoft Excel and the statistical package SPSS Statistics Version 19.

RESULTS AND DISCUSSIONS

The results from the research of the pomological characteristics of fruits of peach cultivar 'Cresthaven' are presented in tables (Table 1) and graphics (Figure 2).
The results represent mean values obtained from analysis of the fruits used in each production year (2011, 2012, 2013).
Analyzed fruits of peach variety of 'Cresthaven' (Figure 1) are characterized with features inherent for the type and variety (Bulatović-Danilović, 2007).
The fruit is big, solid and round shape. The surface is yellow, complemented by red color. The mesocarp is yellow with redness around the pit, which can be easily split (Vuletić, 2016). The smell is pleasant and the taste is sweet.

Figure 1. Peach cultivar 'Cresthaven'

In Table 1 are shown the obtained mean values for mass and dimensions for the peach fruits from the variety 'Cresthaven', and the obtained mean values for the mass of the mesocrap and of the pit and the yield calculated.
The mass of the fruit is one of the major pomological features, which largely affects the yield (Nikolić et al., 2013). Based on the obtained values for the average mass of the fruits in the three year period (from 169.08 ± 26.99 to 174.24 ± 44.39 g), it can be concluded that this variety is characterized with large fruits (150-200 g) (according to Mratinić, 2012 quoted Vuletić, 2016). The high standard deviation value of mass clears good selection of samples.

Table 1. Pomological characteristics of fruits of peach cultivar 'Cresthaven' in 2011, 2012, 2013

Year	Analyzed parameters	n	$\bar{x}$	SD	p- value		
					2011	2012	2013
2011	Mass of the fruit (g)	50	174.24	44.39		0.586	0.489
	Height of the fruit (cm)	50	6.30	0.55		0.521	0.592
	Width of the fruit (cm)	50	6.87	0.60		0.942	0.686
	Thickness of the fruit (cm)	50	6,68	0.61		0.091	0.930
	Mass of the mesocarp (g)	50	166.24	42.79		0.456	0.413
	Mass of the pit (g)	50	5.68	1.73		0.001**	0.001**
	Yield (%)	50	95.35	0.46		0.000**	0.000**
2012	Mass of the fruit (g)	50	170.18	38.17	0.586		0.883
	Height of the fruit (cm)	50	6.36	0.44	0,521		0.915
	Width of the fruit (cm)	50	6.88	0.54	0.942		0.634
	Thickness of the fruit (cm)	50	6.87	0.59	0.091		0.109
	Mass of the mesocarp (g)	50	160.90	36.40	0.456		0.941
	Mass of the pit (g)	50	6.74	1.83	0.001**		0.940
	Yield (%)	50	94.52	0.30	0.000**		0.000**
2013	Mass of the fruit (g)	50	169.08	26.99	0.489	0.883	
	Height of the fruit (cm)	50	6.35	0.43	0.592	0,915	
	Width of the fruit (cm)	50	6.83	0.41	0.686	0.634	
	Thickness of the fruit (cm)	50	6.69	0.49	0.930	0.109	
	Mass of the mesocarp (g)	50	160.37	25.88	0.413	0.941	
	Mass of the pit (g)	50	6.76	1.22	0.001**	0.940	
	Yield (%)	50	94.83	0.33	0.000**	0.000**	

n - number of examined fruits; $\bar{x}$ - average value; SD - standard deviation; p - statistical significance, * Significant differences at the significance level of 0.05 (p<0.05); ** Significant differences at the significance level of 0.01 (p<0.01).

Lower values for mass of the peach fruits from the cultivar 'Cresthaven', cultivated in different locations in Serbia in the period 2000-2003 were found by Zec et al. (2003). The authors found that the mass of the fruits of this variety cultivated at Padinska Skela has average values ranged from 59.90 g to 110.60 g, and at Grocka location from 57.30 g to 126.00 g.

The following mean values expressed in cm for height, width and thickness of the fruit in each of the years vary from 6.30 ± 0.55 to 6.36 ± 0.44, from 6.83 ± 0.41 to 6.88 ± 0.54 and from 6.68 ± 0.61 to 6.87 ± 0.59. From Table 1 can also be determined that the mass of mesocrap is from 160.37 ± 25.88 to 166.24 ± 42.79 g, and the pit mass from 5.68 ± 1.73 to 6.76 ± 1.22 g. The calculated values of the yield are ranging from 94.52 ± 0.30 to 95.35 ± 0.45%.

A comparison of the results from the examinations of the peach fruits from the cultivar 'Cresthaven', by years, indicates the existence of certain differences in the calculated values of the examined parameters. With statistical analysis it has been determined their significance (Table 1).

In terms of mass of fruits in each of the years under examination, it can be concluded that in 2012 and 2013 were obtained lower values (170.18 g and 169.08 g) compared to 2011 (174.24 g). With the statistical analysis of the data it has been determined that these differences are not statistically significant (p>0.05).

In terms of heigh (6.30 cm, 6.36 cm, 6.35 cm) and width (6.87 cm, 6.88 cm, 6.83 cm) of fruits it was found that the differences between the values obtained for this period are not statistically significant (p>0.05).

Small differences in the values obtained have been determined for the thickness of the fruit in 2011, 2012 and 2013 (6.68 cm, 6.87 cm, 6.69 cm), also confirmed by a statistical standpoint (p>0.05). Also, there were identified some variabilities in values obtained for mass of mesocrap, which proved that there was no statistically significant difference (p>0.05).

For pit mass of the fruits have been obtained approximations in 2012 and 2013 (6.74 g, and 6.76 g). The little difference which occurs in the values obtained for this parameter is not statistically significant (p>0.05). In terms of pit mass of the fruits in 2011 it has been determined that it is smaller (5.68 g) than the mass of pit in 2012 and 2013. Statistical analysis of the data has shown that differences which occur between the values of pit mass in 2011 and 2012 and between 2011 and 2013 are statistically significant (p<0.01).

The yield is closely connected with the mechanical composition of raw materials (Niketić-Aleksić, 1994). The differences which appeared in the pit mass of fruits affected the obtained values of the yield. Statistical analysis of the results for the yield of fruits has shown statistically significant differences (p<0.01).

From Figure 2 it can be concluded that the average mass of the fruits from the cultivar 'Cresthaven' for the three year period is 171.17 g, height 6.34 cm, width 6.86 cm, and thickness of 6.75 cm. The average mass of the fleshy interior is 162.50 g, the average pit mass is 6.39 g, and the yield 94.90%.

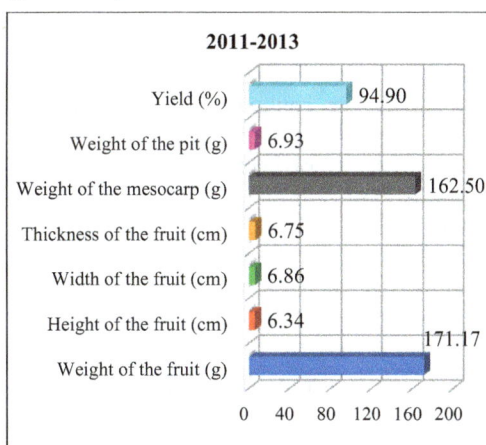

Figure 2. Pomological characteristics of fruits of peach cultivar 'Cresthaven' for all three years

Compared with the results obtained from our research, lower values for the same parameters are presented by Zec et al. (2003). Namely, the three-year research period of peach fruits from the cultivar 'Cresthaven' it has been determined that in average they are characterized by a mass of 92.8 g, a height of 5.41 cm, a width of 5.67 cm and a thickness of 5.37 cm. Higher average mass value (117.00 g) is determined by the same author in 2005 in a two-year examination of the peach fruits from the cultivar Cresthevn cultivated at the locations of Mislodzhin and Padinska Skela, but this average value is lower than the value obtained from our determinations.

Contrary to previous results Nenadović-Mratinić et al. (2003) for the tested parameters have evidenced slightly higher average values of the values obtained from our research. In seven years period of examining peach fruits from the cultivar 'Cresthaven' they have concluded that this variety of peach fruits is featured with a mass of fruit of 184.00 g, a length of 6.75 cm, a width of 7.41 cm, thickness of 7.56 cm, mass of pit 8.09 g and yield of 95.60%.

Differences between the results obtained from the research and the previously mentioned bibliography data are expected, because, although it is a fruit of the same variety of peach, the conditions under which it has cultivated are different.

Heavy metals are defined as metals whose specific density is greater than 5 g/cm^3. They affect human health through pollution of the environment and through the food, mainly through the fruits and vegetables cultivated in soil that has been contaminated. These include: arsenic, cadmium, chromium, copper, lead, nickel, zinc, molybdenum and vanadium (Islam, 2013).

During the research it has been conducted an analysis for the presence of Pb, As and Cd in the fruits of the cultivar 'Cresthaven'. The obtained results of the conducted analyzes in the three years of testing are shown in Table 2.

Table 2. Content of Pb, As and Cd in peach fruits of the cultivar 'Cresthaven' (2011, 2012, 2013)

Heavy metals Year	Pb mg/kg (L)	As mg/kg (L)	Cd mg/kg (L)
2011*	0.008	0.000	0.000
2012*	0.021	0.000	0.006
2013**	0.023	<0.001	0.002

*According to Regulation on general requirements for food safety (Official Gazette of RM No.118/2005).
** According to Regulation on the general requirements for food safety in relation to the maximum levels of certain contaminants (Official Gazette of RM No.102/2013).

From the submitted results it can be found that in the three years of examining the content of Pb, As and Cd in the peach fruits from the cultivar 'Cresthaven' is in accordance with the prescribed standards for food safety (Regulation on general requirements for food safety, Official Gazette of RM No.118/2005; Regulation on the general requirements for food safety in relation to the maximum levels of certain contaminants, Official Gazette of RM No.102/2013).

CONCLUSIONS

Based on three-year research period for the peach fruits of the cultivar 'Cresthaven' cultivated in Rosoman, Macedonia can be concluded that they were characterized by features inherent for their type and variety.

From a statistical point of view between the calculated values of the tested parameters there are statistically significant differences between the years of testing, in terms of pit mass and the yield.

From the calculated values for the content of Pb, As and Cd in fruits in the three years of testing it can be concluded that they are in accordance with the prescribed standards for food safety and are safe for consumption.

It is expected that the results obtained will find applications for the primary producers and direct consumers, which will increase the placement of this kind of fruit in the domestic and foreign market.

REFERENCES

Bulatović-Danilović M., 2007. Peach. Manual for the production of peaches for fresh use. Ministry of Agriculture, Forestry and Water Economy of the Republic of Serbia, Belgrade.

Chandorkar S., Deota P., 2013. Heavy metal content of foods and health risk assessment in the study population of Vadodara. Current World Environment, 8(2): 291-297.

Crisosto H.C., Valero D., 2008. Preharvest factors affecting peach quality. In: Layne D. R., Bassi D. (Eds.), The peach, Botany, Production and uses, (Edited by), London, UK: CAB International, 536-550.

Elbagermi A. M., Edwards M. G.H., Alajtal I. A., 2012. Monitoring of heavy metal content in fruits and vegetables collected from production and market sites in the Misurata Area of Libya. International Scholarly Research Network, ISRN Analytical Chemistry, Volume 2012, Article ID 827645, 5 pages, doi:10.5402/2012/827645.

Ghazanfarirad N., Dehghan K., Fakhernia M., Rahmanpour F., Bolouki M., Zeynali F., Asadzadeh J., Bahmani M., 2014. Determination of lead, cadmium and arsenic metals in imported rice into the West Azerbaijan Province, Northwest of Iran. J Nov. Appl Sci., 3(5): 452-456.

Hajilou J., Fakhimrezaei S., Dehghan G., 2013. Fruit quality, bioactive compounds and antioxidant capacity of 6 Iranian peach cultivars, Research in plant biology, 3(6): 06-16.

Islam R., 2013. Consumption of unsafe foods: Evidence from heavy metal, mineral and trace element contamination. Department of Soil Science Bangladesh Agricultural University, Mymensingh.

Jašić M., 2007. Technology of fruit and vegetables. Part 1. University of Tuzla, Faculty of Technology, Tuzla.

Kantoci D., 2008. Stone fruit. Bulletin of plant protection, 31(5): 6-13.

Karakašova Lj., 2011. Processing of fruits and vegetables. University St. Cyril and Methodius - Skopje, Faculty for Agricultural Sciences and Food, Skopje.

Karakašova Lj., Babanovska-Milenkovska F., 2012. Processing of fruits and vegetables – Handbook. University St. Cyril and Methodius - Skopje, Faculty for Agricultural Sciences and Food, Skopje.

Matei N., Popescu A., Munteanu M., Gabriel Lucian, Radu L. G., 2013. The assessment of Cd, Zn, Pb, Cu and vitamin C in peaches. U.P.B. Sci. Bull., Series B, 75(1): 73-80.

Milošević T., Milošević N., 2011. Quantitative analysis of the main biological and fruit quality traits of F1 plum genotypes (*Prunus domestica* L.), Acta Sci. Pol., Hortorum Cultus 10(2): 95-107.

Mratinić E., 2000. Peach - Nectarine. Agricultural library Draganć, Belgrade.

Nenadović-Mratinić E., Milatović D., Đurović D., 2003. Biological – economic characteristics of peach cultivars of mid-season to late ripening time, Proceedings of scientific papers, 9: 169-175.

Niketić-Aleksić G., 1994. Technology of fruit and vegetables. III Edition. Faculty of Agriculture, Belgrade.

Nikolić D., Rakonjac V., Fotirić-Akšić M., Radović A., 2013. Characteristics of peach hybrids from the crossing combination Flaminia × Hale Tardiva Spadoni, Proceedings of the IV Conference "Innovations in Fruit Growing", Belgrade, 197-207.

Regulation on general requirements for food safety, Official Gazette of RM No.118/2005.

Regulation on the general requirements for food safety in relation to the maximum levels of certain contaminants, Official Gazette of RM No.102/2013.

Sobukola P. O., Adeniran M. O., Odedairo A. A., Kajihausa E. O., 2010. Heavy metal levels of some fruits and leafy vegetables from selected markets in Lagos, Nigeria. African Journal of Food Science, 4(2): 389 – 393.

Stojmilov A., Apostolovska-Toševska B., 2016. Socio-economic geography of the Republic of Macedonia. University St. Cyril and Methodius – Skopje, Faculty of Natural Sciences and Mathematics, Skopje.

Vuletić T., 2016. Biological characteristics of peach (*Prunus persica* (L.) Batsch) in modern farming. Final work. University of Zadar, Department of ecology, agronomy and aquaculture, Zadar.

Zec G., Čolić S., Mišić P., Todorović R., Marinković D., 2003. Cv Gročanka – A new late peach cultivar, Jugosl. Voćar, 37 (143-144): 105-112.

8

AN *IN VITRO* STUDY OF COMMERCIAL FUNGICIDE EFFECTS ON POLLEN GERMINATION IN APPLE

Sultan Filiz GUCLU[1], Fatma KOYUNCU[2]

[1]Süleyman Demirel University, Atabey Vocational School, Isparta, Turkey
[2]Süleyman Demirel University, Agricultural Faculty, Isparta, Turkey
Corresponding author email: sultanguclu@sdu.edu.tr

Abstract

In this study some fungicides effects (Captan, azoxystrobin, mycolobutanil, thiophanate - methyl, maneb) on pollen viability tests and in vitro pollen germination investigated in Red Chief's pollens. Pollen viability test was made by TTC (2, 3, 5-triphenyl tetrazolium chloride) 15% +0.5% agar-agar+5 ppm boric acid at 25⁰C medium as control medium. Pollen germination was conducted at three concentrations: the recommended fields rate (100% RFR), 10% RFR and 1% RFR of each fungicide. 'Agar in plate' method was used for pollen germination tests. Statistical analyses performed with GLM models Using SPSS. Pollen germination rate was inhibited by increasing doses of fungicides when compared with control medium. Captan and azoxystrobin were most inhibitory. Germination was not significantly affected by mycolobutanil. Also Thiophanate –methyl was found inhibitory.

Key words: pollen germination, pesticide, 'Red Chief'.

INTRODUCTION

Rapidly increasing world population is regarded as one of the most important problems for mankind in the decades to come. A decline in the agricultural areas plays an important role in this direction and lead to certain restrictions in feeding the growing population. Therefore, an evaluation of the available land for obtaining the maximum yield has become a major goal. One of the applications in this connection has been the use of pesticides in plants against harmful organism for improving agricultural productivity. However, in addition to the benefits they provide, an excessive use of these chemicals, wrong applications as well as their side-effects have started creating serious environmental problems as well as posing toxicity threat to living organisms.
Inhibitory effects of pesticidal sprays during pollination would be of particular concern in areas where pollination and fertilization are limiting factors in fruit production. Recently, there has been a lot of press related to pollinator health, and some troubling information indicates that certain fungicides, when used during bloom, can negatively affect the health of honey bees. This is a complicated problem with the solutions relying on understanding the detailed relationships among

chemicals, pollinators and pest management needs. The objective of this study was to evaluate the effects of selected fungicides from different class.

MATERIALS AND METHODS

Plant material

Red Chief's pollens were used for pollen tests. Pollens were obtained from flowers of the above mentioned at balloon stage. The flowers were transferred to the laboratory immediately. Anthers were removed and placed into the dark-colored bottle to promote dehiscence at room temperature. The fungicides, which were commonly used for apple in Isparta selected for tests.

Table 1. Used fungicides

Active ingredient	Class	Trade name	Formulation*
Captan	Dicarboximide	Captan	WP
Azoxystrobin	Strobilurin	Quatris	WG
Mycolobutanil	Azole	Rally	WSP
Thio phanate-methyl	Benzimidazole	Topsin M	WSP
Maneb	Dithiocarbomate	Manex	Flowable

*WP=wettable powder, WG=water-dispersable granule, WSP=water-soluble pouches

The class, trade name, formulation and recommended field rate for each of compounds are shown in Table 1. Applies were made according to Yi et al., 2003. Pollen germination and tube growth were conducted at three concentrations: the recommended fiels rate (100% RFR), 10% RFR and 1% RFR of each 'fungicide'. The pollen morphological homogeneity percentages of pollens were assessed with thehaemocytometer (Marienfeld, Germany) slide (Eti, 1990). Imperfectly shaped pollen grains were considered as aborted pollen. The final percentage of morphological homogeneity was defined as:

(number of normal shaped pollen) – (number of aborded pollen) per field

$MH = total\ number\ of\ pollen\ field \times 100$

Pollen viability was determined by TTC (2, 3, 5-triphenyl tetrazolium chloride) stain test. Pollens were scattered onto TTC and stained pollens were counted after 2 hours and 15 minutes, respectively.

To determine the pollen viability, pollens of Red Chief apple (of four different areas) were observed onto two slides under a light microscope ($\times 100$ magnification). The stained pollen was considered as viable in these tests.

Germination medium without fungicide served as control.

For the *in vitro* test, pollen grains were sowed in the medium containing 0.5 agar + 15% sucrose + 5 ppm H_3BO_3 (boric acid) and incubated at the constant temperature of 25°C. The 'agar in plate' method was used to assisgn pollen germination and pollen tube growth (Koyuncu and Güçlü, 2009).

An ocular micrometer was used to measure pollen tube lenght, under a light microscope, at a magnification.

Four Petri dishes were used for germination and pollen tube growth experiments. For each assay, 2mL of medium with or without 'fungicide' was placed into Petri dishes.

Counts were made from 4 different microscope fields (100-150 pollen grains per field for each Petri dishes) (Hedly et al., 2004; Koyuncu, 2006). A factorial design was used in this study. Each treatment (5 'fungicide's x 3 concentrations plus control) was replicated 4 times.

Statistical analysis was conducted using Duncan's multiple range test within the general linear model procedure of SPSS 16.0.

For germination percentage, data were transformed with arcsine to meet the equal variance assumption.

RESULTS AND DISCUSSIONS

Pollens were evaluated from morphological homogeneity and viability there is no significant effect comperad to the control group. (Table 2).

Table 2. Morphological homegeneity and viability ratios (%)

	Morphological homogeneity (%)	Staining Test TTC (%)
Control (No fungicide)	98	97
Captan	96	96
Maneb	94	92
Mycolobutanil	93	90
Thiophanate-methyl	94	92
Azoxystrobin	92	90

Morphological homogeneity and viability rates were obtained upper 90%. However the lightest colour pollens were observed from Azoxystrobin but they were adopted 'viable'. In this case, the azoxystrobin fungicide may impair the structure of the enzyme when exposed to long-term or ex suggesting potential damage layer.

Different staining tests were carried out different fruit cultivars by different researches. Tosun and Koyuncu 2007, studied on cherry pollens, Koyuncu (2006), studied strawberry pollens using TTC and reported that pollen viability ratios reached 82% (Allstar and Elvira) and 86,5% (Chandler).

There are a lot of studies different fruit species about pollen viability tests. Junqueira, 2016 pointed that the viability of pollen grains was affected by the application of fungicide P + E, regardless of the application time. On the other hand, pollen grain germination was not affected by the fungicide or the stage.

Azoxystrobin and captan remained extremely inhibitory, with germination less than 1% of control. Maneb also inhibitory effect of pollen germination with 13.7%.

The highest pollen germination obtained from medium which was contain mycolobutanil with 87.7%. Also thiophanate-methyl have 68.8% germination. The inhibitory effects of some fungicides have been reported different study. Especially in apple, Yi et al 2003 reported the pollen germination in apples treated with Captan decreases by 20 % as compared to the control. Also captan was reported the inhibit pollen germination in pear (Butt et al., 1985).

The new azole product, mycolobutanil, had little or no inhibition, while the other new product, azoxystrobin, had very toxic effects (Yi et al., 2003b). Parallel these findings from our study captan and azoxystrobin were found severely inhibitor for pollen germination. After then maneb has found inhibitory for pollen germination. Mycolobutanil had little inhibition.

A number of compounds tested included fungicides different chemical class, i.e., dicarbomixide, strobilurin, azole, benzimidazole, dithiocarbomates were evalueted. The benzimidazo (thiophanate-methyl) had no to-intermediate effects. The dithiocarbamate compound (maneb) more severely suppressed pollen germination. Azoxystrobin, a strobilurin, was highly inhibitory. shown Table 3.

Table 3 Germination of apple pollen in presence of selected fungicides

Fungicides	Fungicide conc. (%of RFR)		
	100	10	1
Control (No 'fungicide')	100a^y	100a	100a
Captan	0.0b	0d	0.9e
Maneb	0.0b	0.0d	13.7d
Mycolobutanil	0.0b	46.9b	87.7b
Thiophanate-methyl	0.0b	10.2c	68.8c
Azoxystrobin	0.0b	0d	0.3e

^{z}Germination percentages shown are relative to the control, which is expressed at 100%. Actual pollen germination in the no-'fungicide' control was 77.4%. RFR=recommenden field rate
^{y} Mean values within a column followed by the same letter are not significantly different at p=0.05, Duncan's multiple range test.

With 88.6% pollen germination rate obtained from the control medium (0.5 agar + 15% sucrose + 5 ppm H_3BO_3).

There is no pollen germination was observed in assays incorporatting any of the fungicides at 100% RFR (Recommended field rate).

Pollen grains typically exhibit high sensitivity to chemicals with *in vitro* germination assays where contact with chemicals is intense. Although germination was inhibited severely at 10% RFR, the fungicides showed differential effects as pollen germination was observed in the presence of maneb, captan and azoxytrobin. Germination in thiophanate- methyl was only 10.2%.

The highest relative germination occured in the presence of mycolobutanil which was 46.9% of control. Assays conducted at 1% were effective in delineating differences in polen sensitivity to different fungicides. Found that fungicide sprays caused detrimental effects on stigma morphology and enhanced exudates production in almond flowers.

Percent fruitset was not measured in the study, however increased exudates production was raised as possibly causing inhibition of pollen tube growth and germination. It was also suggested that the increased exudates production may be a stress response which could decrease the period of stigma receptivity (Yi et al., 2003). Cyprodinil promoted a copious increase in exudates secretion and caused the most severe collapse of stigmatic cells of all the fungicides evaluated in the almond study.

Fungicides incorporated into the media, or sprayed on the surface of the medium, reduced pollen germination and pollen tube growth at concentrations lower than those commercially recommended for successful disease control (Heazlewood, 2004).

The mode of action of the fungicide, systemic or contact, is thought to alter the level of damage caused to pollination.

It should be noted that effects of 'fungicide's on pollen under in vio conditions will be affected by additional considerations, such as the persistence of the chemical, whether orr not it is systemic, and how it may interact with the constituents of the stigmatic papilae (Yi et al., 2003).

CONCLUSIONS

Commercial 'fungicide's haven't affected so much to morphological homogeneity and pollen viability. Pollen viability rates changed between 90% and 97%. Pollen germination rates decreased by incresing doses of all 'fungicide's.

All 'fungicide's must have used recommended dose. Azoxystrobin was found the most dangerous.

REFERENCES

Butt D.J., Swait A.A. and Robinson, J. D., 1985. *Effect of fungicides on germination of apple and pear pollen*. Ann. Appl. Biol. 106 Suppl: 110-111.

Eti S., 1990. *Cicek tozu miktarini belirlemede kullanilan pratik bir yontem*. Journal of Agricultural Faculty, Cukurova University, 5: 49–58.

Heazlewood J. E. Wilson. Clark S. R. J and Gracıe A. 2005. *Pollination of Vitis vinifera L. cv. Pinot noir as influenced by Botrytis fungicides*. Vitis 44 (3), 111–115

Hedly A.. Hormaza J.I and Herrero M., 2004. *Effect of temperature on pollen tube kinetics and dynamics in sweet cherry, Prunus avium (Rosaceae)*. Am J Bot 91 (4), 558-564.

Junqueria B.V., Costa C.A., Boff T., Muller C., Mendonça C.A.M., Batista F.P., 2016. *Pollen Viability, Physiology and production of Maize Plants Exposed To Pyrocloctrobin+epoxiconozole*. Pesticide Biochemistry and Physiology, YPEST-03991htpp://dx.doi.org/1016

Koyuncu F., 2006. *Response of in vitro pollen tube growth of strawberry cultivars to temperature*. European Journal of Horticultural Science, 71: 125–128

Koyuncu F. and. Tosun (Guclu) F. 2009. *Evaluation of pollen viability and germinating capacity of some sweet cherry cultivars grown in Isparta, Turkey*. Acta Hort.795 ISHS, (1), 71-75.

Tosun (Güçlü), F. and Koyuncu F., 2007. *Investigations of suitable pollinator for 0900 Ziraat sweet cherry cv.: pollen performance tests, germination tests, germination procedures, in vitro and in vivo pollinations*. HortSci(Prague), 34, (2), 47-53

Yı W., S.E. Law and H.Y. Wetzstein, 2003a . *Pollen tube growth in styles of apple and almond flowers after spraying with pesticides*. J. of Hortic. Sci. and Biotec., 78(6): 842-846.

Yi Weiguang, Law S. E., Wetzstein H., Y., 2003b. *An in vitro study of fungicide effects of pollen germination and tube growth in almond*. HortScience 38 (6):1086-1088.

THE INFLUENCE OF STORAGE IN CONTROLLED ATMOSPHERE ON QUALITY INDICATORS OF THREE BLUEBERRIES VARIETIES

Ioana BEZDADEA CĂTUNEANU[1,2], Liliana BĂDULESCU[1,2], Aurora DOBRIN[2], Andreea STAN[2], Dorel HOZA[1]

[1]University of Agronomic Sciences and Veterinary Medicine of Bucharest, Faculty of Horticulture, 59 Marasti Blvd., District 1, Bucharest, Romania
[2] University of Agronomic Sciences and Veterinary Medicine of Bucharest, Research Center for Studies of Food Quality and Agricultural Products, 59 Marasti Blvd., 011464, Bucharest, Romania

Corresponding author email: ioana.catuneanu@gmail.com

Abstract

The aim of this study was to determine which storage conditions can preserve the blueberry quality (Vaccinium corymbosum L.), stored in three different rooms with controlled atmosphere (CA). For this purpose, three varieties of blueberries, like Coville, Blueray and Chandler were stored and monitored for four months. Quality parameters like: dry matter content (D.M.%), titratable acidity (TA), soluble solids (°Brix), firmness, antioxidant capacity and also content in flavonoids, total polyphenols, total anthocyanins and ascorbic acid was monitored during storage period. The experiment conditions were based on the variation of carbon dioxide (CO_2) as follows: Room 1 (CO_2: 0%, representing the control), Room 2 (CO_2: 5%), and Room 3 (CO_2: 10%). Other common parameters of the experiment were: temperature (t°) 1 °C, oxygen quantity (O_2) 3%, relative humidity (RH) 95%. After four months of storage, observations showed that blueberries from Chandler variety presented better quality parameters compared to blueberries from Coville and Blueray varieties. Moreover, notable differences of physical and biochemical parameters were observed within the same blueberries variety stored in different rooms with controlled atmosphere conditions. Blueberries stored in Room 2 (T: 1°C, O_2: 3%, CO_2: 5%, RH: 95%) and Room 3 (T: 1°C, O_2: 3%, CO_2: 10%, RH: 95%) presented the best quality attributes compared with those stored in the other storage room (control), which would translate to a longer shelf life.

Key words: blueberries, controlled atmosphere, storage, quality.

INTRODUCTION

Since the Neolithic, blueberries (*Vaccinium* spp.) were consumed (Wang et al., 2017) at the beginning due to their wonderful taste sweet and sour and after centuries also for their biochemical composition (Wang et al., 2017) and health benefits (Liato et al.,2016). For this reason the production and consumption of blueberries has increased yearly and in recent years, they became one of the most popular horticultural products all over the world, second only after strawberries (Chen et al.,2015). They are sold fresh, processed, and in frozen form for various applications in food retail markets (Yang et al., 2014). Blueberries are appreciated for their taste, their high antioxidant activities, high and rich bioactive level contain of vitamins (C and E) (Liato et al.,2016), anthocyanins (Xu et al., 2016), polyphenolics (Liato et al.,2016), acids, tannins, mineral elements (Xu et al.,2016), chlorogenic acid, procyanidins (Chen et al., 2014) and flavonols (Wang et al., 2017).

Blueberries have antioxidant, anti-inflammatory, antimicrobial, anti-proliferative actions and they can be used in: type 2 diabetes (Shi et al., 2017), diabetic retinopathy (Song et al.,2016), cardiovascular and neuro-vegetative diseases, cancer (Liato et al.,2016), arthritis and obesity (Shi et al.,2017).

Due the increased production of fresh blueberries (Liato et al, 2017) from all over the world, a very important aspect is assuring and maintaining nutritional quality, and microbiological safety during storage and post-harvest sales (Liato et al.,2016).

Liato (2017) suggested that 95% of the blueberries production exhibit fungal contamination. Yang (2014) noticed that fresh blueberries rapidly deteriorate due to water loss and degradation of the fruit, usually caused by fungi such as: Anthracnose (*Colletotrichum acutatum*), Alternaria (*Alternaria* spp.) and grey mold (*Botrytis cinerea*) (Yang et al., 2014). According to Chen (2015) fresh blueberries are highly perishable and they have between 1 and 8 weeks of shelf life, so it is very important how the methods of harvesting, storage and transport conditions are applied. Varela (2008) studied how long controlled atmosphere storage prolong shelf life of apples, until consumption with the following storage conditions: T=1°C, O_2=2% and CO_2=2%, and the result was 7 months. Also it was observed that the loss of firmness is closely related to changes in cell wall composition and decrease in the total water soluble pectin (Chen et al., 2015). Firmness loss, which seriously reduces the commercial value of blueberries (Xu et al., 2016), can be slowed down by a bioactive, biocompatible and biodegradable polysaccharide such as chitosan (Yang et al., 2014). It acts as a barrier of water vapour to reduce the damage, reducing the loss firmness, decreasing mould growth and extending shelf-life (Yang et al., 2014) of strawberries and blueberries. Fungi can be significantly reduced by limited O_2 and increased CO_2 levels in the package (Yang et al.,2014) and in this way the quality of the texture can be maintained during storage (Chen et al.,2015). However, as Bessemans (2016) observed, the blueberries created storage disorders and an off-flavor by low oxygen content in rooms with controlled atmosphere, but Cortellino (2017) showed that firmness of the apples has been maintained better in low oxigen conditions. Francini (2013) suggests that the content of total phenols in apples had decreased during cold storage.

It was observed that the postharvest preservation technologies are applied to reduce damage, prolong shelf life, and keep the nutritional quality of several fruits and vegetables (Liato et al., 2017). Some of postharvest preservation technologies that can be used are: cold room storage, edible coatings, UV irradiation (Xu et al.,2016), packaging in a modified atmosphere (Yang et al.,2014), ozonation, and fumigation of sulphur dioxide (Yang et al.,2014), chlorine dioxide (Xu et al.,2016) (ClO_2 - strong oxidizing and sterilizing power).

The aim of these study was to determine which of the storage conditions of three different rooms with controlled atmosphere (CA), can better preserve the quality of blueberries (*Vaccinium corymbosum* L.).

MATERIALS AND METHODS

Fruits sampling and preparation
In order to accomplish the aim of this study three blueberry varieties (*Vaccinium corymbosum* L.) were used: Coville, Blueray and Chandler. These were acquired in August 2016 from Bilcesti, (Arges, Romania) and were selected by commercial maturity and the same ripening stage of fruits. Blueberries were packed in 250 g perforated trays that were then stored in three rooms with controlled atmosphere conditions from the Research Center for Studies of Food Quality and Agricultural Products - University of Agronomic Sciences and Veterinary Medicine of Bucharest. Temperature (t°) 1°C, oxygen (O_2) 3% and relative humidity (RH%) 95% was the same for all three rooms, but the CO_2 level was different. Thus, in room 1, which represents the control, the CO_2 concentration was 0%, CO_2 concentration in room 2 was 5%, and 10% in room 3 (Rizzolo et al., 2010). The study was conducted in 4 different moments as follows: initial moment (0), after 2, 3 and 4 months of storage in controlled atmosphere (CA). All samples were performed in duplicates.

For total flavonoid, total polyphenol and antiradical activity, the blueberry samples were extracted in 50% ethanol. For total anthocyanin acidified methanol was used (1.0% (v/v) hydrochloric acid in methanol) and for ascorbic acid the samples (5 g each) were extracted in 50 ml 9% metaphosphoric acid (MPA).

Physico-chemical analysis
The *dry matter and water content* of the samples were determined by oven drying for 24 hours at 105°C using a UN110 Memmert oven, method used also by Moura (2005), Skupień (2006), Delian (2011), Corollaro (2014), Mureşan (2014), Ticha (2015). To determine the fruit firmness an electronic penetrometer

TR was used, and the results were expressed in kg/cm^2 (Chen, 2015).

Soluble solids were determined from blueberry juice (Yoon, 2005; Saei, 2011; Mureşan, 2014; Oltenacu, 2015), with refractive device Kruss DR301-95 (% Brix). The titratable acidity was determined by titration with 0.1N NaOH to pH 8.1 (DeEll, 1992; Yoon, 2005; Skupień, 2006; Saei, 2011).

Titratable acidity calculation was done using the formula: $\frac{F \times C \times a \times b \times 100}{b \times c}$, where F is the factor NaOH solution 0.1 N (1,002), C = coefficient of correction for citric acid (0.0064), a = quantity of 0.1 N NaOH titrated, b = volume of the extraction solution, c = mass of the sample. For titration with 0.1 N NaOH the automatic titrator TitroLine easy was used. The results were expressed in g citric acid/100g.

Total flavonoid content was determined after an aluminium chloride adapted method (Žilić, 2011; Shen, 2016; Li, 2017). 0.25 ml hydro-alcoholic extract was mixed with 1.25 ml of distilled H_2O and 0.075 ml of a 5% $NaNO_2$, after five minutes 0.075 ml of a 10% solution of $AlCl_3$ was added. After another six minutes 0.5 mL of 1M solution of NaOH was added, the final volume being 2.5 ml. The absorbance was read at wavelength $\lambda = 510$ nm. The total flavonoid content was expressed in M/ml in fresh weight.

Total polyphenol content was measured by colorimetric Folin-Ciocalteu method after Skupień (2006), Khanizadeh (2008), Delian (2011), Mureşan (2014) and Drogoudi (2016), with some modification. 25 µl of a hydro-alcoholic extract were made up to 2 ml with distilled H_2O. 125 µl Folin - Ciocalteu and 375 µl of Na_2CO_3 (used for an alkaine environment) was added to the mixture. The final volume was 2.5 ml. The wavelength used for mesurements was $\lambda = 750$ nm. The total polyphenol content was expressed in M/ml in fresh weight.

Total anthocyanins content was measured with spectrophotometric absorbance at wavelength $\lambda = 540$ nm (Bărăscu et al., 2016), after an adapted method. The extracts were filtered under vacuum and completed up to 50 ml volume. The results were calculated using the formula: Total anthocyanins = DO_{540} x F, where DO_{540} is absorbance at wavelength $\lambda = 540$ nm and factor F = 11.16. The total anthocyanins content was expressed in mg/100g in fresh weight.

For evaluation of **antiradical activity** an indirect DPPH-radical scavenging activity spectrophotometric method was used (Khanizadeh, 2008; Mureşan, 2014; Drogoudi, 2016). 0.5 ml hydro-alcoholic extract was mixed with 1 ml of 0.1 mM DPPH solution. The results were calculated using the formula: AA_{DPPH} (%) = $\frac{A\ control - A\ sample}{A\ control}$ x 100, where $A_{control}$ is absorbing control sample (containing all reagents except extract) and A_{sample} is the sample absorbance. The absorbance was measured at wavelength $\lambda = 515$ nm. The evaluation of antiradical activity was expressed in % in fresh weight.

All determinations described above were performed with Specord 210 Plus spectrophotometer.

Ascorbic acid content was determined with HPLC – Agilent Technologies 1200 Series equipment, using an ZORBAX Eclipse XDB-C18 (4.6x50 mm, 1.8µm) column with Rapid Resolution HT and a detector UV-DAD detection wavelength 220/30 nm, reference wavelength 400/100 nm. Mobile phases were A= 99% (ultrapure water with H_2SO_4 up to 2.1 pH) and B= 1% (acetonitrile with 10% A). The samples were filtered through a filter Agilent PTFE 0,2 µm. The injection volume was 2 µl, with 4 min post time, flow rate at 0.5 ml/min at 30 °C in column compartment. The samples were analysed in duplicate and were expressed in mg/100g. Ascorbic acid calculation was done using the formula: $\frac{a \times b \times 100}{c}$, where a=ascorbic acid content in mg/ml , b= solution extraction volume (ml) and c= working mass of the sample taken (g).

RESULTS AND DISCUSSIONS

Blueberries fruit quality is assessed by the following indicators: firmness, dry matter, water content, soluble solids and titratable acidity. The quality indicators were different both at harvest and during storage for all the blueberry varieties.

The dry matter and water content of Coville variety had small variations during storage compared to initial moment (0) (Table 1), observing more fluctuations in the control room (CO_2 - 0%).

The water content and the titratable acidity (TA) had the highest value after three month of storage in the room 2 (CO_2 - 5%).

For Coville variety stored in room 1 (control) the soluble solids content had the maximum value after two months of storage and the dry matter and firmness after three months of storage compared to the other two rooms.

Table 1.Variation of firmness and content of: dry matter, water, TA and soluble solids during storage period in CA for Coville variety

Sample	Time of analysis	Dry matter content(D.M.%)	Water content (%)	Titratable acidity (g acid citric/100 g)	Soluble solids (% Brix)	Firmness (kg/cm^2)
Coville	17.08.2016	14.334	85.666	1.016 ± 0.009	10.420 ± 0.751	0.294 ± 0.032
Coville room1	13.10.2016	13.870	86.130	1.087 ± 0.002	11.670 ± 1.063	0.383 ± 0.103
Coville room2	13.10.2016	14.366	85.634	1.061 ± 0.001	11.160 ± 1.251	0.293 ± 0.071
Coville room3	13.10.2016	14.256	85.744	0.889 ± 0.005	9.920 ± 1.396	0.255 ± 0.042
Coville room1	22.11.2016	15.121	84.879	1.136 ± 0.021	10.820 ± 1.434	0.384 ± 0.062
Coville room2	22.11.2016	13.217	86.783	1.302 ± 0.005	10.660 ± 2.219	0.333 ± 0.063
Coville room3	22.11.2016	13.778	86.222	1.092 ± 0.003	9.440 ± 1.608	0.329 ± 0.092
Coville room 1	12.12.2016	14.907	85.093	0.967 ± 0.022	10.470 ± 1.589	0.341 ± 0.074
Coville room 2	12.12.2016	14.794	85.206	1.161 ± 0.001	10.570 ± 1.455	0.257 ± 0.044
Coville room 3	12.12.2016	13.829	86.171	1.105 ± 0.001	10.280 ± 1.605	0.268 ± 0.090

In table 2, for Blueray variety, it has been observed that the content of dry matter and water had some variation during storage in comparison with the initial moment. Blueray variety stored in rooms 1 and 2 have lost turgidity towards the end of storage period. Titratable acidity (TA) and soluble solids content have different values during storage, compared to initial moment (0). The soluble solids content from room 3 (CO_2 - 10%) and titratable acidity in all 3 rooms registered an increase in comparison with the initial moment (0) (Yang et al., 2014). Also, there was an increase in firmness in room 1 (CO_2 - 0%) after two months of storage. In room 2 and 3 the firmness value was close to the initial moment (0), the maxim firmness of the fruits was recorded after three months of storage.

Table 2.Variation of firmness and content of: dry matter, water, TA and soluble solids during storage period in CA for Blueray variety

Sample	Time of analysis	Dry matter content(D.M.%)	Water content (%)	Titratable acidity (g acid citric/100 g)	Soluble solids (% Brix)	Firmness (kg/cm^2)
Blueray	17.08.2016	11.458	88.542	0.620 ± 0.009	8.400 ± 1.339	0.243± 0.063
Blueray room 1	13.10.2016	12.251	87.749	0.780 ± 0.001	9.370 ± 1.434	0.303± 0.069
Blueray room 2	13.10.2016	11.042	88.958	0.751 ± 0.003	8.430 ± 1.318	0.268± 0.063
Blueray room 3	13.10.2016	11.265	88.735	0.792 ± 0.008	9.120 ± 1.989	0.259± 0.049
Blueray room 1	22.11.2016	13.956	86.044	0.856 ± 0.001	8.688 ± 1.698	0.256± 0.069
Blueray room 2	22.11.2016	13.909	86.091	0.818 ± 0.003	9.040 ± 0.873	0.338 ± 0.075
Blueray room 3	22.11.2016	11.879	88.121	0.705 ± 0.005	10.400 ± 1.268	0.283 ± 0.124

Table 3 shows that the dry matter and water content for Chandler variety had small variations in room 2 and 3 from the initial moment (0) compared with room 1 (control).

For control room (CO_2 - 0%) it have noted more fluctuations. Titratable acidity (TA) values were maintained with small variations during storage (Yang et al., 2014).

Table 3.Variation of firmness and content of: dry matter, water, TA and soluble solids during storage period in CA for Chandler variety

Sample	Time of analysis	Dry matter content(D.M. %)	Water content (%)	Titratable acidity (TA)(g acid citric/100 g)	Soluble solids (% Brix)	Firmness (kg/cm²)
Chandler	17.08.2016	12.693	87.307	0.851± 0.009	7.390±1.480	0.079 ± 0.047
Chandler room 1	13.10.2016	13.297	86.703	0.836 ± 0.002	10.050±1.706	0.322 ± 0.037
Chandler room 2	13.10.2016	11.604	88.396	0.909 ± 0.002	10.370±1.113	0.272 ± 0.049
Chandler room 3	13.10.2016	12.413	87.587	0.889 ± 0.005	10.170±1.501	0.274 ± 0.064
Chandler room 1	22.11.2016	13.381	86.619	0.863 ± 0.014	9.560±1.692	0.280 ± 0.035
Chandler room 2	22.11.2016	12.626	87.374	0.833 ± 0.001	10.725±1.320	0.340 ± 0.062
Chandler room 3	22.11.2016	11.352	88.648	0.858 ± 0.003	9.440±1.665	0.315 ± 0.036
Chandler room 1	12.12.2016	11.216	88.784	0.820 ± 0.003	8.750±1.925	0.320 ± 0.049
Chandler room 2	12.12.2016	13.554	86.446	0.857 ± 0.001	9.830±0.953	0.333 ± 0.053
Chandler room 3	12.12.2016	12.686	87.314	0.759 ± 0.001	9.750±0.977	0.278 ± 0.049

A noticeable increase in soluble solids content in all rooms throughout storage can be observed. For the firmness of Chandler variety fruits, an important increase in all rooms throughout the storage period from the initial moment (0) has been observed.

The ascorbic acid content of Coville variety (Figure 1), has declined during storage. Between the three type of storage, there were small differences.

Also at Bluray variety (Figure 2) low values during storage were determined when compared with the initial moment.

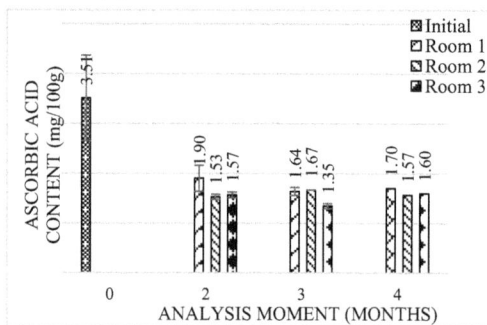

Figure 1. Variation of ascorbic acid content (mg/100g) during storage period in CA for Coville variety where: 0-initial moment, 2 - analyses after 2 months of storage, 3 - analyses after 3 months of storage, 4 - analyses after 4 months of storage

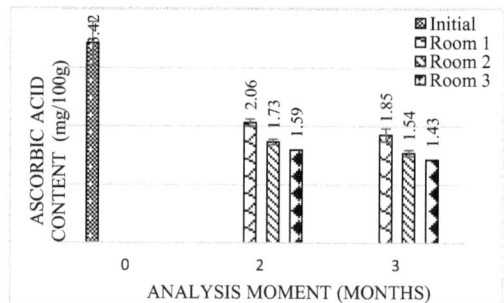

Figure 2. Variation of ascorbic acid content (mg/100g) during storage period in CA for Blueray variety where: 0 - initial moment, 2 – analyses after 2 months of storage, 3- analyses after 3 months of storage

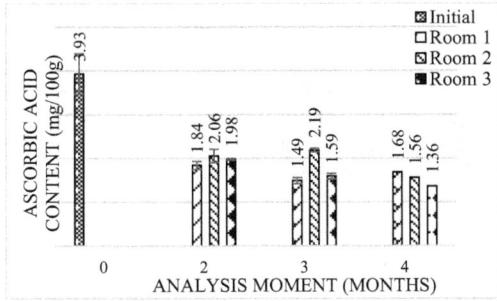

Figure 3. Variation of ascorbic acid content (mg/100g) during storage period in CA for Chandler variety where: 0 - initial moment, 2 - analyses after 2 months of storage, 3 - analyses after 3 months of storage, 4 - analyses after 4 months of storage

In room 1 (CO_2 - 0%) higher values from all the storage variants were present. At Chandler variety (Figure 3) higher values of ascorbic acid content during the storage in the case of room 2 (CO_2 - 5%) were observed.

Koyuncu (2010) showed that the ascorbic acid content of the fruits progressively drops during storage in cold rooms with T: 1 °C and RH: 95%.

Anthocyanins are water-soluble pigments and belong to flavonoid group. They have a very high antioxidant capacity and are responsible for the color in red-purple fruits.

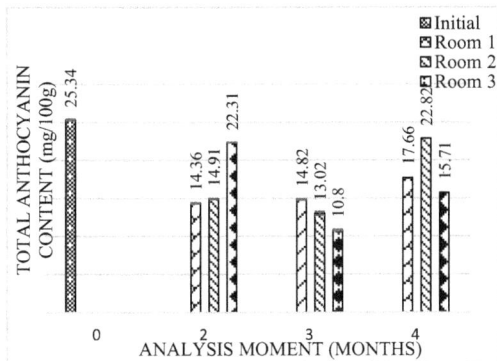

Figure 4. Variation of total anthocyanin content (mg/100g) during storage period in CA for Coville variety where: 0 - initial moment, 2 - analyses after 2 months of storage, 3 - analyses after 3 months of storage, 4 - analyses after 4 months of storage

The fruits content of anthocyanins can also correlate with the antioxidant capacity (Matityahu, 2016).

Coville variety had the best results regarding total anthocyanins content after four months of storage in the room 2 (CO2 - 5%), while for the fruits stored two months the decrease in anthocyanins was at half comparative to the initial moment (Figure 4).

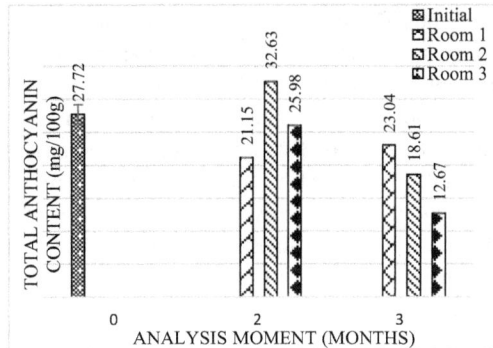

Figure 5. Variation of total anthocyanin content (mg/100g) during storage period in CA for Blueray variety where: 0 - initial moment, 2 - analyses after 2 months of storage, 3 - analyses after 3 months of storage

After two months of storage in controlled atmosphere from room 2, the total anthocyanins content for Blueray variety registered an increase with 17% compared to the initial moment. After three months of storage, total anthocyanins content value was lower compared to the registered value from the initial moment (Figure 5).

At Chandler variety (Figure 6), the total anthocyanins content increased during storage in room 1 (CO_2 - 0%) having the highest value after two months of storage.

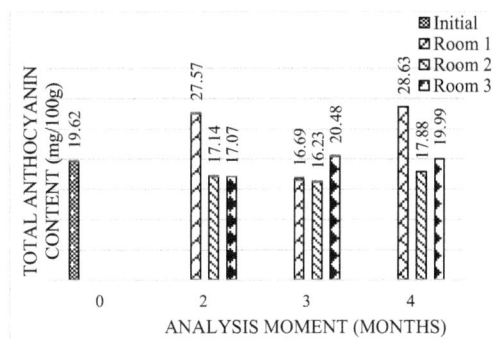

Figure 6. Variation of total anthocyanin content (mg/100g) during storage period in CA for Chandler variety where: 0 - initial moment, 2 - analyses after 2 months of storage, 3 - analyses after 3 months of storage, 4 - analyses after 4 months of storage

The values of anthocyanins content of the fruits from room 3 were comparable to the ones from the initial moment.

Higher values of the total anthocyanins content towards initial moment, indicate that at low temperatures post-ripening process of fruits continues (Matityahu, 2016). The fruits from rooms with CO_2 had total anthocyanins values content close to the initial moment.

The content of total polyphenols in the blueberries of the 3 varieties studied, increased gradually towards the initial moment throughout storage. At the Coville variety (Figure 7) lower values were observed at the end of the storage in room 3 (CO_2 - 10%) compared to other rooms.

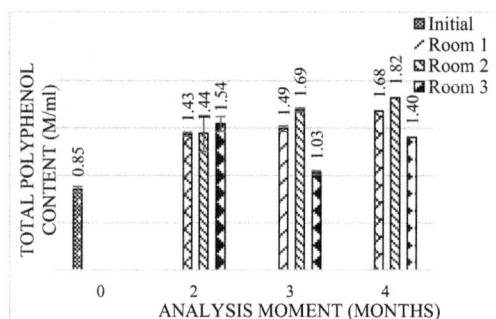

Figure 7. Variation of total polyphenol content during storage period in CA for Coville variety where: 0 - initial moment, 2 - analyses after 2 months of storage, 3 - analyses after 3 months of storage, 4 - analyses after 4 months of storage

Blueray variety (Figure 8) recorded a lower total polyphenol content in rooms 2 and 3 (with CO_2) compared with the blueberries from room 1 (without CO_2) after 2 and 3 months of storage.

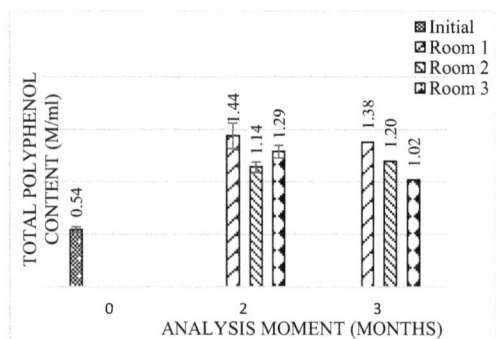

Figure 8. Variation of total polyphenol content during storage periodin CA for Blueray variety where: 0 - initial moment, 2 - analyses after 2 months of storage, 3 - analyses after 3 months of storage

For Chandler variety (Figure 9) in room 1 (without CO_2) there was a progressive increase of total polyphenol content. Yang G. (2014) suggests that this increase in polyphenol content during storage is a process of maturing fruits which were picked when they were not at fully maturity for consumption.

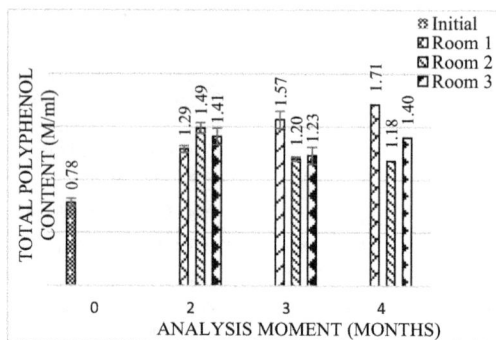

Figure 9. Variation of total polyphenol content during storage period in CA for Chandler variety where: 0 -initial moment, 2 - analyses after 2 months of storage, 3 - analyses after 3 months of storage, 4 - analyses after 4 months of storage

The total flavonoid content of Coville variety (Figure 10) has recorded lower values for all three rooms in 2, 3 and 4 months compared with the value from the initial moment.

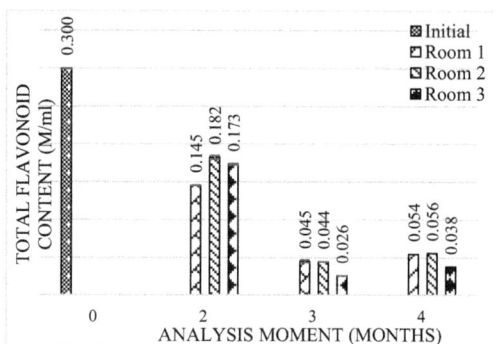

Figure 10. Variation of total flavonoid content during storage period in CA for Coville variety where: 0 - initial moment, 2 - analyses after 2 months of storage, 3 - analyses after 3 months of storage, 4 - analyses after 4 months of storage

The highest content in flavonoids for Blueray variety (Figure 11) was noticed in room 1 (without CO_2) after two months of storage.

The Chandler variety (Figure 12) maintained the flavonoids content after two months of storage in room 1 ($CO2$ - 0%) and 2 ($CO2$ - 5%), and in room 3 ($CO2$ - 10%) a small decrease of the values was recorded.

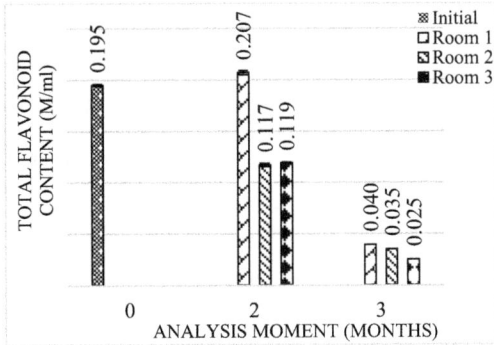

Figure 11. Variation of total flavonoid content during storage period in CA for Blueray variety where: 0 - initial moment, 2 - analyses after 2 months of storage, 3 - analyses after 3 months of storage

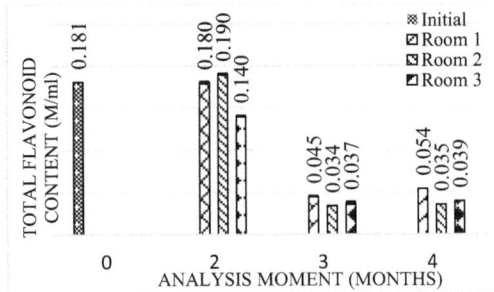

Figure 12. Variation of total flavonoid content during storage period in CA for Chandler variety where: 0 -initial moment, 2 - analyses after 2 months of storage, 3 - analyses after 3 months of storage, 4 - analyses after 4 months of storage

All varieties studied showed an important antioxidant capacity. Coville variety (Figure 13) recorded lower values for antioxidant activity in room 1 (CO_2 - 0%) and room 2 (CO_2 - 5%) after 2 months of storage, compared to the value of the initial moment.

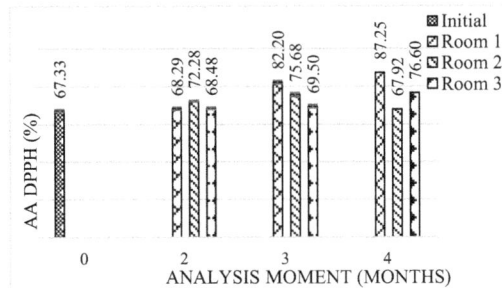

Figure 13. Variation of AA DPPH(%) content during storage period in CA for Coville variety where: 0 - initial moment, 2 - analyses after 2 months of storage, 3 - analyses after 3 months of storage, 4 - analyses after 4 months of storage

Blueray variety (Figure 14) recorded lower values for the antioxidant activity in all three

rooms at 2 and 3 months compared to the value of the initial moment.

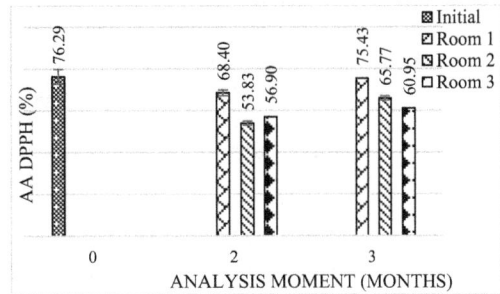

Figure 14. Variation of AA DPPH(%) content during storage period in CA for Blueray variety where: 0 - initial moment, 2 - analyses after 2 months of storage, 3 - analyses after 3 months of storage

Chandler variety (Figure 15) had the strongest antioxidant activity after four months of storage in the room 1 (without CO_2).

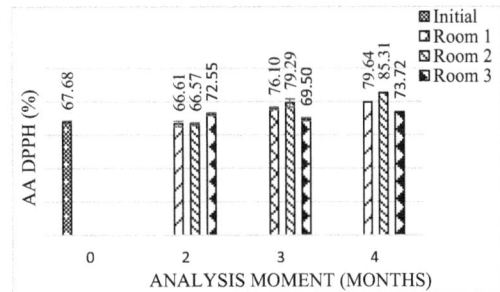

Figure 15. Variation of AA DPPH(%) content during storage period in CA for Chandler variety where: 0 -initial moment, 2 - analyses after 2 months of storage, 3analyses after 3 months of storage, 4 - analyses after 4 months of storage

CONCLUSIONS

The storage in rooms with controlled atmosphere influenced, as expected, fruit quality parameters. It can be noted that all varieties of samples did not have the same level of maturity at harvest, Coville and Blueray varieties being collected at the end of the harvest, while Chandler variety was collected midterm harvest. Because of that, Chandler variety behaved much better during the four months of storage compared to the other two varieties, maintaining much better visual, organoleptic and economical properties. In the case of antioxidant capacity of Coville and Chandler it was observed that room 3 (CO_2 - 10%) had a slight increase towards the end of the storage period suggesting that metabolic processes in fruit were slowed down due to higher CO_2 content.

The varieties behaved differently, observing for example the content of ascorbic acid. The Coville and Blueray varietes behaved better in the room without CO_2 while for the variety Chandler observations concluded that it maintained a higher quantity of ascorbic acid in room 2 (CO_2 - 5%).

Following the obtained results, we can specify that the varieties of the same species (blueberry - *Vaccinium corymbosum* L.) requires different storage conditions depending on crop technology applied and harvesting moment.

The post ripening continued in all rooms but was slowed down in rooms 2 and 3 (with CO_2) compared to room 1 (without CO_2). Small differences were observed in flavonoids and ascorbic acid content of the blueberries between the two rooms with CO_2 (room 2: 5% and room 3: 10%).

REFERENCES

Bărăscu R., Hoza D., Bezdadea-Cătuneanu I., Năftănăilă M., Albulescu A., 2016. Preliminary research regarding the grafting interstock and soil maintenance influence on fruit quality for Pinova Variety. Agriculture and Agricultural Science Procedia 10: 167-171.

Bessemans N., Verboven P., Verlinden B.E., Nicolaï B.M., 2016. A novel type of dynamic controlled atmosphere storage based on the respiratory quotient (RQ-DCA). Postharvest Biology and Technology 115: 91-102

Chen H., Cao S., Fang X., Mu H., Yang H., Wang X., Xu Q.,Gao H.,2015. Changes in fruit firmness, cell wall composition and cell wall degrading enzymes in postharvest blueberries during storage. ScientiaHorticulturae188: 44-48.

Corollaro M. L., Aprea E., Endrizzi I., Betta E., Demattè M. L., Charles M., Bergamaschi M., Costa F., Biasioli F., Grappadelli L. C., Gasperi F., 2014. A combined sensory-instrumental tool for apple quality evaluation. Postharvest Biology and Technology 96: 135-144

Cortellino G., Piazza L., Spinelli L., Torricelli A., Rizzolo A., 2017. Influence of maturity degree, modified atmosphere and anti-browning dipping on texture changes kinetics of fresh-cut apples. Postharvest Biology and Technology 124: 137-146

DeEll J., Prange R., 1992. Postharvest quality and sensory attributes of organically and conventionally grown apples. HortScience 27(10): 1096-1099.

Delian E., Petre V., Burzo I., Bădulescu L., Hoza D., 2011. Total phenols and nutrients composition aspects of some apple cultivars and new studied breeding creations lines grown in Voineşti area – Romania. Romanian Biotechnological Letters vol. 16, no. 6

Drogoudi P., Pantelidis G., Goulas V., Manganaris G., Ziogas V., Manganaris A., 2016. The appraisal of qualitative parameters and antioxidant contents during postharvest peach fruit ripening underlines the genotype significance. Postharvest Biology and Technology 115: 142-150

Francini A., Sebastiani L., 2013. Phenolic Compounds in Apple (*Malus domestica* Borkh.): Compounds Characterization and Stability during Postharvest and after Processing. Antioxidants, 2(3), 181–193

Khanizadeh S., Tsao R., Rekika D., Yang R.,Charles M., Rupasinghe V., 2008. Polyphenol composition and total antioxidant capacity of selected apple genotypes for processing. Journal of Food Composition and Analysis 21: 396-401

Koyuncu M., Dilmaçünal T., 2010. Determination of Vitamin C and Organic Acid Changes in Strawberry by HPLC During Cold Sorage. Notulae Botanicae Horti Agrobotanici, 38(3): 95-98

Li D., Li B., Ma Y., Sun X., Lin Y., Meng X., 2017. Polyphenols, anthocyanins, and flavonoids contents and the antioxidant capacity of various cultivars of highbush and half-high blueberries. Journal of Food Composition and Analysis, ISSN 0889-1575, http://dx.doi.org/10.1016/j.jfca.2017.03.006.

Liato V., Hammami R., Aïder M., 2017. Influence of electro-activated solutions of weak organic acid salts on microbial quality and overall appearance of blueberries during storage. Food Microbiology, vol 64: 56-64

Matityahu I., Marciano P., Holland D., Ben-Arie R., Amir R., 2016. Differential effects of regular and controlled atmosphere storage on the quality of three cultivars of pomegranate (*Punica granatum* L.). Postharvest Biology and Technology 115: 132-141

Moura C., Masson M., Yamamoto C., 2005. Effect of osmotic dehydration in the apple (*Pyrus malus*) varieties Gala, Gold and Fuji. Thermal Engineering, vol 4: 46-49

Mureşan E., Muste S., Borşa A., Vlaic R., Mureşan V., 2014. Evalution of physical-chemical indexes, sugars, pigments and phenolic compounds of fruits from three apple varieties at the end of storage period. Bulletin UASVM Food Science and Technology 71(1)

Oltenacu N., Lascăr E., 2015. Capacity of maintaining the apples quality, in fresh condition-case study. Scientific Papers Series Management, Economic Engineering in Agriculture and Rural Development vol. 15: 331-335

Rizzolo A., Vanoli M., Spinelli L., Torricelli A., 2010. Sensory characteristics, quality and optical properties measured by time-resolved reflectance spectroscopy in stored apples. Postharvest Biology and Technology 58: 1-12

Saei A., Tustin D., Zamani Z., Talaie A., Hall A., 2011. Cropping effects on the loss of apple fruit firmness during storage: The relationship between texture retention and fruit dry matter concentration. Scientia Horticulturae 130: 256-265

Shen Y., Zhang H., Cheng L., Wang L., Qian H., Qi X., 2016. In vitro and *in vivo* antioxidant activity of polyphenols extracted from black highland barley. Food Chemistry 194: 1003–1012

Shi M., Loftus H., McAinch A.J., Su X.Q., 2017. Blueberry as a source of bioactive compunds for the treatment of obesity, type 2 diabetes and chronic inflammation. Journal of Functional Foods 30: 16-29

Skupień K., 2006. Chemical composition of selected cultivars of highbush blueberry fruit (*Vaccinium corymbosum* L.). Folia Horticulturae 18/2: 47-56

Song Y., Huang L., Yu J., 2016. Effects of blueberry anthocyanins on retinal oxidative stress and inflammation in diabetes through Nrf/HO-1 signaling. Journal of Neuroimmunology 301: 1-6

Ticha A., Salejda A., Hyšpler R., Matejicek A., Paprstein F., Zadak Z., 2015. Sugar composition of apple cultivars and its relationship to sensory evaluation. Nauka. Technologia. Jakość 4(101): 137-150

Varela P., Salvador A., Fiszman S., 2008. Shelf-life estimation of "Fuji" apples II. The behaviour of recently harvested fruit during storage at ambient conditions. Postharvest Biology and Technology 50: 64-69

Wang H., Guo X., Hu X., Li T., Fu X., Liu R.H., 2017. Comparison of phytochemical profiles, antioxidant and cellular antioxidant activities of different varieties of blueberry (*Vaccinium* spp.). Food Chemistry 217: 773-781

Xu F., Wang S., Xu J., Liu S., Li G., 2016. Effects of combined aqueous chlorine dioxide and UV-C on shelf-life quality of blueberries. Postharvest Biology and Technology, 117: 125-131

Yang G., Yue J., Gong X., Qian B., Wang H., Deng Y., 2014. Blueberry leaf extracts incorporated chitosan coatings for preserving postharvest quality of fresh blueberries. Postharvest Biology and Technology, 92: 46-53

Yoon K. Y., Woodams E. E., Hang Y.D., 2005. Relationship of acid phosphatase activity and Brix/acid ratio in apples. Lebensm.-Wiss. u.-Technol, 38: 181-183

Žilić S., Šukalović V. H. T., Dodig D., Maksimović V., Maksimović M., BasićZorica, 2011. Antioxidant activity of small grain cereals caused by phenolics and lipid soluble antioxidants. Journal of Cereal Science 54: 417-424.

10

CANOPY MANAGEMENT PRACTICES IN MODERN PLUM (*PRUNUS DOMESTICA* L.) PRODUCTION ON VIGOROUS ROOTSTOCKS

Miljan CVETKOVIĆ[1], Gordana ĐURIĆ[1,2], Nikola MICIC[1,2]

[1]University of Banja Luka, Faculty of Agriculture, Bulevar Vojvode Petra Bojovića 1A, 78000 Banja Luka, Bosnia and Herzegovina
[2]University of Banja Luka, Genetic Resources Institute, Bulevar Vojvode Petra Bojovića 1A, 78000 Banja Luka, Bosnia and Herzegovina

Corresponding author email: miljan.cvetkovic@agrofabl.org

Abstract

Intensive high-density plantings (HDP) of plum trees in the Republika Srpska involve the use of Myrobalan (Prunus cerasifera Ehrh.) seedling as the predominant and, in most cases, the only rootstock. Using Myrobalan as a vigorous rootstock is a serious challenge in growing plums at higher planting densities. Although Myrobalan seedling rootstock increases the vigour of grafted cultivars, plum trees trained to the spindle system on Myrobalan rootstock can also be grown at very high plant densities ranging from 1,000 to 1,800 trees per hectare, depending on the cultivar/rootstock combination and central-leader inclination. The most common training system for plums in high density plantings is the slender spindle or the spindle bush system. Successful training and maintenance of spindle systems in intensive production on high-vigour rootstocks is not possible without the consistent use of canopy management practices, particularly during the first three years after planting, when these practices are most intensive for proper training of both the central leader and main lateral branches. Canopy management practices require a professional attitude and substantial manual labour. Particular importance in training spindle systems for plums as well as in maintaining the training system (replacement of spur-bearing branches) is given to the following specific canopy management practices: notching, shoot bending, shoot twisting, undercutting and replacement of spur-bearing branches. This paper outlines some important canopy management practices and their effect on plum growth and development, focusing on cultivar-specific responses to treatments.

Key words: plum, canopy management practices, cultivar.

INTRODUCTION

On an annual level, plum production in Bosnia and Herzegovina (during 2003 – 2013) showed an increase in total land area (+7.37%) and production (+4.80%) and a decline in average yield (–2.39%) (FAOSTAT, 2016). Although plum is the leading fruit crop in BiH (Statistics Agency, BiH), the intensity of production is rather low, which may be associated with the use of growing methods (Mićić et al., 2005) unadapted to the tendency to introduce new cultivars into production.

The most common training system for plum trees in high density plantings (HDPs) is the slender spindle or spindle bush (Grzyb and Rozpara, 1998; Hrotko et al., 1998; Meland, 2001; Čmelik et al., 2002; Gavrilescu et al., 2004) which uses low-vigour rootstocks. Establishing new highly intensive (high density) plum plantings in Bosnia and Herzegovina characteristically involves the use

of Myrobolan (*Prunus cerasifera* Ehrh.) seedling as the predominant, and in most cases the only rootstock available for plum grafting. In modern highly intensive plantings under high density planting systems which use higher vigour rootstocks, practices designed for canopy management during dormancy and timely summer pruning operations are the preconditions for successful plum production (Mićić et al., 2005; Milošević et al., 2008; Glišić, 2012; Cvetković et al., 2015).

Shoot management operations aimed at creating the best crotch angle possible such as shoot bending (during the first part of the growing season) can have an important effect on generative bud differentiation and, hence, facilitate the control of growth and development processes.

The objective of this study was to analyze shoot bending in plum trees grown on vigorous rootstocks in highly intensive plum production systems.

MATERIALS AND METHODS

The analysis of shoot bending practices in plum trees was conducted in a plum planting at Gunjevci (44°35'33"; 18°56'38") near Kozarska Dubica (BiH) at an altitude of 155 m.

The planting was established in autumn 2009. The experimental plot has a north-western exposure, with a slight inclination (2%). Total land area of the planting is 1 ha.

Four cultivars were planted: 'Stanley', 'Čačanska Lepotica', 'Čačanska Najbolja' and 'Čačanska Rana'. All cultivars were grafted on Myrobalan (*P. cerasifera* Ehrh.) seedling rootstock.

Trees were trained to the spindle system using all necessary canopy management practices. Spacing for all cultivars was 1.5 m within the row and 4.0 m between rows.

The planting received standard cultural practices. Soil management systems were grass mulch for the inter-row space, and bare fallow combined with herbicide band for the within-row space.

The research was carried out in 2010 - 2013. The trees of the tested cultivars were subjected to shoot bending, as follows:

a) initial spreading of the shoots to keep them at an angle subordinate to the central leader – at the beginning of the growing season, and

b) bending of the shoots to retain the desired position relative to the central leader – in the middle of the growing season.

Shoot management treatments were applied in the entire planting, but 30 trees per cultivar subjected to shoot bending operations were randomly selected for detailed analysis.

The shoots were initially spread out by wooden toothpicks, whereas in the second part of the growing season they were bent by plastic hooks, twine (aluminium wire and plastic wire spreaders) and twisting.

After the three–year experimental period, the shoot bending techniques used in the study were analysed for visual shown and integrated evaluation of their efficiency and effectiveness.

RESULTS AND DISCUSSIONS

Spreading of the shoots. The initial spreading of the shoots along the central leader by toothpicks in the tested cultivars (Figures 1a

and 1b) favoured the formation of a proper crotch angle (about 90°) of the newly formed shoots which generally provide the basis for the spindle structure in plum trees (Lučić et al. 1996; Mićić et al., 2005; Gonda, 2006; Glišić, 2012).

Figure 1. Spreading of shoots by toothpicks - variety 'Čačanska Lepotica' (a) and 'Čačanska Rana' (b)

The spreading operation in the tested cultivars led to reduced apical dominance (Wilson, 2000). Importantly, spreading by toothpicks should be done successively in accordance with the shoot growth dynamics.

The shoots subjected to spreading exhibited a higher rate of generative bud differentiation (Mićić et al., 1998). The horizontal position of the shoot resulted in higher percent activation of growth points along the shoot during the same or following growing season, with shorter growth formed and much of this growth developing into fruiting wood. Such a response was effective in reducing growth vigour and producing a more favourable ratio of vegetative to generative growth on the tree, which is of particular importance in the first years of plum production on higher vigour rootstocks.

The spreading efficiency of toothpicks is dependent on the optimum time to use them. Spreading should be performed when the shoot has reached a length of 15 – 20 cm, which is, in part, a cultivar-specific trait.

In 'Čačanska Najbolja' and 'Čačanska Rana', spreading should be applied to shorter lengths of shoots. 'Čačanska Najbolja' exhibits intensive growth, and its shoots longer than 20 cm generally have a greater base diameter and are not spreadable. 'Čačanska Rana' showed a tendency to develop shorter shoots with a higher rate of lignification at the base, thus potentially creating spreading problems. 'Stanley' and, particularly, 'Čačanska Lepotica' show a positive response to shoot spreading by toothpicks even at a later developmental stage. Mitrović et al. (2005) and Glišić et al. (2007) agree that shoot bending is a mandatory practice in establishing dense plum plantings on vigorous rootstocks, but no precise shoot bending times are defined in their studies. Toothpicks should be used in succession – only shoots that have reached the required developmental stage are to be spread out.

Previous experience has shown a very positive effect of toothpicks (Mićić et al., 2005; Glišić et al. 2007; Glišić, 2012). If the shoots are bent in the second part of the growing season without being previously spread at this early stage, the so–called "knee" is formed at the shoot base.

The use of toothpicks has a number of advantages: a high installation efficiency rate (a large number of toothpicks installed per unit time); good spreading performance; simple installation; causing minimum damage to the tissue which shows a high healing rate; the natural material they are made of has no negative effect on plants; they are available on the market and very affordable.

Problems with toothpick use generally include their post-installation instability, which may be due to low toothpick quality, improper installation or adverse weather conditions (heavy rain, wind) after installation. If improperly installed, toothpicks soon fall off, and the treatment must be repeated for satisfactory spreading performance.

Spreading requires intensive manual labour continuously throughout the growing season (until mid-July), which may be a constraint to the use of this practice in large plantations. Upon use of toothpicks to spread out shoots during the initial stage of shoot development and crotch angle formation in the tested cultivars, most of the shoots exhibiting high growth vigour continue their intensive growth in the middle of the growing season. In order to hold the shoots in position, they were further spread by plastic hooks (hereinafter referred to as hooks), aluminium wire and plastic twine shortly before shoot lignification at the base (Figure 2 a,b,c).

The use of hooks to spread shoots during the growing season showed a range of practical advantages: relatively easy installation; hooks can also be used to spread shoots that show strong growth; relatively easy removal after use; hooks can remain on the tree for use in the following growing season (which might cause their partial deformation); they cause no damage to either the leader or the shoot.

Although they increase production costs to some extent, these types of hooks have also been manufactured in the domestic market in the last years at a relatively affordable price. When purchasing hooks, it is advisable to pay attention to their resistance to UV radiation.

The use of aluminium wire (and plastic twine) for shoot bending in this research showed the following disadvantages: relatively low spreading efficiency per unit time (the wire cannot be cut to a required length far in advance of the spreading operation, but rather shortly before the treatment; this slows down the operation and the process of finding a suitable position on both the shoot and the leader or some other type of growth to which the wire is to be fastened – this was a problem especially with plastic twine which was fastened to two positions; a crotch angle of 90° is difficult to establish by wiring and twining in shoots being spread at the base (unless previously done by toothpicks); wires can often fall off after installation, especially at the point of contact with the shoot, and the operation requires correction (which was not the case with the plastic twine); during the growing season, wires must be removed from vigorous shoots to prevent them from cutting into the shoot tissue (in more extreme situations, this can lead to breakage of the shoot and, later on, the year–old branch), which is also a problem with plastic twine.

Wires are given priority in spreading strong shoots that are unspreadable by hooks, if no twisting is used.

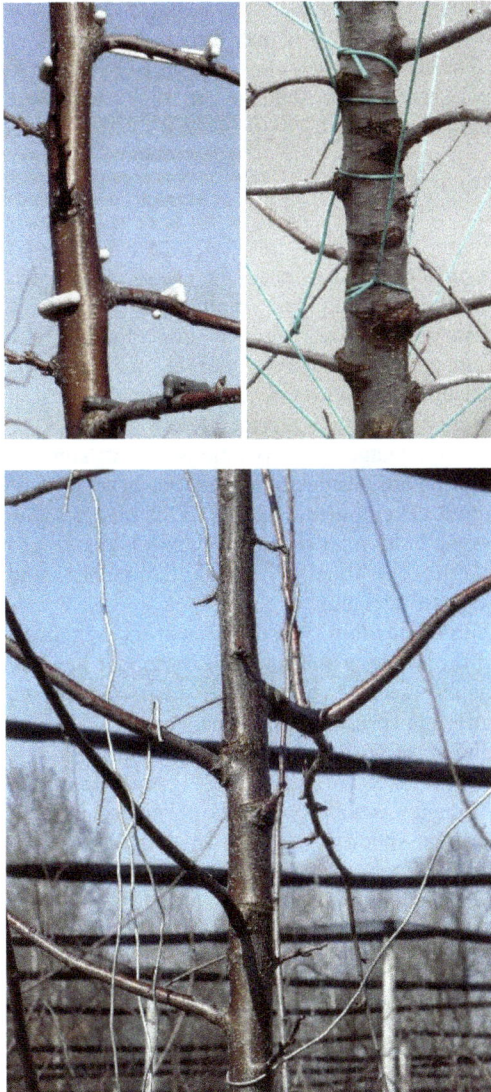

Figure 2. Spreading of shoots by plastic hook (a -variety 'Čačanska Lepotica'), plastic twine (b - variety 'Stanley') and aluminium wire (c - variety 'Čačanska Rodna')

Shoots that have been timely brought to a proper position in relation to the central leader will by the end of the growing season acquire the character of the fruiting wood i.e. mixed–type fruiting branch or, rarely, long vegetative growth.

Shoot twisting. The vigour of a plum tree, especially when Myrobolan is used as a rootstock, very often induces intensive shoot growth in the second part of the growing season, showing the tendency for apical growth, even if hooks or wire spreaders are previously used. In order to keep the shoots subordinate to the central leader and prevent apical dominance, shoot twisting was performed on the tested cultivars in early July.

The operation was aimed at reducing the vegetative growth activity of the shoots and inducing flower bud differentiation. Shoot twisting was done in accordance with the principles set down by Mićić et al. (2005), which ensure satisfactory operation and adequate performance: at twisting, the shoots had active tip growth; twisting involved holding the shoot base in one hand, while twisting the shoot over a wide area with the other hand, to prevent breakage; twisting was repeated if it failed to produce the desired effect in the first twisting treatment. In shoots exhibiting lower growth vigour and smaller base diameter, twisting entails smaller spacing between the hands i.e. the portion of the shoot subjected to the twisting pressure is shorter.

When twisting stronger shoots with greater base diameter, the spacing between the hands is wider, and the portion of the shoot subject to the twisting pressure is longer. This is common in shoots subjected to twisting later than considered optimum (degree of lignification of the shoot base) or which are not spread out in initial developmental stages (by either toothpicks or hooks). The twisting pressure was applied until creases appeared on the bark, and in certain cases until the bark was damaged or cracked. So far, research on the use of twisting has shown that bark damage due to twisting does not have an adverse effect on further growth and development (resin flow, shoot dieback) although this may be an open question from the viewpoint of the ecological and health situation of the trees (Figure 3 a,b,c).

The present research has confirmed that the damaged portion of the shoot shows very good healing in biological terms, and does not differ from the rest of the shoot. The position of the shoot when released speaks of the twisting performance.

The return of the shoot to its original position is a reliable sign of improper twisting. Strong twisting pressure which brings down the shoot completely results in complete crushing of growth vigour and fruit bud formation along the shoot.

Figure 3. Results of shoot twisting (a - variety 'Čačanska Lepotica', b - variety 'Stanley') and aluminium wire (c - variety 'Čačanska Rodna').

This is in favour of the conclusion that greater efficiency is always achieved by strong twisting. Twisting plays an important role in controlling vegetative growth and initiating cropping in initial years of spindle formation. Glišić (2012) observed that the combined use of canopy management practices, with twisting included, leads to trees coming into bearing very early and reaching full productivity at an earlier date in a considerable number of plum cultivars grown under high density system.

Twisting in other stone fruit crops is not so efficient (sour cherry) as in plums or it leads to excessive shoot damage or dieback (sweet cherry).

CONCLUSIONS

Plum production on Myrobalan (*Prunus cerasifera* Ehr.) seedling rootstock under spindle training and high density planting systems requires proper and timely tree management practices during the growing season, along with the need to follow the rest of intensive production principles.

Shoot management operations during summer pruning are designed to bring the shoots at their initial developmental stages to the best position possible in relation to the central leader and enable crushing of apical dominance and vegetative vigour.

Initial spreading by toothpicks at the beginning of the growing season and subsequent bending in the second part of the season provide optimum spreading performance.

Under intensive growth conditions in the second part of the growing season, high efficiency is achieved by shoot bending as well. Cultivar-specific response to these treatments can influence the date, effectiveness and efficiency of the treatments.

REFERENCES

Cvetković M., Mićić N., Đurić G., Bosančić B., 2015. Leader management techniques to induce vegetative bud development in plum. In: A. Papachatzis and Damiano Avanzato (eds) Book of abstracts. 3rd Edition. Present constains of Plum Growing in Europe. Skopelos: 28.

Čmelik Z., Duralija B., Bencic D., Druzic J., 2002. Influence of rootstocks and planting density on performances of plum trees. Acta Horticulturae, 577: 307-310.

Gavrilescu E., Cosmulescu S., Baciu A., Botu M., 2004. The influence of cultivar-rootstock combinaton on dinamical physiological proces in prune species. In: 8th International Symposium on Plum and Prune Genetics, Breeding and Pomology. Book of Abstracts, 8(112E).

Glišić I., Milošević T., Glišić S.I., Milošević N., 2007. Određivanje optimalnog termina za izvođenje zahvata rovašenja kod nekih sorti šljive. Zbornik naučnih radova 22. Savetovanja Unapređenje proizvodnje voća i grožđa, Grocka, 13(5): 41-46.

Glišić I., 2012. Pomotehničke specifičnosti šljive gajene u gustoj sadnji. Thesis. Faculty of Agronomy, Čačak.

Gonda I. 2006. The role of pruning in the intensification of plum production. International Journal of Horticultural Science, 12(3): 83-86

Grzyb Z.S., Rozpara E., 1998. Plum production in Poland. Acta Horticulturae, 478: 19-24.

Hrotkó K., Magyar L., Simon G., Klenyán T., 1998. Effect of rootstocks on growth of plum cultivars in a young orchard. Acta Horticulturae, 478: 95-98

Lučić P., Đurić G., Mićić N., 1996. Voćarstvo I. Institut za istraživanja u poljoprivredi Srbija: Nolit-Partenon.

Meland M., 2001. Early performance of european plum high density production systems. Acta Horticulturae, 557: 265-274.

Mićić N., Đurić G., Cvetković M. 2005. Sistemi gajenja i rezidba šljive. Ministarstvo poljoprivrede, šumarstva i vodoprivrede Republike Srbije, Beograd, 1-60.

Mika A., 1992. Trends in fruit tree training and pruning systems in Europe. Acta Horticulturae, 322: 29-36

Milosevic T, Zornic B, Glisic I., 2008. A comparison of low-density plum plantings for differences in establishment and management costs, and in returns over the first three growing seasons - a mini-review. Journal of Horticultural Science and Biotechnology, 83:539-542

Mitrović M., Blagojević M., Karaklajić-Stajić Ž., Rakićević M., 2005. Sistem guste sadnje u savremenoj tehnologiji gajenja šljive. Traktori i pogonske mašine, 10(2): 103-107.

Wilson BF., 2000. Apical control of branch growth and angle in woody plants. American Journal of Botany, 87: 601-607

http://faostat3.fao.org/browse/Q/QC/E (site visited 20.06.2016)

http://www.bhas.ba/saopstenja/2016/AGR_2015_001_01 -BS.pdf (site visited 20.06.2016)

11

GROWTH AND FRUITING POTENTIAL OF SOME APPLE VARIETIES WITH GENETIC RESISTANCE TO DISEASE, GROWN IN HIGH DENSITY SYSTEM

Gheorghe PETRE[1], Daniel Nicolae COMĂNESCU[1], Adrian ASĂNICĂ[2]

[1]Research and Development Station for Fruit Growing Voinesti,
387 Main Street, 137525, Dambovita, Romania
[2]University of Agronomic Sciences and Veterinary Medicine of Bucharest,
59 Marasti Blvd., District 1, Bucharest, Romania
Corresponding author email: statiuneavoinesti@gmail.com

Abstract

The research conducted at the Research and Development Station for Fruit Growing Voinesti in the period 2014-2016 highlights the growth potential of apple varieties with genetic resistance to diseases such as: 'Ariwa', 'Golden Lasa', 'Goldrush', 'Enterprise', 'Inedit', 'Iris', 'Luca', 'Real', 'Rebra', 'Redix', 'Remar', 'Saturn', 'Voinicel', all grafted on the M9 dwarfing rootstock. Orchard is defined as high density system with 2,500 trees/ha. The growth vigor of the ten years old trees, when the growth potential is well defined, indicates that between apple varieties with genetic resistance to the diseases under study, were significant differences noticed in the thickness of the trunk, size of the trees and crown volume. The production potential of apple varieties ranged from 28 to 44 t/ha in trees aged of 8-10 years. Most varieties yield over 30 t/ha, considered as apple varieties with high production potential. The level of costs for phytosanitary treatments in apple varieties with genetic resistance to disease is lower with more than 55% compared to the sensitive varieties like 'Jonathan', 'Golden Delicious', 'Starkrimson'.

Key words: Vf resistance, productivity, economic efficiency, environment.

INTRODUCTION

Apple's assortment has seen a significant change over the last decades, with varieties dedicated to meet growing consumer demands (Petre et al., 2005). A special situation is the promotion of apple varieties with genetic resistance to diseases, which for the new plantations are the key to the efficient economic technology with immediate effect by the total or partial elimination of the fungicide treatments (Comanescu et al., 2012).

The gradual change of the assortment through the promotion of varieties of apple with genetic resistance to diseases depends on certain characteristics that they can exhibit (Comanescu, 2002). The superiority of new varieties should mainly address the requirements of the producer, oriented to economic efficiency, vigour of growth, high production potential, fruit appearance, consumption period, as well as the taste of the consumer (Petre et al., 2014, 2015).

The important advances made in the research of apple varieties with genetic resistance to

diseases at RDSFG Voineşti allowed study of 13 scab resistant varieties behaviour in the orchard, with 'Jonathan', as sensitive to diseases control variety.

MATERIALS AND METHODS

The research was organized in an apple orchard belonging to RDSFG Voineşti, Experimental Base no. 1, between 2014 and 2016, and highlights the potential for growing and fruiting of the apple varieties with genetic resistance to diseases, representing the main factor in obtaining of an adapted production to the requirements of European quality standards. We have studied 13 varieties with genetic resistance to diseases, namely 'Ariwa', 'Golden Lasa', 'Goldrush', 'Enterprise', 'Inedit', 'Iris', 'Luca', 'Real', 'Rebra', 'Redix', 'Remar', 'Saturn', 'Voinicel' compared to sensitive apple variety 'Jonathan'. All apple varieties were grafted on M 9 rootstocks. The trees were planted in 2007 at a distance of 4 x 1m (2,500 trees / ha), trained as slender spindle (Figure 1).

Figure 1. Experimental field with scab resistant varieties trained as slender spindle

Observations and determinations have been made regarding the growth vigour expressed in in trunk thickness and tree crown size, production record, and average fruit weight.

The orchard's soil is brown eumesobasic, pseudogley, with a pH ranged between 5.7 - 5.9. Humus content is medium in the upper part (2.0 - 2.9%), medium supplied with nitrogen and poorly supplied in phosphorus and potassium.

The climatic conditions were favourable for growing and fruiting of trees, characterized by an average annual temperature higher than 1.9⁰C, with an annual precipitation amount of 755 mm.

In the orchard, natural grass was maintained mowed between the rows and clean along the tree rows. To control pests 6 to 8 treatments were applied only with insecticides.

RESULTS AND DISCUSSIONS

Trees Growth

The vegetative growth of trees is determined by a number of biological factors such as variety, rootstock, disease and pest resistance, but also technological: fruit load, optimal provision of technological measures, as well as nutrition and water conditions necessary for the development of the physiological processes.

The vigour of apple varieties trees with genetic resistance to diseases is manifested in quantitative terms by the annually volume of vegetative growth, trunk size, height and size of the tree crown. These parameters are depending on the vigour of the varieties, corroborated with the degree of fertility of the soil, planting distance etc.

According to the diameter of the tree trunk, 10 years after planting, the scab resistance apple varieties grown in a high density system and grafted on the M9 rootstock rank as follows (Table 1):

- vigorous varieties with trunk diameters above 80 mm: 'Luca' (96.18 mm), 'Golden Lasa' (80.40 mm), 'Enterprise' (87.90 mm), 'Rebra' (81.62 mm), 'Redix' (87.23 mm), 'Remar' (85.90 mm);

- medium vigorous varieties, with trunk diameters between 65-80 mm: 'Iris' (67.35 mm), 'Inedit' (70.50 mm), 'Voinicel' (67.60 mm), 'Real' (67.50 mm), 'Ariwa' (68.03 mm);

- low vigour varieties with trunk diameters below 65 mm: 'Saturn' (62.39 mm), 'Goldrush' (55.44 mm).

Table 1. Vegetative growth of apple varieties trees cultivated in high density system (2,500 trees / ha)

No	Variety / M9	Diameter (mm)		Tree size (cm)		Crown volume		Dif.
		value	Average growth increase	Height	Thickness of the fruiting fence	mc/tree	mc/ha	± control
1	Ionathan (control)	66.38	5.79	245	130	2.53	6,325	-
2	Golden Lasa	80.40	7.17	240	140	2.66	6,650	+325
3	Ariwa	68,03	6.35	250	135	2.70	6,750	+425
4	Goldrush	55.44	4.74	230	120	2.16	5,400	-925
5	Enterprise	87.90	6.57	270	150	3.30	8,250	+1,925
6	Inedit	70.50	6.20	230	120	2.16	5,400	-925
7	Iris	67.35	6.50	205	135	2.09	5,225	-1,100
8	Luca	96.18	8.59	270	145	3.19	7,975	+1,650
9	Real	67.50	6.60	250	130	2.60	6,500	+175
10	Rebra	81.62	7.43	275	140	3.15	7,875	+1,550
11	Redix	87.23	7.53	270	140	3.08	7,700	+1,375
12	Remar	85.90	7.80	260	135	2.83	7,075	+750
13	Saturn	62.39	4.43	230	130	2.34	5,850	-475
14	Voinicel	67.60	5.30	230	130	2.34	5,850	-475

The average growth of the trunk, in the 10th year since planting, exceeded 7 mm in the varieties: 'Golden Lasa', 'Luca', 'Rebra', 'Redix' and 'Remar', and below 7 mm varieties such as 'Inedit', 'Iris', 'Ariwa', 'Enterprise', 'Saturn', 'Real', 'Voinicel' and 'Goldrush', which recorded an increase in growth between 4.43 - 6.60 mm.

The tree crown volume provides the skeleton to support the branches, leaves and fruits. Depending on the type of orchard, crown volume differs, both at the tree level and at the surface unit level. Through its structure, the crown's volume must ensure that the light penetrates all the tree elements, a prerequisite for maintaining its garnish with fruit branches to produce as much productive volume as possible.

The volume of the crown depends on the size of the trees, which is influenced to a large extent by the vigour of the variety. Based on the vigour of the trees, the crown volume was calculated at the tree level and at surface unit.

Regarding the crown volume recorded in the 10th year of planting at the tree level, it oscillated with quite large variation between varieties, from 2.09 cm / tree to 'Iris', 'Goldrush' and 'Inedit' variety 2.16 cm / tree, up to 3.30 cm / tree in the 'Enterprise' variety. Larger volumes were remarked at 'Enterprise', 'Luca', 'Ariwa', 'Real', 'Golden Lasa', 'Rebra', 'Remar', 'Redix', with the crown volume on the tree of more than 2.5 cubic meters.

The volume of the crown calculated on the surface unit follows the same grading of varietal vigour, given by the planting density of 2500 trees/ha for all varieties. Thus, the largest crown was recorded in the 'Enterprise' variety - 8,250 mc / ha and the lowest in 'Iris' varieties with 5,225 and 'Goldrush' with Inedit 5,400 mc / ha respectively.

Lower values of the crown volume at the surface unit were also recorded for the apple varieties 'Saturn' and 'Voinicel' with values of 5,850 mc / ha.

In terms of vegetative growth, small and medium-sized varieties are suitable for high-density orchards at 4 x 1 m planting distances with 2,500 trees / ha.

Trees Productivity

One of the main goal of the research is the evaluation of apples production capacity, a very strong reason for new modern orchards.

The high productive potential associated with the superior quality of fruit expresses the highest degree of genetically-resembled ability of scab apple varieties in the ecological conditions of the area in which they are grown.

The productivity of apple varieties with genetic disease resistance, which is the subject of this study, is a complex, genetically determined hereditary base from which it originates, but is influenced by the interaction between the variety and the climatic conditions of the area of culture. Other factors contributing to the mapping of this attribute of genetically resistant apple varieties are related to the precocity of the fruit, the type of fructification, the applied technology, the resistance to diseases and pests, the compatibility to grafting and pollination, the density of planting and the rootstock used.

For the correct analysis of the harvest quantity, it was envisaged to record the fruit production each year and to determine the level of production, the self-regulation or the intermittent fructification trend.

The annual record of apple production at the variety level shows that there are differences in production levels.

For the appreciation of the productivity of the varieties, the production recorded in the 8-10 years from planting was taken into account (Table 2).

The yield obtained in the 8th - 10th years from planting show the outstanding performance of the apple-high-density system, which can be expanded into well-established fruit-growing areas, only with the most productive varieties, which give quality fruit adapted to market demands.

In the 8th leaf, the largest productions were recorded for the varieties 'Real', 'Remar', 'Iris', 'Luca', ranging from 38.8 to 46.8 t/ha. Most varieties recorded over 30 t/ha, including Jonathan.

In the 9th year after planting, the highest yields were obtained from 'Ariwa', 'Iris', 'Real', 'Remar' and 'Saturn', ranging from 34 to 38 t/ha. The other varieties recorded yields ranging from 24.7 to 29.5 t/ha, compared to the Jonathan, which was recorded 22.5 t / ha.

Table 2: Production of apple varieties with genetic resistance to diseases, cultivated in a high density system (2,500 trees / ha), in the 8-10 years leaf

No	Variety	Yield (t/ha)				Dif. ± Control
		8th year / 2014	9th year / 2015	10th year / 2016	Average	
1	Ionathan (control)	30.3	22.5	24.3	25.7	-
2	Golden Lasa	31.3	29.0	29.8	30.0	+ 4.3
3	Ariwa	30,3	34.0	28.5	30.9	+ 5.2
4	Goldrush	36.8	24.7	28.8	30.1	+ 4.4
5	Enterprise	36.5	38.2	28.5	34.4	+ 8.7
6	Inedit	35.6	28.7	30.3	31.5	+ 5.8
7	Iris	46.8	38.2	48.3	44.4	+ 18.7
8	Luca	39.5	25.2	24.5	29.7	+ 4.0
9	Real	38.8	35.5	29.3	34.5	+ 8.8
10	Rebra	28.0	25.7	30.6	28.1	+ 2.4
11	Redix	29.3	29.5	28.3	29.0	+ 3.3
12	Remar	43.5	38.5	40.3	40.8	+ 15.1
13	Saturn	32.8	37.2	29.3	33.1	+ 7.4
14	Voinicel	33.6	28.5	29.3	30.5	+ 4.8

In the year 2016 (10th year after planting), the largest production was recorded in the varieties 'Golden Lasa', 'Real', 'Rebra', 'Saturn', 'Voinicel', 'Inedit', 'Iris', 'Remar' ranging from 29.3 to 48.3 t/ha. In the other varieties, yields ranging from 24.5 to 28.8 t/ha were recorded, compared to the variety Jonathan, which was registered in 10th year of planting 24.3 t/ha.

Analysing the average of the yields obtained in the three years of study, the yields of the varieties: 'Real' (34.5 t/ha), 'Saturn' (33.1 t/ha), 'Enterprise' (34.4 t/ha), 'Remar' (40.8 t/ha) and 'Iris' (44.4 t/ha). Most varieties have produced over 30 t/ha, being included as branded varieties with high production potential.

Production quality

Apples for fresh consumption or for processing must be healthy, ripen for commercial or consumption purposes and to have organoleptic properties specific to the variety. In order to sell apples at higher prices, they must be at market-standardized levels and maintain their quality after harvesting and during storage and delivery time.

The quality of the fruits is a genetic trait influenced by the variety, the degree of maturation, the action of the environmental factors, as well as of the technological factors. The effects of the interaction of these factors materialize by obtaining fruits with special qualities or they can be negatively influenced.

Low-vigour vegetative rootstocks significantly contribute to the enhancement of apple quality if the varieties grown under appropriate ecological conditions and some technological performance measures are applied.

Fruit Quality Parameters

Average weight and size are important elements in assessing the commercial quality of fruits. These are characteristics of the variety and can be influenced to a greater or lesser extent by the amount of production, the age of the tree, the applied crop technology and the climatic conditions of the year.

The interest in obtaining large fruit with high commercial value is one of the major objectives in the pomological appreciation of the variety. It is known that in European standards, apple size is far superior to that obtained in Romania. Under these circumstances, we must be demanding to choose varieties to promote them in commercial crops and to apply appropriate technology to achieve this goal.

The size of the fruit, combined with an intense and uniform colour (red or golden type) and a symmetrical shape, gives the fruit the appearance and attractiveness required. The study of apple varieties with genetic resistance to diseases grown in high density system reveals a genetic variability in fruit size. A fruit size variability among the varieties studied was found.

It is appreciated that an average fruit size of 170 - 180 g is appropriate for a modern apple variety. Most apple varieties in the study have a fruit size that corresponds to and competes with modern varieties, to the extent that varieties of genetic scab resistance grown in a high density system are promoted in culture in order to obtain some organic production, increasingly demanded by consumers.

The size of the fruit in the 9th year after planting had fairly large amplitude, from 140 g to 'Ariwa', 'Inedit', 'Saturn', to 180 g in the 'Real' variety. Fruit over 160 g were obtained in the varieties 'Enterprise', 'Luca', 'Rebra', 'Redix', 'Remar' and smaller in the varieties 'Goldrush', 'Inedit', 'Iris', 'Ariwa', 'Golden Lasa', 'Saturn' and 'Voinicel', with 140 - 155 g; the differences are due to the genetic factors, respectively the varieties (Table 3).

Table 3: Fruits quality parameters, from apple scab resistant varieties at the harvest date (2015 - 2016)

No	Variety	Fruit weight (g)		Flesh firmness (kgf/cm^2)		Dry substance (%)	
		2015	2016	2015	2016	2015	2016
1	Ionathan	145	148	7.6	8.2	14.2	14.6
2	Golden Lasa	145	165	10.6	10.2	13.8	15.8
3	Ariwa	140	155	10.2	10.5	13.8	15.2
4	Goldrush	145	148	11.4	10.8	14.1	15.0
5	Enterprise	175	165	9.8	10.7	14.0	15.9
6	Inedit	140	158	7.4	8.2	14.2	16.0
7	Iris	145	153	7.8	7.2	13.8	13.6
8	Luca	160	162	10.8	8.7	15.2	15.7
9	Real	180	175	9.6	10.2	13.5	13.0
10	Rebra/M9	170	171	9.1	8.1	14.1	13.4
11	Redix	175	165	10.2	8.6	14.1	15.5
12	Remar	165	172	8.8	8.6	14.0	13.8
13	Saturn	140	160	9.4	9.0	15.4	15.5
14	Voinicel	145	155	7.0	7.5	14.2	14.7

In the 10th leaf, the fruit biomass oscillated from 148 grams at the 'Goldrush' variety (figure 2) and 175 grams at the 'Real' variety. Fruit with biomass over 160 g was obtained in most of the varieties with genetic resistance to the diseases studied: 'Golden Lasa', 'Enterprise', 'Redix' with 165 g, 'Rebra', 'Remar', 'Real' with biomass fruits ranging from 171 to 175 g. The smallest fruits were recorded in the 'Goldrush', 'Iris' and 'Voinicel' varieties with biomass ranging from 148 to 155g.

Figure 2. 'Goldrush' fruit size in overloaded trees

The average value of fruit variety at the level of the variety shows that the varieties 'Enterprise',

'Luca', 'Real', 'Rebra', 'Redix' and 'Remar' have the potential to ensure the large size of the fruit to compete on the market, and the other varieties fall into the middle fruit size group.

The firmness of the fruit pulp is considered to be appropriate when it is compact, crisp or fine, with uniform colouring. A poor firmness is considered to be inappropriate, and is generally found in over-ripen fruits. Also, a rough, coarse consistency with glassy areas is considered negative.

Under the conditions of 2015, the firmness of the fruit pulp at harvest time ranged from 7.0 kgf/cm^2 to the 'Voinicel' variety and 11.4 kgf/cm^2 in the 'Goldrush' variety.

From the data recorded in 2015, it appears that there are varieties producing firm fruit such as: 'Goldrush', 'Ariwa', 'Golden Lasa', 'Enterprise', 'Luca', 'Real', 'Rebra', 'Saturn', 'Redix' and 'Remar', with average penetration resistance of 8.60 kgf/cm^2 and 11.4 kgf/cm^2 and varieties such as 'Voinicel', 'Inedit', 'Iris' and 'Jonathan' as control, whose average fruit penetration resistance is much lower than 7 - 7.8 kgf/cm^2.

In 2016 the firmness of the pulp was between 7.2 kgf/cm^2 in the 'Iris' variety and 10.8 kgf/cm^2 in the 'Goldrush' variety.

The dry matter content of apple varieties with genetic resistance to disease in the years 2015-2016 was 13.0 - 13.5% for 'Real' apple variety, being the earliest variety with a dry matter content of 13,4 - 16,0%. For the other varieties, dry matter content accumulated according to the variety and the ripening period of the fruits.

Varietal conveyer of scab apple varieties

Studies and research undertaken at RDSFG Voineşti presents for apple growers, especially for those who promote modern apple culture systems, a group of varieties with different maturation ages covering a long period of consumption with apples of resistant varieties.

Now, the basic range includes the varieties with genetic resistance to diseases: 'Romus1', 'Romus 3', 'Romus 4', 'Prima', 'Pionier', 'Voinea', 'Ciprian', 'Florina' which are propagated in the fruit trees nurseries in the country. Some of the varieties mentioned, even if they currently meet the requirements of fruit resistance, productivity and quality, can be replaced as new varieties become more valuable, both in terms of production and quality of fruit.

The apple varieties that have been studied can cover much better the consumer season, along with some genetically scab resistant varieties already known and appreciated on the market by consumers.

Table 4 indicate scab apple varieties that were the subject of the study during the period 2014 - 2016 and the way they fit among the valuable varieties with genetic resistance to diseases from the current apple assortment.

Most of the varieties presented are suitable for growing in high-density orchards that will represent future orchards for apple growers.

Depending on the period of maturation and consumption of fruits, the apple varieties studied, which have been distinguished by genetic resistance to disease, productivity and quality of fruit, fall differently in the variety conveyor for Dâmboviţa fruit area.

Thus, according to the varieties of apple 'Romus 1', 'Romus 3', 'Romus 4', 'Irisem', the 'Real' variety can be introduced, thanks to appetizing fruits, ripen in the last decade of August - first decade of September.

The varieties of apple such as 'Saturn', 'Remar', 'Golden Lasa' fall between the varieties 'Voinea' and 'Pionier' with the goal of replacing 'Voinea' variety by 'Remar' because of his superior qualities, taste and fruit coloring.

Table 4: The apple varieties consumption period, in the frame of the scab resistant apple varietal conveyer, cultivated in the Voinesti area

Variety	VII			VIII			IX			X			XI			XII			I			II			III		
	1	2	3	1	2	3	1	2	3	1	2	3	1	2	3	1	2	3	1	2	3	1	2	3	1	2	3
Romus1(Vf)		•	•																								
Romus3(Vf)				•	•																						
Romus4(Vf)							•	•																			
Irisem (Vf)							•	•																			
Real (Vf)								•	•	•	•	•	•														
Prima (Vf)								•	•	•	•	•	•														
Voinea (Vf)									•	•	•	•	•														
Saturn (Vf)									•	•	•	•	•														
Remar (Vf)									•	•	•	•	•														
Golden Lasa (Vf)									•	•	•	•	•														
Pionier (Vf)									•	•	•	•	•	•	•	•											
Voinicel (Vf)									•	•	•	•	•	•	•	•											
Iris (Vf)									•	•	•	•	•	•	•	•											
Ciprian (Vf)								•	•	•	•	•	•	•	•												
Ariwa (Vf)									•	•	•	•	•	•	•	•	•	•	•	•	•						
Luca (Vf)									•	•	•	•	•	•	•	•	•	•	•	•	•						
Rebra (Vf)									•	•	•	•	•	•	•	•	•	•	•	•	•	•	•				
Redix (Vf)									•	•	•	•	•	•	•	•	•	•	•	•	•	•	•				
Inedit (Vf)										•	•	•	•	•	•	•	•	•	•	•	•	•	•	•	•	•	•
Enterprise (Vf)										•	•	•	•	•	•	•	•	•	•	•	•	•	•	•	•	•	•
Florina (Vf)										•	•	•	•	•	•	•	•	•	•	•	•	•	•	•	•	•	•
Goldrush (Vf)										•	•	•	•	•	•	•	•	•	•	•	•	•	•	•	•	•	•

Also for autumn fresh consumption, the 'Iris' and 'Voinicel' varieties can be up-scaled, and multiplied along with the 'Pionier' variety, which should be replaced.

The 'Ariwa', 'Luca', 'Rebra' and 'Redix' varieties fill up the consumption period that extends until January and February.

The varieties 'Inedit', 'Enterprise' and 'Goldrush' exceed Florina's consumption period.

Economic and environmental impact by promoting scab resistant apple varieties

Apple's cultivar assortment is very dynamic due to a competitive market. Is very important to stay economic efficient and therefore, total or partial elimination of the fungicide treatments has a significant role in it.

It can be appreciated that the orientation towards apple varieties with genetic resistance to diseases will gradually be imposed also in high density orchards, not only for the economic efficiency, but also for obtaining ecological productions.

By promoting the cultivation of apple varieties with genetic resistance to diseases adapted to the ecological conditions in our country, we increase the quantity of apples on the market with low pesticide levels, with beneficial influences on the consumers and the environment too.

Due to the reduction of the number of phytosanitary treatments and the quantities of pesticides, the costs are reduced by more than 55%.

The research and development center for fruit growing in Voinesti has been and remains the promoter of the disease-resistant assortment and the high-density apple-tree system in our country.

The economic and environmental effects are highlighted by the costs of phytosanitary treatments per one hectare of orchard.

From the data presented in Table 5, it results that between the two cultivated varieties there are significant differences in the total number of warnings requested during the vegetation period, the quantities of pesticides, the consumption of diesel fuel and the related costs.

Thus, in the orchard with sensitive varieties in the years 2014 - 2016 it was performed 14

sprays; while in the orchard with resistant varieties only 7. The savings made in the orchard with resistant varieties, by the elimination of fungicides in the proportion of 90% and 81% reduction in insecticides and acaricides, represents 66% of susceptible varieties, which means that in orchards with resistant varieties apply 50% fewer sprays and their value is 2 times lower than orchards with a classically sensitive range. Diesel fuel consumption is reduced by 53%.

Table 5. Economic efficiency of some scab resistant apples varieties vs susceptible varieties (2014 - 2016)

Item	Susceptible varieties	Scab resistant varieties	Economic effect (%)
No of sprays	14	7	50
Insect-fungicide consumption (kg,l):	122	54	56
- fungicide (kg,l)	63	6	90
- insecticide – acaricide (kg,l)	59	48	81
Costs (ron):	11808	4464	62
- phytosanitary products	8400	2856	66
- labour	1200	600	50
- mechanic works	2208	1008	54
Diesel consumption (l)	90	42	53
- value (lei)	540	252	53

In addition to the beneficial economic effects, we must add pollution reduction, faster recovery of predators and natural parasites, and maintaining the quality standard of fruit.

CONCLUSIONS

Depending on the diameter of the trunk of the trees, the scab resistant varieties were grouped as follows:

- vigorous varieties: 'Luca', 'Golden Lasa', 'Enterprise', 'Rebra', 'Redix', 'Remar';
- medium-vigorous varieties: 'Iris', 'Inedit', 'Voinicel', 'Real', 'Ariwa';
- small vigorous: 'Saturn', 'Goldrush'.

The volume of the tree crown recorded in the 10th year since planting indicate 'Iris', 'Goldrush' and 'Inedit' as less vigorous varieties than 'Enterprise' variety.

The volume of crown calculated per unit area oscillated between 5,100 cubic meters/ha for 'Goldrush' and 'Inedit' varieties, up to 8,250 mc/ha in the 'Enterprise' variety.

Varieties with small vegetative growth are suitable for expansion in high-density orchards at distances of 4 x 1 m at a density of 2,500 trees / ha.

The highest yield was achieved by 'Real' 'Saturn', 'Enterprise', 'Remar' and 'Iris'

The size of the fruit was influenced by the size of the crop load with fairly large amplitude, from 148 g ('Goldrush') to 175 g ('Real' variety).

'Goldrush', 'Ariwa', 'Golden Lasa', 'Enterprise', 'Luca', 'Real', 'Rebra', 'Saturn', 'Redix' and 'Remar' varieties produce firm fruits with average penetration resistance between 8.60 and 10.8 kgf/cm^2.

The lowest content in dry substance was recorded in the 'Real' variety.

The high-density apple system with genetically resistant varieties is recommended for fruit-growing areas in our country due to the high economic efficiency and continuously need for better cultivars.

The elimination of fungicidal products and use of highly selective insecticides, correlated with the quality and productivity of new resistant varieties, support the environmental effects.

REFERENCES

Comănescu D.N, 2002. Cercetări privind sistemul de mare densitate la măr, în scopul obţinerii de producţii adaptate la cerinţele de comercializare, Teză de doctorat.

Comănescu D., Petre G., Petre V., 2012, The behaviour of some apple tree varieties with genetic disease-resistance in the high density system. Scientific Papers. Series B. Horticulture, Vol. LVI, 63-68.

Petre Gh., Petre V., Neagu I.O., 2005. Particularităţile de creştere şi rodire şi efecte ale unor secvenţe tehnologice specifice soiurilor de măr cu rezistenţă genetică la boli, Lucrări ştiinţifice ICPP Piteşti-Mărăcineni, Vol.XXII, Ed. Pământul, Piteşti, 157 – 163.

Petre Gh., Andreieş N., Petre Valeria, 2005. Tehnologia obţinerii unor producţii de mere competitive, Ed. Pildner, Târgovişte.

Petre V., Petre G., Asănică A. 2015, Research on the use of some apple genitors in the breeding process for genetic resistance to disease and fruit quality, Scientific Papers. Series B, Horticulture, Volume LIX, 75-80.

Petre G., Comănescu D. N., Petre V. 2014, Peculiarities of growth and fruitfulness of apple cultivars with genetic resistance to diseases grown under high density system. Scientific Papers. Series B, Horticulture, Volume LVIII, 75-80.

12

THE IMPROVEMENT OF THE ROMANIAN APPLE ASSORTMENT HERITAGE WITH NEW VARIETIES WITH GENETIC RESISTANCE TO DISEASE - REVIDAR, CEZAR AND VALERY

Valeria PETRE[1], Gheorghe PETRE[1], Adrian ASĂNICĂ[2]

[1]Research and Development Station for Fruit Growing Voinesti,
387 Main Street, 137525, Dambovita, Romania
[2]University of Agronomic Sciences and Veterinary Medicine of Bucharest,
59 Marasti Blvd., District 1, Bucharest, Romania

Corresponding author email: statiuneavoinesti@gmail.com

Abstract

Between 2014-2016, at the Research and Development Station for Fruit Growing Voinesti it was evaluated the performance of three brand new scab apple varieties: 'Valery', 'Cezar' and 'Revidar', all patented by ISTIS in 2016. The growth vigor of trees in the 9th year, grafted on the M9 rootstock, is well defined by the trunk circumference, which records between 14.9 cm in the 'Caesar' variety and 16.5 cm in the 'Valery' variety. The crown volume calculated at the surface unit and for a density of 2,857 trees/ha was of 8,200 mc/ha for the 'Caesar' and 'Revidar' and 8,900 mc/ha for the 'Valery' variety. Productivity of trees in the ages of 7-9 years was between 35-40 t/ha in the 'Caesar' and 'Valery' varieties and 28.30 t/ha at the 'Revidar' variety. The fruit weight ranged from 160 to 190 g, smaller fruits being recorded in the 'Revidar' variety and higher in the 'Caesar' and 'Valery' varieties, which correspond to market requirements. The Remarkable quality of the varieties fruits recommends them for an increasingly demanding fruit market, meeting the current quality requirements and consumer's needs.

Key words: Vf resistance, yield, fruit quality, variety.

INTRODUCTION

Apple's assortment has seen a significant change in the past decades, promoting varieties that primarily target the producer's requirements, sensitive to economic efficiency, high production potential, fruit appearance, ripening time, etc., as well as consumer tastes. These requirements are satisfied by expanding the cultivars of apple varieties with genetic resistance to diseases, which for the new plantations are linked of the efficient economic technology, with immediate effect by the total or partial elimination of the fungicide treatments (Petre V., 2009, 2014).

In the promotion of varieties of apple with genetic resistance to diseases, an important role was played by the Research and Development Station for Fruit Growing Voinesti, either by own creation or by studying foreign varieties, managing to greatly change the vision of the fruit farmers and the gradual change of the assortment.

The continuous breeding process has allowed the creation and patenting of 18 varieties of apple with genetic resistance to diseases (CociuV. et al, 1999), among them 'Valery', 'Cezar' and 'Revidar', were approved in 2016, valuable varieties which will surely meet the growing demands of consumers.

MATERIALS AND METHODS

The complex genetic base existing at the Research and Development Station for Fruit Growing Voinesti, consisting of selection fields, hybrid nursery and competition microcultures, was the main source of selection of valuable apple tree elites and registered with ISTIS for testing for new patenting varieties.

Elite apple H 1/16-90, H 1/78-90, H 4/ 37-04, existing in the microculture set up in 2009 for compete, corresponded in terms of fruit productivity and quality, being registered at ISTIS To be tested for approval from 2014, it being reviewed for 2 years (2014-2015) and analysed according to the DUS and VAT test criteria and techniques required for approval. These became varieties from 2016 under the names 'Revidar', 'Caesar' and 'Valery'.

In order to highlight the performance traits of the three varieties still existing in the competition microculture set up in 2009, the crop technology was applied correctly in order not to affect the production capacity and the quality of the fruits.

The researches carried out during the period 2014-1016 highlight the growth and fruiting potential of new varieties of apple-resistant apple, 'Revidar', 'Caesar 'and 'Valery', where the trees were planted at a distance of 3.5 x 1 m (2,857 trees / ha), grafted on the M9 rootstock, with the spindle shape of a crown.

During the study years, observations and determinations have been made regarding the growth in trunk thickness, tree crown size, production record and fruit quality, by their biomass and dry matter content.

The orchard soil is brown eumesobasic, slightly pseudogley, with a acidic pH (5.7 - 5.9), the humus content is medium to surface (2.0-2.9%), medium supplied with nitrogen, poorly supplied in phosphorus and potassium.

The climatic conditions were favourable for growing and fruiting of trees, characterized by an average annual temperature higher than 1.0°C, normal for the area of 8.8°C, with an annual rainfall of 755 mm.

In the orchard, the soil was maintained as permanent grass cover and weedy clean interval on the row of trees. To control pests, 6 to 8 treatments were applied with insecticides only. The other works were executed according to the technology specific to the high-density orchards.

RESULTS AND DISCUSSIONS

In the process of developing performances in fruit production, varieties must be promoted with a genetic basis that allows:
- increasing the production potential and the quality of the fruit;
- natural increase of resistance to diseases and pests in order to protect the environment;
- suitability in the application of high performance technologies.

An essential condition is compliance with the crop technology in order not to affect the production capacity of the variety and the quality of production. Proper application of technologies preserves or improves varieties' performance.

Obtaining varieties of apple is a long-lasting activity and a great deal of complexity, especially when it comes to obtaining varieties of genetic resistance to disease, irrespective of the research method used.

A new variety, in addition to the productivity, superior fruit quality, genetic resistance to diseases, depending on the area of culture, has to meet other attributes that are added to the essential conditions, namely:
- degree of adaptability to climatic conditions;
- destination of the production, depending on the degree of knowledge of the variety;
- market requirements of the production obtained;
- safety of the production and delivery source of the fruit propagating material;
- economic efficiency of cultural technology.

In the competition microculture, an agrotechnical scheme was applied to the varieties of genetically resistant apple cultivars, including varieties patented in 2016 (mowing the grass between the rows, herbicidation along the row of trees, phytosanitary treatments, only insecticide) and monitoring the vegetation condition of the biological material.

In the 9 years of vegetation, apple elites / apple varieties approved in 2016, 'Valery', 'Cezar' and 'Revidar' have grown properly, fruiting since the 3rd leaf, proving superior quantitative and qualitative production, resistance to scab and mildew.

In order to highlight the characteristics of apple varieties approved in 2016, the growing and fruiting traits recorded during the exam period, namely 2014 - 1016, are forward reproduced.

The vigour of trees in quantitative terms is given by the volume of vegetative growth accumulated annually, expressed by the size of the trunk, the height and size of the tree crown, these being determined by the vigour of the variety, the stable factor being the rootstock.

Growth vigour in tree of nine years old, when the growth potential is well defined, indicates that there are significant differences in the growth in trunk thickness, the height and crown dimensions between the apple varieties with genetic resistance to the diseases under study (Table 1).

Table 1. Growing and fruiting particularities of the new varieties with genetic resistance to diseases patented in 2016

Item	Variety			
	'Valery'	'Cezar'	'Revidar'	Jonathan (control)
I. Growth vigour				
Trunk circumference - cm	16.5	14.9	15.0	14.2
Trees height - cm	290	270	270	280
Fruit tree fence thickness - cm	130	130	130	130
Crown volume - mc/ha	8,900	8,200	8,200	8,500
II. Fruiting				
Phenophases				
Start of blossom	24 – 29.04	22 – 28.04	22 – 28.04	20 – 26.04
End of blossom	01 – 06.05	30.04 – 05.05	30.04 – 05.05	28.04 – 03.05
Blooming time	7 – 9	7 – 9	7 – 9	8 – 9
Date of the fruit ripening	25.09 – 01.10	15 – 20.09	25 – 31.08	20.09 – 30.09
Consumption period	oct. - martie	oct. – decembr.	septembrie	oct. – ianuarie
Storability (days)	145 – 150	80 – 86	30 – 35	125 – 135
Yield (t/ha)	35 – 40	35 – 40	28 – 30	28 – 30
Production quality				
Fruit weight (g)	185	190	160	155
Dry substance content (%)	16.5	13.8	13.0	13.5

The trunk circumference of the three varieties of apple, grafted on the M9 rootstock, recorded values between 14.9 cm in the 'Caesar' variety and 16.5 cm in the 'Valery' variety, compared to 14.2 cm as recorded to the variety 'Jonathan' assigned as a control.

The volume of the crown provides the structure for supporting the branches, the leaves and the fruits and ranges according to the size of the trees. The height of the trees in year 9 records values between 270 cm and 290 cm comparing to the 'Jonathan' variety where the height of the trees was of 280 cm. The thickness of the fruit tree fence has values corresponding to the three varieties of apple taken in the study, including the 130 cm of the control variety.

The volume of the crown calculated at the surface unit at the planting density of 2,857 trees / ha was of 8,200 mc / ha for the 'Caesar' and 'Revidar' varieties and 8,900 mc / ha for the 'Valery' variety, with the medium values at the 'Jonathan' variety as a control with 8,500 mc / ha.

Based on vigour values, the present varieties studied have been included into middle-strength varieties ('Valery') and small-medium vigorous varieties ('Caesar', 'Revidar').

For the evaluation of the rhythm of the fructification phenophases, the data related to: the beginning and the end of the flowering, its duration, but also the maturation period of the fruits were recorded.

From the dates recorded between 2014 and 2016, the first flowers were opened in 'Caesar' and 'Revidar' varieties from April 22 to 28, followed by the 'Valery' variety between 24 and 29.04. The end of the bloom took place between 30.04 and 06.05 so that the three varieties studied are enclosed into varieties with a medium flowering period and a flowering period of 7 to 9 days, depending on the evolution of climatic conditions.

The varieties taken in the study overlapped totally or partially during flowering, allowing mutual pollination.

Fruit maturation took place between 25th and 31st of August in the 'Revidar' variety, being considered as summer variety, 15-20.09 in the 'Caesar' variety and 25.09-01.10 in the 'Valery' variety, which has a longer storage of the fruits.

The duration of fruit storage was 30-35 days for the 'Revidar' variety, 80-86 days for the 'Caesar' variety and 145-150 days for the 'Valery' variety, compared to the 'Jonathan' variety as control, where the storage duration of the apples in the cooling space is 125 - 135 days.

One of the priority objectives of the study is the evaluation of the production capacity, being the most important characteristic in promoting of the varieties for the establishment of the new commercial plantations. The high productive potential, associated with superior fruit quality, expresses to the highest degree the capacity of apple varieties with genetic resistance to diseases, taken in the study to assimilate and capitalize the ecological conditions of the area in which they are grown.

Productivity of apple varieties is a complex attribute, genetically determined by the hereditary base from which it originates, but is influenced by the interaction between the variety and the climatic conditions of the area of culture. Other factors contributing to the shaping of this attribute are related to the yielding precocity, the type of fructification, applied technology, disease resistance, grafting and pollination compatibility, planting density and rootstock used.

The production potential at 7 to 9 trees year old in the experimented varieties of apple cultivated in a high-density system was between 35 to 40 t / ha for the 'Caesar' and 'Valery' varieties and of 28 to 30 tons / ha in the 'Revidar' variety.

The quality of the fruits expressed in their biomass was between 160 g and 190 g, smaller fruits registered in the 'Revidar' variety and higher in the 'Caesar' and 'Valery' varieties. Dry matter content ranged between 13% and 16.5%.

Apple varieties with genetic resistance to diseases, patented in 2016, have other valuable features, as follows:

- 'Valery' (sin. H 4/37 - 04) - fruits are conic-truncated, yellow-orange on the sunny side, with a yellowish-yellow pulp, crisp at harvest, with a sweet and very good taste (Figure 1). It is resistant to scab and poorly attacked by *Podosphaera leocotricha*. It is characterized by precocity, outstanding fruit quality and constant production.

Figure 1. 'Valery' fruits

- 'Cezar' (sin H 1/79 - 90), present conical-globular fruits, covered with red on almost the entire surface (Figure 2). White, sweet, slightly acidified pulp is very tasty. It is resistant to *Venturia inaequalis* and *Podosphaera leocotricha*. It is distinguished by small-medium vigour, precocity, fruit quality and constant production.

Figure 2. 'Cezar' fruits

- 'Revidar' (sin H 1/16 - 90), have conical-shaped fruits, covered in red on 2/3 of the surface (Figure 3). The pulp is white, acidified and juicy, with good taste. It is resistant to scab and mildew. It is distinguished by medium-small growth, precocity, productivity and constant yielding trait.

Figure 3. 'Revidar' fruits

The studies and researches undertaken at RDSFG Voineşti present for apple growers a group of varieties with different ripening range covering a long period of consumption with apples from genetically resistant varieties.

Nowadays, the basic assortment includes varieties with genetic resistance to diseases such as: 'Romus 1', 'Romus 3', 'Romus 4',

'Prima', 'Pionier', 'Voinea', 'Ciprian', 'Florina', which are propagated in the Romanian fruit nurseries. Some of the varieties mentioned, even if they currently meet the requirements in terms of fruit resistance, productivity and quality, they can be replaced by new varieties more valuable in this respect.

Table 2 displays apple varieties with genetic resistance to diseases, 'Revidar', 'Caesar' and 'Valery', which were the subject of the study between varieties of apple with genetic resistance to diseases from the current assortment of apple.

Table 2. Consumption time and position of brand new varieties inside the assortment chart of the scab resistant varieties in the Voineşti area

Variety	Month																										
	VII			VIII			IX			X			XI			XII			I			II			III		
	1	2	3	1	2	3	1	2	3	1	2	3	1	2	3	1	2	3	1	2	3	1	2	3	1	2	3
Romus 1 (Vf)	•	•																									
Romus 2 (Vf)	•	•																									
Romus 3 (Vf)					•	•																					
Irisem (Vf)							•	•																			
Real (Vf)							•	•																			
Prima (Vf)							•	•	•																		
Revidar (Vf)							•	•	•	•																	
Remar (Vf)								•	•	•	•	•															
Iris (Vf)								•	•	•	•	•															
Voinea (Vf)									•	•	•	•															
Voinicel (Vf)									•	•	•	•	•	•	•												
Pionier (Vf)									•	•	•	•	•	•	•												
Cezar (Vf)										•	•	•	•	•	•	•	•	•									
Inedit (Vf)										•	•	•	•	•	•	•	•	•	•	•	•	•					
Generos										•	•	•	•	•	•	•	•	•	•	•	•	•					
Redix (Vf)										•	•	•	•	•	•	•	•	•	•	•	•	•		•			
Valery (Vf)										•	•	•	•	•	•	•	•	•	•	•	•	•	•	•	•		
Florina (Vf)										•	•	•	•	•	•	•	•	•	•	•	•	•	•	•	•	•	•

Depending on the period of maturation and consumption of fruits, the apple varieties studied, which have been distinguished by genetic resistance to disease, productivity and quality of fruit, fall differently in the variety conveyor for Dâmboviţa fruit basin.

Therefore, after the apple varieties as 'Romus 1', 'Romus 3', 'Romus 4', it can be introduced 'Irisem' and 'Real', varieties with ripening time in the last decade of August to first decade of September. After 'Prima' variety, the apple variety 'Revidar', which has a consumption period from 25 to 31 August until 1 October is very welcomed.

'Remar' apple variety is situated between the varieties 'Voinea' and 'Pionier', with the perspective of replacing 'Voinea' by the 'Remar' variety, because it has some superior qualities, both for taste and for fruit coloring.

For autumn season, the 'Iris' and 'Voinicel' varieties represent also the alternative for 'Pionier' variety.

After the 'Pionier' variety, the 'Caesar' variety filling gaps with the fruits harvested between September 15 - 20 and consumed until the end of December.

'Redix' variety completes a period of consumption that extends until December to January.

'Inedit' variety exceeds 'Redix' consumption period, closer to that of the 'Florina' variety. During the 'Florina' variety season, the 'Valery' variety arises, with Golden delicious fruit type.

The orientation towards apple varieties with genetic resistance to diseases in Romania will gradually be imposed not only as a result of economic efficiency, but also for the fact that they are the main factor in obtaining organic production.

CONCLUSIONS

- The new varieties of apple 'Valery', 'Cezar' and 'Revidar', patented in 2016, meet the requirements of the producers, oriented to economic efficiency, high production potential, quality fruits, and consumers demands.
- The growth vigour of the trees in the 9th year after planting, represented by the circumference of the trunk and trees height, rank 'Valery' variety as more vigorous than "Revidar" and "Cezar" which are similar in vigour.
- The volume of the crown calculated at the surface unit, at the density of 2,857 trees / ha, oscillated between 8,200 mc / ha in the 'Caesar' and 'Revidar' varieties, and 8,900 mc / ha in the 'Valery' variety.
- The highest production potential was achieved in the 7th - 9th years by 'Caesar' and 'Valery' varieties with yields of 35-40 t / ha and 28-30 t / ha by the 'Revidar' variety.
- Apples varieties created and patented by RDSFG Voineşti in 2016 cover a large

period of the consumption season alongside other varieties with genetic resistance to diseases already known and appreciated on the market by consumers, completing the conveyer recommended for Dambovita area.
- By promoting the varieties of apple with genetic resistance to diseases, there are beneficial economic outcomes for producers, environmental protection and apple production with low pesticide residues.

REFERENCES

Cociu V., Botu I., Şerboiu L., 1999. Progrese în ameliorarea plantelor horticole din România. Vol.I., Pomicultura. Edit. Ceres, Bucureşti.

Collectively by authors, 2000. Staţiunea de cercetare şi producţie pomicolă Voineşti la aniversarea a 50 de ani de cercetare ştiinţifică şi dezvoltare (1950 -2000). Ed. Domino, Târgovişte.

Collectively by authors, 2014. Genotipuri pomicole tolerante la stress termic, hidric şi biotic, pretabile sistemelor tehnologice specifice agriculturii durabile. Subcap. 4.1. Ameliorarea mărului şi părului la SCDP Voineşti. Ghid elaborat în cadrul proiectului sectorial ADER 1.1.9., finanţat de MADR.

Petre V., Petre Gh., 2014. Contributions regarding the apple trees genetic variability increase in the process of obtaining improving biological material. Scientific papers. Series B. Horticulture., Vol LVIII, 71-74.

Petre V., 2009. Tehnica obţinerii soiurilor de măr cu rezistenţă genetică la boli prin mutageneză. Ed. Moroşan, Bucureşti.

13

EFFECT OF AUXIGER GROWTH REGULATOR ON FRUITS DEVELOPMENT, PRODUCTION AND CHRACING INDEX OF 'REGINA' CHERRY VARIETY

Ananie PEŞTEANU, Valerian BĂLAN, Igor IVANOV

State Agrarian University of Moldova, 44 Mircesti Street, MD-2049,
Chisinau, Republic of Moldova

Corresponding author email: a.pesteanu@uasm.md

Abstract

Fruit size and yield can be improved with the application of growth regulators, such as synthetic auxins and may be effective in enhancing fruit growth, when applied during the second stage of fruit development. The aim was to evaluate the influence of Auxiger growth regulator on average weight of fruits, fruit production, fruit size, period of maturation and cracking index. The study subject of the experience was 'Regina' cherry variety, grafted on Gisela 6. The trees were trained as spindle system. The distance of plantation is 4.0 x 2.0 m. The experimental plot it was placed in the orchard „Vindex-Agro" Ltd. founded in 2012 year. The research was conducted during the period of 2016 year. To study average weight of fruits, fruit production, fruit size, period of maturation and cracking index were experimented the following variants of treatment: 1.Control – without treatment; 2. Auxiger, 0.5 l/ha; 3. Auxiger, 0.7 l/ha. Active ingredient of Auxiger is NAD – 1.5 g/l + ANA – 0.6 g/l. Growth regulator Auxiger were sprayed one time, during the period of intensive fruit growing, when the fruits diameter was 12-13 mm (26.05.16). During the research, it was studied the average of fruits, tree production and their quality, period of maturation and cracking index. During the analyzed period, it was established that the average weight of fruits, the plantation productivity, fruit size, period of maturation increase when treating with Auxiger growth regulators in dose of 0.7 l/ha and reduced the cracking index when the diameter of the fruits was 12 - 13 mm.

Key words: *growth regulator, fruit size, production, quality, maturation.*

INTRODUCTION

Cherries are the first fresh fruit of the year. Cherry fruits have a significant food value (Asanică, 2012; Cimpoieş, 2002). The food value, as well as other qualities of the fruits, such as size, color and firmness of flesh, are generally the main criteria by which the destination to exploit the cherries is determined (Asanică et al., 2013; Balan, 2015).

Small fruit size is one of the limiting factors in marketing cherry fruit (Sansavini and Lugli, 2005; Whiting and Ophardt, 2005).

As consumers prefer large cherries, fruit size is a very important marketing consideration, and the economic benefits of treatments capable of improving average fruit size are potentially very high.

Several techniques have been used to improve fruit production and fruit size of cherry (Balan, 2012; Budan and Grădinăriu, 2000; Long et al., 2014; Whiting and Lang, 2004). Fruit size can be improved with the application of growth regulators, such as synthetic auxins and gibberellins (Zeman et al., 2013).

Synthetic auxins may be effective in enhancing fruit growth, when applied during the second stage of fruit development (Faust, 1989).

The effectiveness of synthetic auxins in increasing fruit size is affected by the type of auxins, its concentration and the fruit crop. Some synthetic auxins are effective in increasing fruit production and fruit size of sweet cherry (Stern et al., 2007), though others, such as CPA, showed no effect (Zhang and Whiting, 2011).

In the fruit growing practice, the growth regulators are used in small amounts, but their effect is quite striking, if applied in recommended phases in active physiological concentrations, allowing be easily absorbing and transporting to the reaction.

NAA applied alone or in combination 30 - 35 days before the harvest decrease cracking index (Demirsoy and Bilgener, 1998). Pre-harvest spray of NAA has also been reported to reduce the field cracking and cracking index and increase the firmness of two cherry varieties (Anonymous, 1994; Yamamoto et al., 1992).

A combination of auxins gives better results than the application of single compound (Long et al., 2014; Stern et al., 2007; Zeman et al., 2013).

The objective of this study was to evaluate the effect of growth regulator Auxiger (NAD and NAA) on fruit development, fruit size, cracking, maturation, quality and yield in Regina sweet cherry.

MATERIALS AND METHODS

The research was conducted during the year of 2016, in the cherry orchard founded near the village Malaesti, Orhei district, during the spring of 2012 in the „Vindex-Agro" Ltd., with one-year-old trees shaped as a rod. The subject of the experience was 'Regina' cherry variety grafted on rootstock Gisela 6. The crowns it conducted by thin spindle system. The planting distance was 4.0 x 2.0 m.

To establish the influence of Auxiger growth regulator on the plantation production, cherry fruits quality and they cracking were tested the following variants (Table 1).

Table 1. Scheme experiments to determine the effectiveness of Auxiger growth regulator on tree production and their quality

Variants	Active ingredient	Application
Control	-	-
Auxiger, 0.5 l/ha	NAD - 1.5 g/l + NAA - 0.6 g/l	Spraying during the period of intensive fruit growing
Auxiger, 0.7 l/ha		

In the second and third variant the treatment date was 26.05.16, when was registered an intensive cherry growing. Location of plots made into blocks, each variant having four replicates. Each replication has 7 trees. At the border between the rehearsals and experimental plots were left one untreated tree to avoid the duplication of variants or repetitions while performing treatments.

Trees treatment was performed with portable sprinklers in the morning hours. The amount of the solution was 0.8 liters tree, based on the number of trees per unit area and the amount of water recommended of 1000 l/ha.

The number of fruits, the average weight of a fruit, the production from a tree and a unit area

were settled during the harvest. The establishment of harvest for each variant was performed by individual weighing of the fruits on 28 trees. The average weight of the fruits was determined by weighing a sample of 1 kg of cherries from each repetition and counting them.

The fruit diameter was determined during the harvesting period using the template recommended for sorting cherries by holes of 26, 28, 30, 32, 34 and 36 mm.

The height of the fruits was determined by the measuring and it is the distance between the base and the top. The large and small diameter of a fruit was measured at the equatorial area. The evaluation of mentioned parameters was carried out using calipers at time of harvest gathering 20 fruits in the row from each repetition.

The average weight of the seed was determined by the method of weighing, an indicator which was obtained as a result of removing the pulp from the seed. The ratio of the seed in the fruit is the ratio between the weight of 20 seeds and the weight of these fruits in each repetition reported in percent.

To have a more real index of cracking of cherry fruits, it resorted to setting cracking index of cracking natural and artificial. Natural cracking index was determined by the counting method at harvest time. After collecting 100 fruits in a row from the tree crown, it was counted the number of cracked fruits, then, using the correlation was established the index of cracking. Theoretically, the cracking index was determined by the method described by Christensen (1972).

Fruit harvesting was carried out in two rounds based on their maturation. The share of fruits harvested in the first half and the second one was determined by the method of weighing and counting on specific trees out of each variant.

The significance of differences men values of investigated parameters was determined by using the LSD test for the likelihood of 0.05.

The soil between the rows consisted of grass silage, and the strips between the trees per row with a width of 1.2 m are worked with the mechanical milling FA 086. The irrigation in the plantation is carried out by drip irrigation. The fertilization system calculated according to soil fertility and scheduled crop.

RESULTS AND DISCUSSIONS

The fruit production is the final index which indicates how all agro-technical measures were performed in the cherry plantation 'Regina' variety.

Investigations conducted proved that the number of fruit in the trees crown included in the research were not different in the studied variants (Table 2). This is explained by the fact that to create identical conditions for fruit development was necessary to leave a constant number of fruits in the trees crown. To maintain this number of fruit in the trees crown after the fall of ovaries in June, the load of fruit was corrected by manual thinning, leaving a number as precisely as possible of fruits.

Table 2. The influence of the Auxiger growth regulator on the amount of fruits, the average weight of fruits and the production of 'Regina' cherries variety

Variants	Number of fruits, pcs/tree	Average weight, g	The production of fruit		In %, compared to control variant
			kg /tree	t/ha	
Control	495	10.07	4.98	6.23	100.0
Auxiger, 0.5 l/ha	491	10.68	5.24	6.55	105.1
Auxiger, 0.7 l/ha	498	10.83	5.39	6.74	108.2
LDS 0.05	18.4	0.33	0.14	0.18	-

The lowest average weight of fruits was recorded in the control variant, without treatment being 10.07 g. Followed, in the ascendant order by the variant treated with Auxiger in dose of 0.5 l/ha with an average fruit weight of 10.68 g and the variant treated with Auxiger in dose of 0.7 l/ha, where the studied index was 10.83 g, or an increase of 0.76 g compared with the control variant. This difference in average weight between control variant and variants 2 and 3 was recorded due to treatment with growth regulator Auxiger.

Analyzing the influence of the dose treatment, it was recorded that with the increase of the dose quantity from 0.5 to 0.7 l/ha, the average fruit weight increased, but not as much as it increased between the control variant and the treated variants. If the difference between the variant treated with Auxiger in dose of 0.5 l/ha and 0.7 l/ha was 0.15 g, then between the control variant and the treated variant with the growth regulator Auxiger in dose of 0.5 l/ha was 0.61g. This results were proven statistically too.

The production of fruits on a tree and a surface unit is in direct correlation with the number of fruits and their average weight. The lowest fruit production was recorded in the control variant being 4.98 kg/tree or 6.23 t/ha.

In the variant treated with Auxiger in dose of 0.5 l/ha, the fruit production was 5.24 kg/tree or 6.55 t/ha, or it increased with 5.1% compared with the control variant.

The highest fruit production was registered in the variant treated with Auxiger in dose of 0.7 l/ha being 5.39 kg/tree or 6.74 t/ha, or and increase with 8.2% compared with the control variant.

Studying the influence of treatment dose on fruit production showed that with increasing the amount of product administered from 0.5 to 0.7 l/ha, the studied index increased, but not as essentially as between control variant and tested variants. If the difference between the treated variant with Auxiger in dose of 0.5 l/ha and 0.7 l/ha was 3.1%, then between the control variant and the variant treated with Auxiger in dose of 0.5 l/ha was 5.1%.

The insignificant difference between the variant treated with Auxiger in dose of 0.5 l/ha and Auxiger 0.7 l/ha was proven statistically too.

Statistical data about the production of fruit from a tree and a unit area showed a statistical difference between the control variant and variants treated with Auxiger.

Currently, in the modern research conducted on cherry plantations in order to increase the average weight of the fruits and their quality parameters (height, width, thickness, seed weight) are widely used treatments with growth regulators from auxin group.

While studying the fruit size 'Regina' cherry variety, we recorded higher values on their large diameter (d_1), and then in descendent order was the height and lastly the small diameter (d_2). If the large diameter during the research was 30.7 - 31.5 mm, then the height index and the small diameter was respectively 28.0 - 29.0 and 27.1 - 28.3 mm (Table 3).

Between the studied variants, the lowest height of a fruit was recorded in control variant, being 28.0 mm. In ascendant order is placed the variant treated with Auxiger in dose of 0.5 l/ha, with the studied index being 28.7 mm. Followed by the variant treated with Auxiger in dose of 0.7 l/ha, where the height of a fruit was

29.0 mm, or it increased with 3.6% compared with the control variant.

Table 3. Influence of Auxiger growth regulator on the quality of cherry fruits of 'Regina' variety

Variants	Size , mm			H/D	Average seed weight, g	% of seed
	Height (h)	Large diameter (d₁)	Small diameter (d₂)			
Control	28.0	30.7	27.1	0.91	0.58	5.7
Auxiger, 0.5 l/ha	28.7	3.3	27.9	0.92	0.59	5.5
Auxiger, 0.7 l/ha	29.0	3.5	28.3	0.92	0.59	5.4

Analyzing the influence of the treatment dose on the fruit height, it was noticed that once the treatment dose increased the height of the fruit increased too. If the difference between the variant treated with Auxiger in dose of 0.5 l/ha and dose of 0.7 l/ha was 1.1%, then between control variant and the variant treated with the Auxiger in dose of 0.5 l/ha was 2.5%.

At harvest time, the smallest value of the large diameter on cherry fruits was recorded in the control variant, being 30.7 mm. When treatment was applied with Auxiger, an increase in the studied index was noticed being 31.3 - 31.5 mm, so it increased with 0.6 - 0.8 mm compared with the control variant. The increase in the treatment dose did not influence significantly the large diameter index on Regina cherry variety. The same thing is valid and for the small diameter perhaps with small deviations between the variants.

The treatments made with Auxiger also influenced on the ratio between the height and the large diameter on the fruits. The smallest value of this ratio was registered in the control variant, being 0.91. In the variants treated with Auxiger, the ratio height/large diameter of fruits was 0.92.

The size of the seed is an important index for the quality of the fruits and productivity. On different varieties of cherries the seed ratio stands around 7.0%, but cherries varieties are quite different (Donica Il et al., 2005).

The conducted researches showed at the average seed weight in the control variant, was the smallest being 0.58g, but when treatment was made with Auxiger, its value was 0.59 g.

The seed ratio in the fruit is influenced by the average seed weight and the average fruit weight. Conducted research highlighted the variants treated with Auxiger where the seed weight was 5.4 to 5.5%. In the control variant,

the above index was higher, being 5.7%. Therefore, the treatments made with had a positive influence both on height, width and thickness of the fruit, and also on the fruit and the seed weight.

Effectuated research showed that there is a direct influence between the fruit weight and their diameter. The results from table 4 show that the fruit production obtained in the studied variants differ, registering higher values when treating with the Auxiger growth regulator.

If, in the control variant, the diameter of fruits with 22-26 mm was 24.7%, the fruits with 26-30 mm diameter were 30.1% and the fruits with the diameter larger than 30 mm were 45.2%. Therefore, the fruits with the diameter larger than 26 mm in the control variant were 75.3%.

Table 4. The influence of Auxiger growth regulator on fruits redistribution according to their diameter in the 'Regina' cherry variety

Variants	The share of fruits (%) according to their diameter (mm)		
	22-26	26-30	>30
Control	24.7	30.1	45.2
Auxiger, 0.5 l/ha	14.6	28.7	56.7
Auxiger, 0.7 l/ha	12.8	27.7	59.5

When treatment was made with Auxiger, the cherry fruit quality improved compared to the control variant. When treatment was made with Auxiger growth regulator in dose of 0.5 l/ha, the share of fruits with the diameter 22 - 26 mm decreased in comparison with the control variant being 14.6%, those with diameter 26 - 30 mm were 28.7%. The fruits with a diameter larger than 30 mm increased to 11.5%. This means that the share of fruits with the diameter larger than 26 mm were 85.4% or it increased with 10.1% compared with the control variant.

The same thing was valid and for the variant treated with Auxiger in dose of 0.7 l/ha. The share of fruits with the diameter 22 - 26 mm decreased in comparison with the control variant, being 12.8%, those with diameter 26 - 30 mm were 27.7% and the fruits with a diameter larger than 30 mm - 5.5%. Practically, this variant showed higher values compared with the control variant and the variant treated with Auxiger in dose of 0.5 l/ha.

By studying the influence of the dose treatment on the distribution of cherry fruit by diameter, once the amount of product increased from 0.5 to 0.7 l/ha, the studied index increased too, but not as much as compared with the control variant. If the difference between the fruits with a diameter larger than 2 mm between the variant treated with Auxiger in dose of 0.5 and 0,7 l/ha was 1,%, then between the control variant and the variant treated with Auxiger in dose of 0.5 l/ha was 20.1%.

Cherry fruit cracking is an inherent characteristic of the species and under certain genetic, physiological, chemical conditions can affect up to 90% of the harvest which influences negatively the financial situation of companies (Demirsoy and Bilgener, 1998).

The factors that may promote the phenomenon of cracking of the cherry fruits can be chemical, technological and genetically. They influence on the maturation of the cherry, the intensity of respiration, the capacity to absorb the water at the root and the skin of fruit, and also the osmotic pressure and turgor potential of the mesocarp cells (Yamamoto et al., 1992).

Cherry fruits are more prone to cracking during the period when they move from the yellow - purple color until they become black which is considered the full maturation and they are ready for consume. During the reference period (10-29.06.2016), the quantity of atmospheric precipitation was 85.9 mm.

These rainfalls affected the natural fruit cracking index on Regina cherry variety. The high'est value of the natural fruit cracking index on 'Regina' cherry variety was recorded after precipitation fallen during their maturation were in the control variant was 2.0%. In the variants treated with growth regulators Auxiger, it did not register fruits cracked naturally despite the precipitation fallen during fruit ripening.

To have a more real value of the theoretical fruit cracking index on 'Regina' cherry variety, it was used the method described by Christensen (1972).

Conducted research after two hours of cherries fruits immersion in water, it demonstrated that in the control variant, only one fruit cracked. The number of fruits cracked in the same variant after being immersed in water for four hours was 2 pcs, and after 6 hours - 7 pcs. The obtained results showed that the theoretical index of artificial cracking was 7.2% (Table 5).

In the variants treated with Auxiger, after the fruits were immersed in water for 2 and 4

hours, don't were registered the cracked fruits. If, the period of time that the fruits where in the water increased to 6 hours, it was recorded the quality of cherries improved compared to the control variant. In the variant treated with Auxiger in dose of 0.5 l/ha, the number of fruits artificially cracked was 4 pcs, or it decrease of 5.6% compared with the control variant.

Table 5. The influence of the growth regulator Auxiger on the fruit cracking on cherries of 'Regina' variety

Variants	Index of natural cracking, %	Fruits cracked artificially, psc.			Index of theoretical cracking, %
		After 2 hours	After 4 hours	After 6 hours	
Control	2.0	1	2	7	7.2
Auxiger, 0.5 l/ha	-	-	-	4	1.6
Auxiger, 0.7 l/ha	-	-	-	3	1.2

The same thing happened and in the variant treated with Auxiger in dose of 0.7 l/ha where the number of cracked fruit artificially was 3 psc, or it decreased by 6.0% compared with the control variant.

Analyzing the influence of the treatment dose on the artificially fruit cracking, it was noticed that once the dose treatment increased for 0.5 to 0.7 l/ha, the studied index didn't change as much as it did in the control variant. If the difference between the artificially cracked fruits in the variant treated with Auxiger in dose of 0.5 and 0.7 l/ha was 0.4%, then between the control variant and the variant treated with Auxiger in dose of 0.5 l/ha was 5.6%.

Optimal harvest time is determined by the fruit capitalization way. In this context, it should be borne in mind the gradual maturation of cherries and that after their separation from the tree no longer occur physiological processes to improve quality, as happens in other species. Therefore, cherry fruits are collected in two phases, during the time when they have the highest food value and good taste. The best harvesting time is determined usually empirically based on experience taking into account the color of the fruit, since there is no other index more accurately. Thus, the cherries are harvested when they got the color typical of the variety, the flesh softens and releases easily from the stalk branch.

Conducted research proved that treatments made with growth regulators Auxiger intensified fruit coloring. 'Regina' cherry variety is a late maturing variety which requires for the fruits to be collected in two stages.

The most important index is the share of fruits picked in the first and second stage of harvest. The research showed that in the control variant in the first picking stage (27.06.2016) were collected 48.8% of fruits from the trees crown and in the second stage (30.06.206) were picked the rest 51.2% (Table 6).

The treatments performed with the growth regulator Auxiger which is based on active ingredients NAD and NAA increased the share of fruits pick in the first stage of harvest.

Table 6. The influence of Auxiger growth regulator on the maturation of fruits of 'Regina' cherry variety, %

Variants	Harvest time	
	27.06.2016	30.06.2016
Control	48.8	51.2
Auxiger, 0.5 l/ha	67.5	32.5
Auxiger, 0.7 l/ha	70.4	28.6

When treatments were made with Auxiger in dose of 0.5 l/ha, the share of fruits picked in the first stage of harvest was 67.5% or it increased with 18.7% compared with the control variant. Once, the treatment dose increased to 0.7 l/ha, the studied index increased to 70.4%, which increased by 21.6% compared with the control variant and a 2.9% increase compare with the variant where the treatment dose was 0.5 l/ha.

CONCLUSIONS

Treatments made with Auxiger in dose of 0.7 l/ha increased the average weight of fruits, the plantation productivity, fruit size, period of maturation and reduced the cracking index.

The results presented here, indicate that the effect of the application of plant growth regulator Auxiger in dose of 0.7 l/ha during the intensive cherry growth, when the fruits reach a diameter was 12 - 13 mm, improve the physiological processes of the plant and increase the fruit development, quality and yield of 'Regina' sweet cherries variety.

REFERENCES

Anonymous , 1994. All about cherry cracking. Tree Fruit Leader, 3: 1-11.

Asanică A., 2012. Cireşul în plantaţiile moderne. Bucureşti.

Asanică A., Petre Gh., Petre V., 2013. Înfiinţarea şi exploatarea livezilor de cireş şi vişin. Bucureşti.

Balan V., 2012. Perspective în cultura cireşului. Pomicultura, Viticultura şi Vinificaţia Moldovei. Chişinău, 2: 7.

Balan V., 2015. Tehnologii pentru intensificarea culturii mărului şi cireşului. Akademos, Chişinău, 3: 82-87.

Budan V., Grădinăriu G., 2000. Cireşul. Iaşi.

Christensen J. V., 1972. Cracking in cherries. I. Fluctuation and rate of water absorption in relation to cracking susceptibility. Tidsskr. Planteavl, 76: 1-5.

Cimpoieş Gh., 2002. Pomicultura specială. Chişinau.

Demirsoy L., Bilgener S., 1998. The effect of preharvest chemical applications on cracking and fruit quality in 0900 Zirat, Lambert and Van sweet cherry varieties. Acta Hort., 468: 663-670.

Donica Il., Ceban E., Rapcea M., Donică A., 2005. Cultura cireşului. Chişinău.

Faust M., 1989. Physiology of Temperate Zone Fruit Trees. Wiley, New York, USA, 169-234.

Long L. E., Lond M., Peşteanu A., Gudumac E., 2014. Producerea cireşelor. Manual tehnologic. Chişinău.

Sansavini S., Lugli S., 2005. Trends in sweet cherry cultivars and breeding in Europe and Asia. Proc. 5th Int. Cherry Symposium. Turkey, (Abstract), 1.

Stern R.A., Flaishman M., Applebau, S., Ben-Arie R., 2007. Effect of synthetic auxins on fruit development of "Bing" cherry (Prunus avium L.). Sci. Hortic. Amsterdam, 114: 275-280.

Yamamoto T., Satoh H., Watanabe S., 1992. The effects of calcium and naphthalene acetic acid Sprays on cracking index and natural rain cracking in sweet cherry fruits. J. Japan. Soc. Hort. Sci., 61: 507-511.

Whiting M.D., Lang G.A., 2004. "Bing" sweet cherry on the dwarfing rootstock Gisela 5: thinning affects tree growth and fruit yield and quality but not net CO_2 exchange. J. Am. Soc. Hort. Sci., 129: 407-415.

Whiting M.D., Ophardt, D., 2005. Comparing novel sweet cherry crop load management strategies. HortScience, 40; 1271-1275.

Zeman S., Jemrić T., Čmelik Z., Fruk G., Bujan M., Tompić T. 2013. The effect of climatic conditions on sweet cherry fruit treated with plant growth regulators. Journal of Food, Agriculture and Environment, 11(2): 524-528.

Zhang C., Whiting M., 2011. Pre-harvest foliar application of prohexadione-Ca and gibberellins modify canopy source-sink relations and improve quality and shelf-life of "Bing" sweet cherry. Plant Growth Regul., 65: 145-156.

14

EMERGING PESTS OF *ZIZIPHUS JUJUBA* CROP IN ROMANIA

Roxana CICEOI[1], Ionela DOBRIN[2], Elena Ştefania MARDARE[1], Elena Diana DICIANU[2], Florin STĂNICĂ[2]

[1]University of Agronomic Sciences and Veterinary Medicine of Bucharest, Laboratory of Diagnosis and Plant Protection of Research Center for Studies of Food Quality and Agricultural Products, 59 Marasti Blvd, District 1, Bucharest, Romania
[2]University of Agronomic Sciences and Veterinary Medicine of Bucharest, 59 Marasti Blvd, District 1, Bucharest, Romania

Corresponding author email: roxana.ciceoi@gmail.com

Abstract

The jujube crop (Ziziphus jujuba Mill.) is one of the oldest crops in China and in the world. Written indications about its technology (grafting, pruning, pest control etc) are older than 1500 years, when the jujube was one of the top five most important fruits in China, together with peach, apricot, plum and chestnut. Their nutraceutical qualities were highly appreciated, but also the beauty of the tree and the flower perfume, which legends says, made people fall in love. Nowadays, the crop is expanding, both in China and on the other continents, due to the high content of bioactive compounds that can help in lowering the blood pressure, in liver diseases, anaemia and also inhibit the growth of tumor cells etc. In Romania, the species grows sub spontaneous since the antique times and until the present moment no important pests had been reported. Still, giving the climatic changes in the last years, many pests became invasive and enlarge their host plant spectrum. On the USAMV Bucharest experimental field, we monitor the pests incidence and we identify few species that might become a threat: Halyomorpha halys (Stal) (Heteroptera: Pentatomidae), Metcalfa pruinosa Say (Homoptera: Flatidae), Ceratitis capitata (Wiedemann) (Diptera, Tephritidae) and Nezara viridula (L.) (Heteroptera: Pentatomidae). The other potential polyphagous pests, like weevils, fruit borers and moths, made insignificant damages on the fruits, but they should still be under careful observation. Data about the pests' biology, damage and control are given.

Key words: Zizipus jujuba, Halyomorpha halys, invasive alien species, polyphagous pests, quarantine pests.

INTRODUCTION

The jujube fruits are nowadays worldwide appreciated as an outstanding source of biologically active compounds, with high nutraceutical value (Liu et al., 2014, Preeti et Tripathi, 2014). According to Gupta, 2004, jujube, along with date palms and grapes, start to be domesticated on the Indian subcontinent around the year 9000 BC, together with wheat and barley, which were cultivated from the very beginning of agriculture. In China, many written sources date back the jujube fruits 4000 years ago (Li et al., 2007, Liu et al., 2009) while in Romania it seems the tree was brought via the Silk road 2000 years ago, (Stănică, 2009). Jujube had been used both as crude and dried in the Chinese Traditional Medicine, for its antitussive, palliative, analeptic and nutraceutical properties, but also as a food and food flavourant for thousands of years (Li et al., 2007). It is cultivated also in Australia (Johnstone, 2014) and in Europe, research being done mainly in Romania, Italy and Macedonia (Cossio et Bassi, 2011, 2013; Markovski et Velkoska-Markovska, 2015; Stănică, 2002). Jujube leaves proved to have an insecticide action against one of the major pest in the world, *Helicoverpa armigera*, by inhibiting the digestive and mitochondrial enzymes which lead to growth retarding of the larvae (Varghese et Patil, 2005) and also against *Tribolium confusum* (Vasudha, 2012).

Around the world, except its native environment, the Chinese date seems to be very resistant to pests and no major damage had been registered until now outside the Asian continent borders (Yao, 2013).

In 2013, Balikai et al. mention almost 130 pest species recorded on *Ziziphus* crop in India and specify that 177 species of insect and non-insect jujube pests were recorded around the world. Only a few of them cause substantial economic damage (Balikai et al., 2013). In

India, in an IPM governmental meeting in 2015, 10 pests are cited as pests of national significance: the fruit flies *Carpomyia vesuviana* Costa, *B. zonata* Saunders, *B. dorsalis* (Diptera: *Tephritidae*), the fruit borers *Meridarchis scyrodes* Meyr (Lepidoptera: *Carposinidae*), the green slug caterpillars *Thosea sp.* (Lepidoptera: *Limacodidae*), the grey hairy caterpillars *Thiacidas postica* Walker (Lepidoptera: *Noctuidae*), the mites *Larvacarus transitans* Ewing (Tetranychoidea: *Tenuipalpidae*), the ber beetles *Adoretus pallens* Blanchard (Coleoptera: *Scarabaeoidea*), the grape mealybugs *Maconellicoccus hirsutus* (Green) (Hemiptera: *Pseudococcidae*), the ber mealybugs *Perissopneumon tamarindus* (Green) (Hemiptera: *Pseudococcidae*), the thrips *Scirtothrips dorsalis* Hood (Thysanoptera: *Thripidae*) and the termites *Odontotermes obesus* (Isoptera: *Termatidae*) (Satyagopal, 2015). Four of these ten categories were also cited by Azam-Ali et al. in 2006, as major pests for Asia: the fruitfly, the fruit borers, the ber beetles, the mites. The same authours included in their list the bark eating caterpillars *Indarbela quadrinotata*, *I. watsoni* and *I. tetraonis* (Coleoptera: *Cerambycidae*), the hairy caterpillars *Dasychira mendosa*, *Euproctis fraterna* (Lepidoptera: *Lamantriidae*), *Thiacidas postica* (Lepidoptera: *Noctuidae*) and the lac insect *Kerria lacca* and *K. sindica* (Hemiptera: *Keridae*). Acording to Balikai, 2013, in Europe, the following jujube pests were mentioned: *Carpomyia vesuviana* Costa (by Tominic, in 1954, in Yugoslavia), *Carpomyia incompleta* (Becker) (by Monastero, in 1970, in Italy), *Bactrocera zonata* (Saunders), (by Anonimus, in 2010, in the EPPO member countries, mentioned by Sarwar, 2006 as jujube pest), *Ceratitis capitata* (Wiedemann) (by Martinez et al., in 2006, in Amposta, Spain), *Hispa sp.* (by Jolivet, in 1989, in Turkey), *Grammadera clara* Brunner von Wattenwyl, (by Liebermann, in 1970, in Spain).

In Romania, Stănică, 1997, mentions that the most important pests of *Ziziphus jujube* are *Carpomya vesuviana* (Costa) and *C. incompleta* (Beck), dipterous that lay their eggs in July under the fruit epiderma and *Carposina sasakii* (Mats.), one Lepidopterous which in

China destroys 15-20% of fruits, other minor pests being *Ceratitis capitata*, *Cydia molesta* and *Polycrosis botrana*.

No major research programs have been started until now regarding the resistance to pests, except some mentioned by Azam-Ali et al., 2006, against the fruitfly *Carpomyia vesuviana* and fruit borer *Meridarchis scyrodes*.

During the last 20 years, since we first introduced the jujube tree in the experimental fields of University of Agronomic Science and Veterinary Medicine Bucharest, no pests were observed in the field, except *Ceratitis capitata* fly, starting with 2013 and no chemical applications were needed. The year 2016 was totally exceptional from the point of view of climatic conditions and pests evolution, especially for the new invasive species. Four species were damaging the crop, namely *Metcalfa pruinosa* Say, *Ceratitis capitata* (Wiedemann), *Nezara viridula* (L.) and *Halyomorpha halys* (Stal), while another 2 species producing damages were not yet identified. Using online databases, we estimate the possible risk raised by other recorded pests of jujube crop in Romania.

MATERIALS AND METHODS

The present article includes a comprehensive literature review and also presents our own research results. The literature review is based on the online and offline bibliographical references that we found on the WEB and in the University library. We used the following international databases: Web of Science - Core Collection (Journal Citation Reports, Derwent Innovations Index, Thomson Reuters), SpringerLink Journals (Springer), Scopus (Elsevier), ScienceDirect Freedom Collection (Elsevier), PROQUEST Central, Oxford Journals, CAB Abstracts, Google Scholar, Agris Fao, simple google research.

Our observations were made in the experimental field of USAMV Bucharest, in the new fruit species testing field of the Faculty of Horticulture. Regular monitoring visits were done weekly, starting from August 2016. For the presence of *Ceratitis capitata* adults, Tephri traps were used by the researchers of Research and Development Institute for Plant Protection, Bucharest, Romania and the results will be presented by them (Chireceanu et al., 2013). In laboratory, during the chinese dates morphometric determinations, we identified *C. capitata* larvae and apreciate the number of affected fruits, without precise counting. This preliminary observation will be follow-up in 2017.

To check the scientifically accepted name, spelling, alternative names and geographical distribution of pest species, we used the online Catalog of life databases, which holds essential information on the names, relationships and distributions of over 1.6 million species and the number is continuously rising. To verify the presence of a specific pest species or genus in Romania, we used the ARTHropod Ecology, Molecular Identification and Systematics database, belonging to INRA, France.

For the new non-native species *Halyomorpha halys*, as we consider it a serious threat, our laboratory launch its first citizen science action in July 2016. Warning leaflets were given directly or sent by mail to over 4000 citizens, including agricultural producers and also shared on social networks, as LinkedIn and Facebook.

RESULTS AND DISCUSSIONS

The pests of jujube crop could be grouped into four main categories, as illustrated in fig. 1.

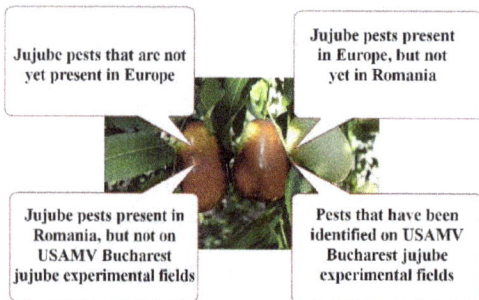

Figure 1. The four possible pests groups associated to *Ziziphus jujuba*

We present the four groups of pests in the order of their threat on jujube crop in Romania.

I. Pests that have been identified on USAMV Bucharest jujube experimental fields
The mediterranean fruit fly (MFF) - *Ceratitis capitata (Wiedemann) (Diptera, Tephritidae),* EPPO code CERTCA, A2 no. 105.

Ceratitis capitata is a highly invasive species, worldwide established, extremely polyphagous, having a high economic impact, affecting production, rising the pests control costs and limiting the international trade access (EPPO, 1997). According to EPPO, 2011, *C. capitata* is the most serious pest of some fruits, including citrus, especially in the countries with a warm, Mediterranean, tropical or subtropical climate, but the species is spreading to the north and the recent climate changes makes a larger area suitable for its establishment. Its presence, even as temporary population, leads to severe constraints regarding the export of fruits to uninfested areas and make *C. capitata* one of the most significant quarantine pests for the EPPO region. It originates in tropical Africa, from where it has spread to the Mediterranean area and the world. The larvae develop in fruits, in a very wide range of host plants, in the EPPO region the main hosts belonging to a large number of tree fruit crops, as apples, peaches, plums, cherry, avocados, *Citrus sp.*, figs, kiwifruits, mangoes, pears etc. It has also been recorded on wild hosts. *Ceratitis capitata* is an EPPO A2 quarantine pest (EPPO, 1981), at no. 105 (EPPO, 2015) and is also of quarantine significance throughout the world. The spread to Europe, North and Eastern Africa, the Middle East, Australia and the Americas is the result of accidental transportation during the trade (CABI, 2016a).

The presence of the Mediterranean fruit fly in Romania was first time mentioned by Stanciu, 2007, on *Diospyros kaki* and described by Chireceanu et al. in 2013, following their research on 17 locations in Romania, on apricot, jujube and wild blackberry. The authors found *C. capitata* adults in five locations, on jujube in orchards and in private garden (Bucharest and Moara Domneasca, Ilfov County), on apricot in private garden (Agigea-Constanta County) and on wild blackberry bushes (Bucharest).

Regarding the attack on jujube, Chireceanu et al., 2013, states that in Bucharest the percentage of affected jujube fruits was under 10% and that the adults were collected from September to October, the first individuals being captured just when the jujube fruits start ripening.

During our observations, in 2016, in the laboratory, while weighting the jujube fruits which were previously held in plastic bags, we identified the *C. capitata* larvae, which flex and 'jump' repeatedly up to 2 cm when we pick the fruits from the bag. This behavior was previously described when the surrounding air temperature was warm and the larvae fully grown (EPPO, 2011). The degree of affected fruits was appreciated under 5%.

The morphological identification with a binocular microscope is the recommended diagnostic method and the description of the fly is provided by EPPO Diagnostics, 2011.

Biology

The eggs of *Ceratitis capitata* are laid below the skin of the fruit and hatch within 2-4 days in warm days and 16-18 days in cool weather and the larvae feed for others 6-11 days. The pupa forms in the soil, beneath the host plant and adults emerge after 6-11 warm days or longer in cooler days and adults live for up to 2 months (Christenson et Foote, 1960).

The adult flight and the transport of infested fruits are the major means of movement and dispersal to previously uninfested areas. Scientists suggest that *C. capitata* can fly at least 20 km (EPPO, 1990). Some fruits are only infested when ripe but precautions should be taken and the production should be carefully checked before transportation in another region.

Damage

The *Ceratitis capitata* larvae consume the heart of the fruit, making them inappropriate for consumption or processing. The attacked fruits have puncture marks made by the female's ovipositor and they often fall down before maturation.The fly can act also as a potential vector for *Erwinia amylovora* (Ordax et al., 2015), for other fungi, as *Rhizopus stolonifer* (Cayol et al., 1994), while Sela et al., in 2005 demonstrate that the fly is a potential vector of human pathogens as coliforms and *E. coli*.

Control

When detected, it is important to gather all fallen and infected host fruits and destroy them. The detection of *Ceratitis capitata* is done during monitoring with traps with trimedlure *(t-butyl-4(or5)-chloro-2-methyl-cyclohexane-carboxylate)* and terpinyl acetate, which attracts the males.

Malathion was long time used for chemical spraying and bait sprays (both male and female of *C. capitata* are attracted to a protein source which emanates ammonia). Bait sprays have minimal impact on natural enemies, as only the pests are attracted and killed by the chemical insecticide. Qazzaz et al. in 2015, demonstrate that extracts of entomopathogenic *Beauveria bassiana* significantly reduced the peach infestation.

Regarding the phytosanitary measures, the fruits imported from countries where *Ceratitis capitata* occurs, need to be carefully inspected and those suspected for symptoms of infestation should be cut and controlled for larvae. EPPO recommends treatments of the fruits belonging to the genus *Citrus* and *Prunus* by either cold treatment while in transit either by vapour heat, forced hot air or hot water treatment (EPPO, 1997). Fumigants as ethylene dibromide or methyl bromide are no longer recommended either because of the carcinogen effect, either because of the damaging effect on fruits or fruit shelf-life. Irradiation, a combination of methyl bromide fumigation and cold treatment and wrapping fruits in shrink-wrap film were investigated as possible methods of disinfesting procedures (Ohta et al., 1989, Jang, 1990).

The brown marmorated stink bug (BMSB) - *Halyomorpha halys (Stal) (*Heteroptera: Pentatomidae)*. EPPO code HALYHA, deleted from the Alert list in 2013

Halyomorpha halys is a highly polyphagous pest, attacking mainly fruit trees and vegetables, but also field crops, some ornamentals and weeds. BMSB has become a major nuisance pest in the USA, in mid-Atlantic region and Pacific Northwest, due to its overwintering behavior which leads to unimaginable homes invasions (Inkley, 2012). In the mid-Atlantic region, serious crop losses have been reported for apples, peaches, sweet corn, peppers, tomatoes and row crops such as field maize and soybeans since 2010. Paula et al., 2016 consider this pest a devastating invasive species in the USA, as the direct losses produced only in the mid-Atlantic apples orchards alone, in the year 2010, were about $37 million. In Europe, the damages produced by *H. halys* in Italy, in 2016, were approximated around 1 billion Euro (Fontana, 2016). For humans, allergic reactions and dermatitis have been mentioned (Anderson, 2012).

In Europe, the first specimen was captured in a light trap in 2004 in Liechtenstein at Balzers (EPPO, 2008), afterwards in 2007 in Switzerland (Wermelinger et al., 2008) and its evolution was silent until 2015, when it became invasive and major damages were reported, as in Italy, where despite all control measures applied, pest feeding caused severe damage to fruit crops, especially pear (EPPO, 2016). The pest was mentioned for the first time in 2011 in Germany (Heckmann, 2012) and Greece (Milonas et Partsinevelos, 2014) in 2012 in Italy (EPPO, 2013a) and France (EPPO, 2013b), in 2013 in Hungary (Vétek, 2014), in 2014 in Russia (EPPO 2016a), in 2015 in Austria (EPPO, 2016b) and Serbia (EPPO, 2016c).

In Romania, Macavei et al., 2015 first time described this species, found in Bucharest in September 2014, but they state that its presence could date back to at least 1-2 years ago, due to the fact that individuals were found at several kilometers away. Our observations confirm this theory and leads to the hypothesis that the cohabitation habit between *Nezara viridula* and *Halyomorpha halys* species is the misleading factor in the late identification of *H. halys* specimens in Romania. In 2016, in the USAMV Bucharest corn testing fields, goji testing fields and edible roses testing fields, the two species were always observed in cohabitation and they produced crop losses of 100% to goji crop, starting with the month of august (Ciceoi et al., 2016a,b). On the corn field, the estimated *H. halys* density/m^2 plant was 25.5 individuals on the field border area and 9.75 individuals in the field interior area, which indicates an exponential growth of the marmorated stink bug populations. On jujube experimental field, the adults of *H. halys* were observed since early September and we consider that the main two reasons for this late occurrence are the fact that the goji crop found in the vicinity was more "feeding" attractive and the population was fix until the complete fruit depreciation of the fruits and the fact that the fruits of jujube start to ripen in the first decade of September.

Biology

The biology of BMSB depends on its environment; it has up to five generations/year in southern China, one or two generations in USA and one generation had been reported until now in Europe (CABI, 2016c). Although the overwintering adults become active in April, due to the high demand of temperature, the oviposition peak in Europe starts at beginning of July (Haye et al., 2014) and then continue until the end of September. The eggs are laid in clusters of around 30 eggs (CABI, 2016c), on the inner leaves surfaces and a female may lay in average 79 eggs, maximum observed being 160 eggs (Haye et al., 2014). The species has 5 larval instars and the adults emerge in 60 to 131 days, depending on the temperature fluctuations. In controlled conditions, at 30°C, the cycle from egg to adult only lasted only 33.2 days and below 15 and above 35°C no development was observed (Nielsen et al., 2008, Haye et al., 2014). More than 300 host plants have been recorded, the most severely damaged being the tree fruits and small fruits, the vegetables, especially tomatoes, ornamentals and field crops (Nielsen et al., 2008).

Damage

The attacked fruits present feeding punctures, caused by the plant juices sucking, which lead to suberifications, the formation of necrotic areas and sometimes to deliquescent fruit pulp (Rice et al., 2014). The adults feed on fruits and the nymphs feed on leaves, stems and fruit. In 2016, in Bucharest some hobby tomatoes crops were completely destroyed by the BMSB, while at the USAMV Bucharest experimental field heavy infestations were observed on corn and goji crops, the goji fruits being severely distorted and quickly become blackish and juiceless, unmarketable. In Asia, *H. halys* cause significant damage to soybean, in Northern Japan to apple, in Italy to pear. In addition to plant damage, BMSB is a nuisance to humans because the overwintering adults aggregate in buildings and houses. *H. halys* has a strong flight capacity, including for long distances (more than 5 km/day), particularly in the summer generation, so the spreading capacity is very high.

Control

Monitoring and surveillance of the pest spreading in new areas are capital and should be followed by immediate eradication measures. The IPM programs for this pest are correlated with the number of generations per year and the peak of the last adult population, which is the most damaging stage (Nielsen et al., 2008). The black light traps and baited black pyramid traps (with methyl (2E,4E,6Z)-decatrienoate) have been successfully used to monitor *H. halys* (Leskey et al., 2012). The H. *halys* aggregation pheromone has been identified by Zhang et al., 2013. Leskey et al., 2012, tested the efficacy of 37 insecticide treatments and proved that dimethoate, malathion, bifenthrin, acephate, permethrin, methidathion, endosulfan, methomyl, chlorpyrifos and fenpropathrin have a greater potential for controlling BMSB.

Biological control methods are desirable, but as *H. halys* is a newly established invasive pest, the lack of knowledge combined with the lack of predators and parasites in the newly invaded areas makes the chemical control methods preferable. Some observations mention *Harmonia axyridis* and earwigs being efficient in eggs predation.

Citrus flatid plant-hopper - *Metcalfa pruinosa Say* (Homoptera: *Flatidae*), EPPO code: METFPR, no list

The citrus flatid plant-hopper is an extremely polyphagous pest, more than 300 hosts being mentioned until the present moment (Vlad et Grozea, 2016). It is a major nuisance pest in parks and gardens and produce qualitative depreciation of agricultural products, caused by the feeding with the sap of plants and by the massive productions of waxy secretions and honeydew, when in numerous populations (CABI, 2016d). The species originate from eastern North America and was first time mentioned in Italy in 1979 (Bărbuceanu, 2015, Strauss, 2010) and is now present in at least 15 European counties (Strauss, 2010, EPPO, 2016). Recently was discovered also in the eastern palearctic area, in Korea, where it imedialtely became a quarantine pest (Kim et al. 2011)

In Romania *M. pruinosa* was first time mentioned by Preda et Skolka, 2009, in the city of Constanta, in 2009, then in 2010 in Timisoara (Gogan et al. 2010, Grozea et al., 2011) and in 2011 in Bucharest, (Chireceanu et Gutue, 2011).

Our observation in the jujube experimental field showed low infestations and only on the lower leaves of the trees, having no influence on the fruit quality. Still, considering its evolution on other crops, we recomend a throughout monitoring of the *M. pruinosa* activity all year long.

Biology

According to Mead, 2014, *Metcalfa pruinosa* overwinters as eggs, hatching starting early in the spring. First adults emerge in around 70 days after the hatching and the species has only one generation per year. The species has five larval instars. In Romania, the larvae occur in the second half of May and the first adults can be seen in late July. An overlap of the stages usually happen (Bărbuceanu, 2015).

Damage

Although new and non-native pest for Europe, considering the low economic impact of the species in its area, *M. pruinosa* was not considered quarantine pest by EPPO. Although Mead, 2014 considers the pest does very little damage to plants, in Europe, Strauss, 2010 mentions serious qualitative damage of grapes and quantitative damage in soybean in Italy. For viticulture and fruit growing, the negative impact on quality is due to the *M. pruinosa*'s honeydew secretions, wich stimulate the black sooty mold development Strauss, 2010.

Control

Limiting the spreading of this pest is difficult, due to its very high mobility, very large host plants spectrum and waxy protection of the individuals. Bărbuceanu, 2015, conclude that the insecticide with the highest effectiveness is the mixture of imidacloprid 75 g/l, deltamethrin 10 g/l, lambda cyhalothrin 50 g/l, followed by a treatment with pirimiphos methyl 500 g/l, at 3-5 days after the first one. Still, the chemical treatments should be avoided, as the broad spectrum insecticides kills the natural beneficial species. The biological control with *Neodrynus typhlocybae* is assumed to be advantageous by Strauss, 2012.

Southern Green Stink Bug - *Nezara viridula* (L.) (Heteroptera: Pentatomidae) *EPPO code NEZAVI*

The southern green stink bug originates from Ethiopia and East Africa, although some authors believe it came from southern Asia (Grozea et al, 2012). It is spread now in the tropical and subtropical regions of Europe, Asia, Africa and the Americas (Squitier, 2013). The first reports of pest in Europe date from 1998, in Italy (CABI, 2016b). It is a highly polyphagous species, causing economic damage to legumes, soybean and beans (Portilla et al., 2015) but also fruits and vegetables.

In Romania, qualitative and quantitative depreciations of tomatoes were described by Grozea et al. in 2012, in Timisoara area while in Muntenia area the presence of the pest is stated in 2015 (Kurzeluk et al. 2015). The author mentions that in the Bucharest area *Nezara viridula* was identified on a goji experimental field since 2011.

The adults of *N. viridula* were observed on jujube experimental field since the beginning of September. The biology and ecology of *N. viridula* seem to be closely related to the *H. halys*'s one, as these two species cohabits. The vicinity of the depleted goji trial field and the jujube fruit maturity are considered the two main reasons for the late occurrence of the adults in the testing field. No major damages were noticed following the feeding on fruits, but further research is recommended, as the biology of the pest might change in the long and warm autumns, when the fruit start to ripe.

Biology

In the warm climates, the southern green stink bug may have four generations per year while in Europe just one generation had been mentioned. Overwinters as an adult under the bark of trees or leaf litter. In spring, after the feeding, the oviposition may start. In US, eggs have been found starting the second week of April till December. One female of *Nezara viridula* could maximum lay 260 eggs (Squitier, 2013). Its development includes five larval instars.

Damage

The damages include drop and malformation of the fruits (Panizzi, 2008) on the rostrum inserting points into the fruit or growing shoots tissues. The damages are even higher due to the digestive enzymes introduced while feeding (Grozea et al., 2015). Black spots or suberification may be observed on the mature fruits, while the younger fruits usually drop out of the plant.

Control

Although not considered a major pest, the economic thresholds for different crops were determined in the USA. For example, in soybeans, 36 stink bugs per 100 swings of a net, for cowpea 5000 southern green stink bugs per ha and 3 - 4 stink bugs per 100 swings for cotton are considered the upper limits before chemical interventions. Trap crops are considered having great potential. As biological control means, two parasites were introduced, a tachinid fly, *Trichopoda pennipes*, which parasitizes the adults and the wasp *Trissolcus basalis*, which parasitizes the eggs.

II. Jujube pests that are present in Romania, but have not been identified on USAMV Bucharest jujube experimental fields

Among the pests that have been mentioned in the literature as attacking jujube crop and that exist in Romania, but we did not identify it on the jujube crop yet, two are more important:

a. the leopard moth, *Zeuzera pyrina*, (Lepidoptera, *Cossidae*), EPPO code ZEUZPY, which is included in EPPO Alert list since 2001 and

b. the oriental fruit moth, *Grapholita molesta*, (Lepidoptera, *Tortricidae*), EPPO code LASPMO, formerly on A1/A2 list.

III. Jujube pests that are present in Europe, but not yet in Romania

Among the pests that have been mentioned in the literature as attacking jujube crop and that exist in Europe, but not yet in Romania, the ber fruit flies *Carpomya vesuviana* and *Carpomya incompleta (Diptera, Tephritidae)* EPPO code CARYVE, formerly on A1/A2, are the most dangerous. *C. vesuviana* was found in Bosnia-Herzegovina, Italy and Turkey, while *C. incompleta* was mentioned in South of Europe, Israel, France and Italy. (Vadivelu, 2014, Pollini, 2014, fera, 2016, catalogue of life.org, 2016)

The ber fruit fly, *Carpomya vesuviana* Costa, 1854 (Diptera: Tephritidae) is the most destructive pest of *Ziziphus* crop in its area of distribution, that includes Bangladesh, China, Georgia, India, Indian Ocean Islands, Iran, Mauritius, Oman, Pakistan, Southern Europe, Turkmenistan, Turkey, and Uzbekistan (Amini et al., 2014, Vadivelu 2014).

The bug *Apolygus lucorum* (Meyer-Dur, 1843) (*Miridae: Hemiptera*), is another major pest of jujube crop in Asia. Overwintering *A. lucorum* eggs are primarily laid in the summer pruning wounds of Chinese date trees. It was reported in Europe in Britain, Denmark, Germany, Italy, Macedonia, Poland, Russia, Spain, Sweden, Switzerland, and Turkey, (catalogoflife.org, 2016, Pan et al., 2014). Another jujube pest is *Lygus pratensis* L. (*Miridae: Hemiptera*), which, according to the catalogue of life, was mentioned in Austria, Belgium, Bulgaria, Cyprus, Czech Republic, Denmark, Finland, France, Georgia, Germany, Greece, Iran, Italy, Hungary; Britain; Corsica, Macedonia, Siberia, Sweden and Turkey.

IV. Jujube pests that are not yet present in Europe

The most dangerous pest for jujube crop, but also for the entire fruit production in Europe is the Peach fruit moth - *Carposina sasakii* (Lepidoptera: Carposinidae), EPPO code CARSNI, A1 list: No. 163 (CABI, 2016e)

Also known as *Carposina niponensis*, the peach fruit moth can be easily mistaken as *Cydia pomonella* or *Cydia molesta*. Considered as one of the most important pests of pome fruits in the Far East, the quarantine pest can damage pears up to 100% in some cases, while apples are less heavily infested, in a degree of 40-100% (EPPO QP, 1988). Special phytosanitary measures are imposed for all *Chaenomeles, Crataegus, Cydonia, Eriobotrya, Malus, Prunus, Pyrus* and *Ziziphus* plants with roots imported from Asian country where this pest occurs. Additionally, the consignment should have been kept under conditions which prevent a reinfestation by the organism (EPPO, 1990). The growth of the larvae in the apple fruits was carefully examined by Koizumi, 2010.

CONCLUSIONS

Until the present moment, in Romania, *Ziziphus jujube* keeps its status of crop without major phytosanitary problems. All four identified pests produced a low to inexistent degree of damage.

Major attention should be given to the plants intended for planting imported from Asian countries. Although it might seem a good opportunity, importing plants with a questionable origin or doubtful phytosanitary certificate might have a severe economic impact on the fruit-growing areas of the EPPO region.

Considering that many alien invasive polyphagous pests start producing severe damages in Europe and USA and the recent climate changes, we recommend a thorough monitoring program and a detailed pest biology observation program.

REFERENCES

Amini A. Sadeghi H. Lotfalizadeh H. Notton D., 2014. Parasitoids Hymenoptera: Pteromalidae, Diapriidae) of *Carpomyia vesuviana* Costa (Diptera: Tephritidae) in South Khorasan province of Iran. Biharean Biologist 8(2):122-123.

Anderson B.E., Miller J.J., Adams D.R., 2012. Irritant contact dermatitis to the brown marmorated stink bug, *Halyomorpha Halys*. Dermatitis, 23(4):170-172.

Azam-Ali S., Bonkoungou E., Bowe C., deKock C., Godara A., Williams J.T., 2006. Ber and other jujubes. International Centre for Underutilised Crops, Southampton, UK.

ARTHropod Ecology, Molecular Identification and Systematics *database,* accessed on 09.10.2016.

Balikai R.A., Kotikal Y.K., Prasanna P.M., 2013. Global scenario of insect and non-insect pests of jujube and their management Options. Proc. 2nd International Jujube Symposium Acta Hort. 993:253 - 277.

Bărbuceanu D., Mihăescu C. F., 2015. New data about the control of the planthopper *Metcalfa pruinosa* (Say 1830) (Hemiptera: Flatidae) in the town of Pitești (Argeș County). Annals of the University of Craiova - Agriculture, Montanology, Cadastre. XLV:26-31.

CABI, 2016a. *Ceratitits capitata* (Mediterranean fruit fly), http://www.cabi.org/isc/datasheet/12367

CABI, 2016b. *Nezara viridula* (green stink bug) http://www.cabi.org/isc/datasheet/36282#201631022 87.

CABI, 2016c. *Halyomorpha halys* (Marmorated Stink bug), http://www.cabi.org/isc/datasheet/27377.

CABI, 2016d. *Metcalfa pruinosa* (frosted moth-bug), http://www.cabi.org/isc/datasheet/35054.

CABI, 2016e. *Carposina sasakii* (peach fruit moth), http://www.cabi.org/isc/datasheet/11401

Cayol J.P.; Causse R.; Louis C.; Barthes J., 1994. Medfly *Ceratitis capitata* as a rot vector in laboratory conditions. Journal of AppEnt 117, 338-343.

Ciceoi R., Mardare E. Ș., Teodorescu E., Dobrin I., 2016a. The status of brown marmorated stink bug,

Halyomorpha halys, in Bucharest area. JHFB, 20(4):18-25, DOI: 10.5281/zenodo.267978

Ciceoi R, Mardare E.S., 2016b The risks assessment of *Aceria kuko* (Kishida) and *Halyomorpha halys* (Stal) pests for the Romanian goji growers. Poster. DOI: 10.5281/zenodo.345966

Chireceanu C., Gutue C., 2011. *Metcalfa pruinosa (Say) (Hemiptera: Flatidae)* identified in a new south eastern area of Romania (Bucharest area), Romanian Journal of Plant Protection, 4:28-34.

Chireceanu C., Stănică F., Chiriloaie A. 2013. The Presence of the Mediterranean fruit fly *Ceratitis Capitata* (Wied.), (Diptera: Tephritidae) In Romania, Romanian Journal of Plant Protection. VI:92-97.

Christenson L.D., Foote R.H., 1960. Biology of fruit flies, Annual Review of Entomology 5:171-192.

Cossio F., Bassi G., 2011. Prospettive di diffusione in Italia di varietà cinesi di giuggiolo a frutto grosso, Frutticoltura. 11:58-61.

Cossio F., Bassi G., 2013. Field Performance of Six Chinese Jujube Cultivars Introduced and Tested In Northern Italy. Acta Hortic. 993:21-28.

EPPO, 1981. Data sheets on quarantine organisms No. 105, *Ceratitis capitata*. Bulletin EPPO Bulletin 11 (1).

EPPO, 1990. Specific quarantine requirements. EPPO Technical Documents No. 1008.

EPPO/CABI, 1997. *Ceratitis capitata*. Quarantine Pests for Europe, Wallingford (GB), 2nd edn, 146–152.

EPPO, 2011. Diagnostics. Ceratitis capitata PM 7/104(1). OEPP/EPPO Bulletin 41:340-346.

EPPO Services, EPPO A1 And A2 Lists Of Pests Recommended For Regulation As Quarantine Pests, 2015. EPPO.

EPPO, 2013a. First report of *Halyomorpha halys* in Italy (2013/108), EPPO Reporting Service, 05:10-11.

EPPO, 2013b. *Halyomorpha halys* continues to spread in the EPPO region: first reports in France and Germany (2013/109), EPPO Reporting Service, 05:11.

EPPO, 2016a. First report of *Halyomorpha halys* in Russia (2016/148), EPPO Reporting Service, 08:12.

EPPO, 2016b. First report of *Halyomorpha halys* in Austria (2016/150), EPPO Reporting Service, 08:13.

EPPO, 2016c. First report of *Halyomorpha halys* in Serbia (2016/151), EPPO Reporting Service, 08:14.

Fontana L., 2016. Question for written answer to the Commission. Subject: Damage to Italian agriculture: the case of the brown marmorated stink bug, Parliamentary questions.

Gogan A., Grozea I., Vîrteiu A. M., 2010. *Metcalfa pruinosa* Say (Insecta:Homoptera: Flatidae) –first occurrence in western part of Romania. Res. J. Agric. Sci., 42(4):63-67.

Golmohammadi F., 2013. Medicinal plant of Jujube (*Ziziphus jujuba*) and its indigenous knowledge and economic importance in desert regions in east of Iran: situation and problems, Tech J Engin & App Sci., 3 (6): 493-505.

Grozea I., Gogan A., Virteiu A.M., Grozea A., Stef R., Molnar L., Carabet L., Dinnesen S., 2011. *Metcalfa pruinosa* Say (Insecta: Homoptera: Flatidae): A new pest in Romania, African Journal of Agricultural Research Vol. 6(27), 5870-5877.

Grozea I., Ştef R., Virteiu A.M., Cărăbeţ A., Molnar L., 2012. Southern green stink bugs (*Nezara viridula* L.) a new pest of tomato crops in western Romania. R. Journal of Agricultural Science, 44(2):24-27.

Grozea I., Virteiu A.M., Ştef R., Cărăbeţ A., Molnar L., Florian T., Vlad M., 2015 Trophic Evolution of Southern Green Stink Bugs (*Nezara viridula l.*) in Western Part of Romania, Bulletin UASVM Horticulture 72(2), 371-375.

Gupta A.K., 2004. Origin of agriculture and domestication of plants and animals linked to early Holocene climate amelioration. Current Science. 87:54- 59.

Haye T., Abdallah S., Gariepy T., Wyniger D., 2014, Phenology, life table analysis and temperature requirements of the invasive brown marmorated stink bug, *Halyomorpha halys*, in Europe. J. Pest Sci. 87:407–418.

Heckmann R., 2012. Erster Nachweis von *Halyomorpha halys* (Stål, 1855) (Heteroptera: Pentatomidae) für Deutschland. Heteropteron 36:17-18.

Inkley DB, 2012. Characteristics of home invasion by the brown marmorated stink bug (Hemiptera: Pentatomidae). J. Entomol. Sci. 47:125-130.

Jang, E.B., 1990. Fruit fly disinfestation of tropical fruits using semipermeable shrink-wrap film. Acta Horticulturae No. 269, abstract.

Johnstone R., 2014, Development of the Chinese jujube industry in Australia, Rural Industries Research and Development Corporation, Publication No. 14/001.

Kim Y., Kim M., Hong K-J., Lee S., 2011. Outbreak of an exotic flatid, *Metcalfa pruinosa* (Say) (Hemiptera: Flatidae), in the capital region of Korea. Journal of Asia-Pacific Entomology, 14:473–478.

Koizumi M, Ihara F, Yaginuma K, Kano H, Haishi T., 2010. Observation of the peach fruit moth, *Carposina sasakii*, larvae in young apple fruit by dedicated micro-magnetic resonance imaging. Journal of Insect Science 10:145.

Kurzeluk D.K., Fătu Ac., Dinu Mm, 2015. Confirmation of the presence of the Southern Green Stink Bug, *Nezara viridula* (Linnaeus, 1758) (Hemiptera: Pentatomidae), in Romania, Romanian Journal for Plant Protection, Vol. VIII.

Leskey T.C., Hamilton G.C., Nielsen A.L., Polk D.F., Rodriguez-Saona C., Bergh J.C., Herbert D.A., Kuhar T.P., Pfeiffer D., Dively G.P., Hooks C.R.R., Raupp M.J., Shrewsbury P.M., Krawczyk G., Shearer P.W., Whalen J., Koplinka-Loehr C., Myers E., Inkley D., Hoelmer K.A., Lee D.-H., Wright S.E., 2012, Pest Status of the Brown Marmorated Stink Bug, *Halyomorpha Halys* in the USA. Outlooks Pest Manag. 23:218–226

Li J.W., Fan L. P., Ding S. D., Ding X. L., 2007. Nutritional composition of five cultivars of chinese jujube. Food Chemistry 103:454-460.

Liu M.J., Zhao Z.H., 2009, Germplasm resources and production of Jujube in China, Proc 1st Int'l jujube Symp, Acta Hort 840:25-31.

Liu M.J., Zhao J., Cai Q.L., Liu G.C., Wang J.R., Zhao Z.H., Liu P, Dai L., Yan G, Wang W.J., Li X.S., Chen Y., Sun Y.D., Liu Z.G., Lin M.J., Xiao J., Chen Y.Y., Li X.F., Wu B., Ma Y., Jian J.B., Yang W.,

Yuan Z., Sun X.C., Wei Y.L., Yu L.L., Zhang C., Liao S.G., He R.J., Guang X.M., Wang Z., Zhang Y.Y., Luo L.H., 2014. The complex jujube genome provides insights into fruit tree biology. Nature Comm., 5: 5315

Macavei L.I., Bâețan R, Oltean I, Florian T, Varga M, Costi E, Maistrello L., 2015. First detection of *Halyomorpha halys* Stål, a new invasive species with a high potential of damage on agricultural crops in Romania. Lucrări Științifice seria Agronomie 58(1):105-108.

Markovski A., Velkoska-Markovska L., 2015. Investigation of the morphometric characteristics of Jujube types (*Ziziphus jujuba* Mill.) fruits in Republic of Macedonia. Genetika, Vol. 47(1):33-43.

Mead F.W., 2014. Citrus Flatid Planthopper, *Metcalfa pruinosa (Say) (Insecta: Hemiptera: Flatidae)*, Florida Department of Agriculture and Consumer Service-Division of Plant Industry: Gainesville FL.

Milonas P.G., Partsinevelos G.K. 2014. First report of brown marmorated stink bug *Halyomorpha halys* Stål (Hemiptera: Pentatomidae) in Greece. Bulletin OEPP/EPPO Bulletin 44(2):183-186.

Nielsen A.L., Hamilton G.C., Matadha D., 2008, Developmental Rate Estimation and Life Table Analysis for Halyomorpha halys (Hemiptera: Pentatomidae). Environ. Entomol. 37:348–355.

Ohta A.T.; Kaneshiro K.Y.; Kurihara J.S.; Kanegawa K.M.; Nagamine L.R., 1989. Gamma radiation and cold treatments for the disinfestation of the Mediterranean fruit fly in California-grown oranges and lemons. Pacific Science 43:17-26.

Ordax M., Piquer-Salcedo J.E., Santander R.D., Sabater-Muñoz B., Biosca E.G., López M.M., 2015. Medfly *Ceratitis capitata* as Potential Vector for Fire Blight Pathogen *Erwinia amylovora*: Survival and Transmission. PLoS ONE 10(5):e0127560.

Pan H, Liu B, Lu Y, Desneux N., 2014. Identification of the key weather factors affecting overwintering success of *Apolygus lucorum* eggs in dead host tree branches. PLoS ONE 9(4): e94190.

Panizzi A.R., 2008. Southern green stink bug, *Nezara viridula* (L.) (Hemiptera: Heteroptera: Pentatomidae). Encyclopedia of Entomology pp. 3471.

Paula D.P., Togawa R.C., Costa M.M.C., Grynberg P., Martins N.F., Andow D.A., 2016. Identification and expression profile of odorant-binding proteins in *Halyomorpha halys* (Hemiptera: Pentatomidae), Insect Molecular Biology 1-15.

Preda C., Skolka M., 2009. First record of a new alien invasive species in Constanta - *Metcalfa pruinosa* (Homoptera: Fulgoroidea). In: Paltineanu C. (Ed.) Lucrarile Simpozionului Mediul si Agricultura in regiunile aride: prima editie. Estfalia, 141-146.

Preeti S., Tripathi S., 2014. A phytopharmacological review on "*Ziziphus jujuba*", Int. J. Res. Dev. Pharm. L. Sci., 3(3):959- 966.

Pollini, A., Cravedi, P., 2014 *Carpomya vesuviana* A. Costa (Diptera: Tephritidae Trypetinae Carpomyini) from jujube tree in Emilia-Romagna (Northern Italy), Redia-Giornale Di Zoologia, 97:117-118.

Portilla M., Snodgrass G., Streett D. Luttrell R., 2015. Demographic Parameters of *Nezara viridula* (Heteroptera: Pentatomidae) Reared on Two Diets Developed for *Lygus spp.*, J. Insect Sci. 15(1):165.

Rice K., Bergh C., Bergman E., Biddinger D., Dieckhoff C., Dively G., Fraser H., Gariepy T., Hamilton G., Haye T., Herbert A., Hoelmer K., Hooks C., Jones A., Krawczyk G., Kuhar T., Mitchell W., Nielsen A.L., Pfeiffer D., Raupp M., Rodriguez-Saona C., Shearer P., Shrewsbury P., Venugopal D., Whalen J., Wiman N., Leskey T., Tooker J., 2014. Biology, ecology and management of brown marmorated stink bug (*Halyomorpha halys*) J. Integr. PestManage. 5(3):1-13

Qazzaz F.O., Al-Masri, M.I., Barakat, R.M., 2015. Effectiveness of *Beauveria bassiana* native isolates in the biological control of the Mediterranean fruit fly (*Ceratitis capitata*). Advances in Entomology.3:44-55

Sarwar M., 2006. Incidence of Insect Pests on Ber (*Zizyphus jujube*) Tree. Pakistan J. Zool., vol. 38(4):261-263.

Sela S., Nestel D., Pinto R., Nemny-Lavy E., Bar-Joseph M., 2005. Mediterranean fruit fly as a potential vector of Bacterial Pathogens. Applied and Environmental Microbiology, Vol. 71 (7):4052–4056.

Stanciu I., 2007. Comportarea unor soiuri de kaki în condițiile Câmpiei Române" Ph.D thesis. Faculty of Horticulture, University of Agronomic Sciences and Veterinary Medicine, Bucharest, Romania.

Stănică F., 1997. Curmalul chinezesc (*Ziziphus jujuba* Mill.), specie pomicolă pentru partea de sud a României (Chinese date (*Ziziphus jujuba* Mill.), fruit specie for the southern part of Romania). Hortinform.

Stănică F., 2002. Multiplication of Chinese date (*Ziziphus jujuba* Mill.) using conventional and *in vitro* techniques, XXVIth Congress ISHS, 11-17, Toronto, Canada, 303-304.

Stănică F., 2009. Characterization of Two Romanian Local Biotypes of *Ziziphus jujube*, ISHS Acta Horticulturae 840:259-262.

Strauss G., 2010. Pest risk analysis of *Metcalfa pruinosa* in Austria. Journal of Pest Science, 83:381-390.

Strauss G., 2012: Environmental risk assessment for *Neodryinus typhlocybae*, biological control agent against *Metcalfa pruinosa*, for Austria. European Journal of Environmental Sciences, 2(2):102-109.

Squitier J.M., 2013. Southern Green Stink Bug, *Nezara viridula* (Linnaeus) (Insecta: Hemiptera: Pentatomidae), UF/IFAS Extension Service, University of Florida, EENY-016 (IN142).

Varghese J., Patil M.B., 2005. Successive use of proteinase inhibitor of *Zizyphus Jujuba* leaves for effective inhibition of *Helicoverpa armigera* gut enzymes and larval growth, Biosciences Biotechnology Research Asia 3(2):367-370.

Lingampally V., Solanki V.R., Jayaram V., Kaur, A., Raja S.S., 2012. Betulinic acid: a potent insect growth regulator from *Ziziphus jujuba* against *Tribolium confusum*. Asian J. Plant Sci. Res. 2(2):198-206.

Vétek G., Papp V., Haltrich A., Rédi D., 2014. First record of the brown marmorated stink bug, *Halyomorpha halys* (Hemiptera: Heteroptera:

Pentatomidae), in Hungary, with description of the genitalia of both sexes. Zootaxa 3780(1):194-200.

Vadivelu K., 2014. Biology and management of ber fruit fly, *Carpomyia vesuviana* Costa (Diptera: Tephritidae): A review. African Journal of Agricultural Research 9(16):1310-1317.

Vlad M., Grozea I., 2016. Host Plant Species of the Cicada *Metcalfa Pruinosa* in Romania, Bulletin UASVM series Agriculture 73(1):131-137.

Yao, S. 2013. Past, present and future of jujubes-Chinese dates in the United States. HortScience, 48:672-680.

Wermelinger B, Wyniger D., Forster B., 2008. First records of an invasive bug in europe: *Halyomorpha halys* Stål (Heteroptera: Pentatomidae), a new pest on woody ornamentals and fruit trees? Bulletin De La Société Entomologique Suisse, 81:1–8

***species details: *apolygus lucorum* (meyer-dur, 1843), http://www.catalogueoflife.org/col/details/species/id/7fb6162f963b017e91741263c7764e20

***species details: *carpomya vesuviana* costa, 1854 http://www.catalogueoflife.org/col/details/species/id/ba931ddef181c172c517e6d6644301a5

***species details: lygus *pratensis* (linnaeus, 1758) http://www.catalogueoflife.org/col/details/species/id/1d6b9b22b095fa0df3d3c8e92e62a509

***https://gd.eppo.int/taxon/METFPR
***https://gd.eppo.int/taxon/NEZAVI
***https://gd.eppo.int/taxon/CERTCA

15

STUDY ON THE RELATED CRACKING-RESISTANT GENES IN CHINESE JUJUBE

Yong-Xiang REN[1], Lian-Ying SHEN[1], Xiao-Ling WANG[1], Yao-WANG[1], Chun-Mei YAN[1], Li-Hui MAO[2], Yong-Min MAO[1]

[1]Agricultural University of Hebei, Lingyusi Street, No. 289, 071001, Baoding, China
[2]Renxian Agriculture Bureau, Guangming Road, NO.216, 055151, Xingtai, China
Corresponding author email: mym63@126.cn

Abstract

The problem of fruit cracking for Chinese jujube (Zizyphus jujuba Mill.) causes serious yield losses in China, however, the existing prevention and control measures are hard to solve it. Two groups of high resistant cracking type and easy cracking type for the filial generation of Dong Zao × Linyili Zao were divided in the study. Two allelic genetic pool and the related genes cDNA libraries for the cracking-resistant characters to be specifically expressed were built to conduct the preliminary feasibility study on the genes related to the resistance of fruit-cracking through the differently expressed genes. The sequencing production of the two group transcriptomes were received more than 4.5 Gb and 4.6 Gb data after removing the low quality segments, and conducted 45,401,606 and 46,468,222 times reading, the Q20 (the base ratio no less than 20) of the two libraries were both greater than 96%. After comparing the genetic expression within two groups, 391 items of differently expressed genes (DEGs) were filtered. Genome function annotation on the 391 items DEGs were conducted and the results showed that there were 92 items genes added, 45 items defined, and 299 items gene without annotation. The annotated genes probably involving in jujube cracking phenomenon were Aquaporin PIP, Tubulin, Calreticulin and Calmodulin.

Key words: *Chinese jujube, fruit cracking, cracking-resistant genes, RNA-Seq.*

INTRODUCTION

The phenomenon of fruit cracking on jujube is widespread, which causes serious yield losses, however, the existing prevention and control measures cannot fundamentally solve the problems. The researches showed that there were great differences between different varieties of jujubes on fruit cracking resistance (Mao et al., 1998 ; Wang et al., 2011 ; Yuan et al., 2013). So, it is effective means to solve the fruit cracking by breeding new varieties with high resistant cracking in Chinese jujube, the molecular marker assisted selection is a important methods in plant breeding.

Researches on the genes related to fruit cracking characteristics have been reported, such as MdExp3 gene in apples (Wakasa et al., 2003; Kasai et al., 2008), β-galactosidase gene (TBG6) in tomatoes (Moctezuma et al., 2003), Expansin gene - LcExp1gene and LcExp2 - in litchi pericarp (Wang et al., 2006), which

closely related to cell wall loosening gene and fruit cracking gene. With the rapid development of high throughput sequencing, the function of transcriptome sequencing analysis has opened new avenues for the study of fruit cracking gene. Li used the high throughput sequencing technology to do the transcriptome sequencing analysis on crack resistance and easy to crack pericarps, through the analysis of the differences in gene expression, there were 67 candidate genes about fruit cracking screened out (Li et al., 2014), including 4 water transportation related genes (LcAQP, 1; LcPIP, 1; LcNIP, 1; LcSIP, 1), 5 genes related to Gibberellic Acid (Gibberellic Acid, GA) metabolism (LcKS, 2; LcGA2ox, 2; LcGID1, 1), 21 Abscisic Acid (Abscisic Acid , ABA) metabolism related genes (LcCYP707A, 2; LcGT, 9; Lcβ-Glu, 6; LcPP2C, 2; LcABI1, 1; LcABI5, 1), 13 genes related to calcium transportation (LcTPC, 1; Ca2+/H+ exchanger, 3; Ca2+-ATPase, 4;

LcCDPK, 2; LcCBL, 3), and 24 cell wall metabolism related genes (LcPG, 5; LcEG, 1; LcPE, 3; LcEXP, 5; Lcβ-Gal, 9; LcXET, 1).

The main objectives of this study are (1) building two near-isogenic pools with the hybrid offspring of Dong Zao × Linyili Zao, which have the same genetic background; (2) establishing jujube resistance to cracking fruit traits specific related genes expression cDNA library, and to do the sequencing analysis generating the relevant ESTs sequences; (3) sequencing the differential expression genes, and to do gene function analysis and prediction on the ESTs sequences. This will provide references for the study of new genes and new germplasm about jujube and related plants, and it will have important significance to breed new cracking-resistance varieties through molecular techniques.

MATERIALS AND METHODS

Plant material

The supplied samples were 12 year-old hybrid offspring of Dong Zao × Linyili Zao in Jujube breeding base located in Wangcun town, Daming County, Hebei Province).

Two group of 12 cracking-resistance (cracking fruit rate < 5%, group A) and 12 sensitive-cracking (cracking fruit rate > 80%, group B) hybrid offspring were divided, according to the survey results of consecutive years cracking fruit rate on the hybrid offspring.

During young fruit stage (at the end of the 20 days after flowering), 10 young fruits of each type were picked and covered with silver paper immediately, then put into liquid nitrogen to quick-freeze. The samples were taken back to laboratory and preserved in an icebox (T=-80°C) for subsequent analyses.

RNA extraction

The fruit RNA of the offspring was extracted using the RNeasy plant mini kit (Qiagen); On-column DNase digestion with the RNase-Free DNase set (Qiagen) was performed to remove contaminated DNA.

Twelve mixed Chinese jujube cDNA samples were prepared for each group, and then the RNA samples were sent to prepare library with NEB RNA library prep kit and sequence with Illumina HiSeq2000. The RNA-Seq data were subjected to bioinformatic analysis.

Sequence bioinformatics

The raw reads were filtered with FASTQ_Quality_Filter tool from the FASTX-toolkit. Reads both with more than 35bp and having a quality score higher than 20 were kept. Then all the valid reads of A and B samples were combined to perform de novo splicing by paired-end method with Trinity software (Grabherr et al., 2011). The longest transcript per locus was used as a unigene, and as a result we got 94,984 unigenes.

Several complementary approaches were utilized to annotate the unigenes, which were conducted using the Basic Local Alignment Search Tool (BLAST). The unigenes were compared against the NCBI NR, SWISS-PROT, TrEMBL, Cdd, pfam and KOG databases with an E-value of 1e-5 and Identity of 30%. Functional annotation were operated using gene ontology terms method by Blast2GO software (GO, http://www.geneontology.org) (Ashburner et al., 2000). The Kyoto Encyclopedia of Genes and Genomes (KEGG) pathways were assigned to the sequences using the online KEGG Automatic Annotation Server (KAAS) (http://www.genome.jp/kegg/kaas/). This method was used to obtain KEGG Orthology (KO) assignment. The output of KEGG analysis includes KO assignments and KEGG pathways (Kanehisa, 1997; Kanehisa and Goto, 1999).

Gene expression analysis

Researchers compared the reads with unigenes using single-end mapping method by bowtie2-2.2.2 software. To compare the unigene expression level in the A and B libraries, the transcript level of each expressed unigene was calculated and normalized to the reads per kilobase of exon model per million mapped reads (RPKM) (Mortazavi et al., 2008). Significance of differential unigene expression was determined by using General Chi-squared test and assigned P-values (<0.01). The P-values were adjusted to account for multiple testing by using the false discovery rate (FDR) and assigned error ratio Q-value (<0.05).

The unigenes with an adjusted P-value <0.01 and the absolute value of log2 >1 (expression fold change) were deemed to differently expressed, while the unigenes with an FDR-adjusted P-value <0.01 was considered statistically significant (Audic and Claverie, 1997; Reiner et al., 2003; Simonsen and Mcintyre, 2005).

RESULTS AND DISCUSSIONS

Sequence analysis and assembly

To understand the molecular bases of Zizyphus jujuba and identify the new valid genes, six libraries representing two groups of high resistant cracking fruit and sensitive cracking fruit were constructed. RNA from the two groups was used for Illumina RNA-Seq.

Each sequenced sample yielded 100bp reads from paired-end sequencing of cDNA fragments. After quality assessment and data clearance, 4.5~4.6 billion (G) reads with more than 96% Q20 bases (those with an average base quality greater than 20) were kept as high quality reads for each library and used in the later analysis (Table 1). All the valid reads above were combined to perform de novo splicing by paired-end method with Trinity software (trinityrnaseq_r20131110 version). A total of 94,984 unigenes were obtained, among which 25,287 unigenes were longer than 1kb. An overview of the assembled transcripts and unigenes was presented in Table 2. The length distributions of unigenes were shown in Figure 1. The results demonstrated the effectiveness of Illumina pyrosequencing in rapidly capturing a large portion of the transcriptome.

The distribution of gene expression levels was used to evaluate the normality of the library data. The level of gene expression was determined by calculating the number of unigenes and then normalizes to the RPKM (Mortazavi et al., 2008). As shown in Figure 2, the majority of mRNA was expressed at low levels, whereas a small proportion of mRNA was highly expressed.

The gene expression variations were analysed by compared the two libraries of resistant cracking fruits (group A) and sensitive cracking (group B). A total of 391 differently expressed genes (DEGs) including 218 up-regulated and 173 down-regulated genes were detected.

Table 1 Description of two groups of *Zizyphus jujuba* RNA-Seq libraries

Library name	Total reads	Total bases	Average reads length (bp)	Cycle Q20 (%)
A[a]	45,401,606	4,507,948,150	99.29	96.37
B[b]	46,468,222	4,606,988,891	99.14	96.04
All	91,869,828	9,114,937,041	99.22	96.20

A [a]: cracking-resistant fruit; B [b]: sensitive-cracking fruit;

Table 2 Summary of Illumina transcriptome assembly for *Zizyphus jujuba*

Library name	Total reads	N50[a]	N90[b]	Total Length	Max length	Min length	Average length
Transcript	198,993	2010	589	252,481,032	28,719	201	1268.79
Unigene	94,984	1584	302	79,036,677	28,719	201	832.11

N50 a, sorted the transcripts from long to short, then accumulated bases of transcripts in turn, when the total bases number reached half of total number of bases, the length of transcript, as well as unigenes; N90 b, counted in a similar way.

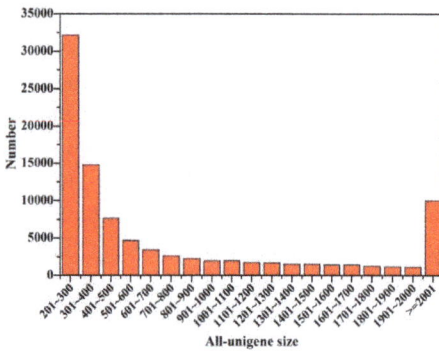

Figure 1. Length distributions of All-Unigenes

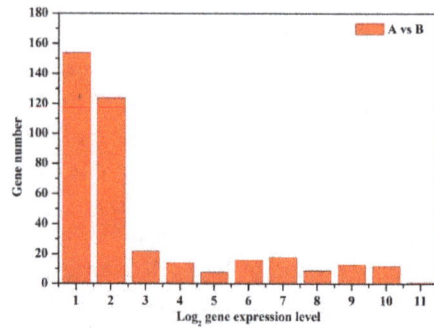

Figure 2. Distribution of gene expression levels

GO enrichment analysis of DEGs

GO enrichment analysis was the function annotations of DEGs. GO term with P-value< 0.05 were defined significantly enriched, and there were 1515 items DEGs involved in it. We mapped the sensitive-cracking higher than cracking-resistant fruit gene expression (UP.A.B) and sensitive-cracking less than cracking-resistant fruit gene expression (DOWN.A.B) DEGs to terms in GO database respectively.

The DEGs up of GO enrichment analysis are shown in Figure 3. There were 467 items UP.A.B DEGs involved in biological metabolism, mainly including metabolic processes, accounting for 30%; cellular processes, 26%; single-organism processes, 15%, and multi-organism processes, 10%; 257 items involved in cellular components, the cell accounted for 17%; cell part, 17%; other organism, 16%, and other organism part, 16%; 255 items involved in molecular function, the catalytic activity took up 60%, and binding was 27%.

The DEGs down of GO enrichment analysis are shown in Figure 4. There were 233 items DOWN.A.B DEGs involved in biological metabolism, mainly including cellular process, taking up 30%; cellular component organization or biogenesis, 15%, and metabolic process, 14%; 256 items involved in cellular component, The cell accounted for 20%; cell part, 20%; macromolecular complex, 14%; organelle, 14% and organelle part, 14%; 73 items involved in molecular function, including binding accounting for 55% and catalytic activity taking up 42%.

We compared GO enrichment analysis between UP.A.B and DOWN.A.B DEGs, there were 11 processes fully upward trend, multicellular organismal process, developmental process, reproductive process, biological adhesion, membrane, membrane part, extracellular matrix, extracellular region part, nucleic acid binding transcription factor activity, molecular function regulator and antioxidant activity. The membrane belongs to cellular component including 24 genes were all up-regulated, membrane component may have important relationship with fruit sensitive cracking.

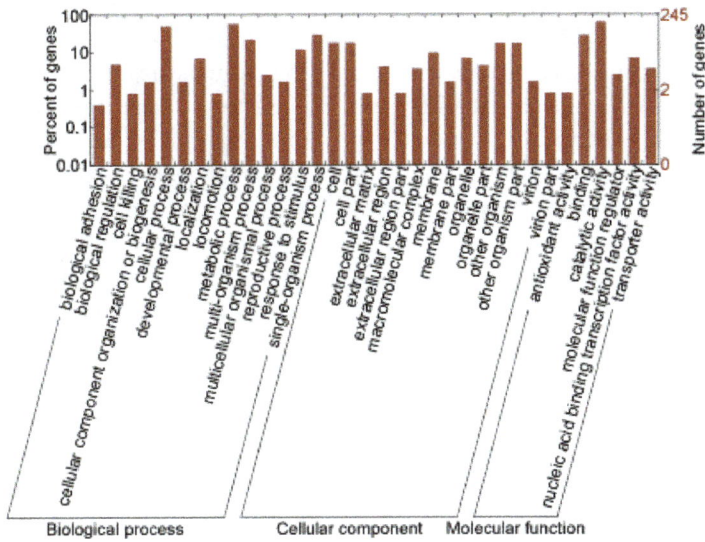

Figure 3 DEGs up of GO enrichment analysis

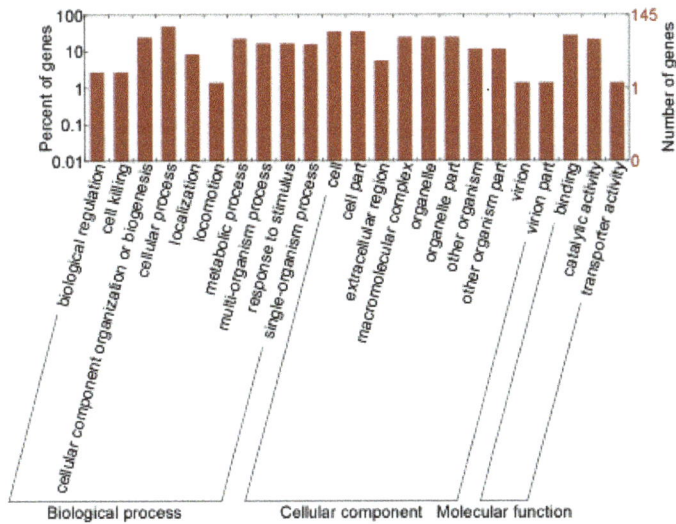

Figure 4. DEGs down of GO enrichment analysis

Pathway enrichment analysis of DEGs

KEGG pathway enrichment analysis was performed to categorize the biological functions of DEGs. (Kanehisa, 1997; Kanehisa and Goto, 1999) We mapped all the genes to terms in KEGG database.

Enrichment analysis of differentially expressed genes by KEGG is shown in Table 3. KEGG with P-value < 0.05 were defined significantly enriched, and there were 246 items DEGs involved in it. Among all the KEGG, there were 10 items significantly enriched KEGG; The DEGs participated in amino acid metabolism and lipid metabolism were more, there were 15 items and 12 items, respectively. The amino acid metabolism primarily referred to cysteine and methionine metabolism, and valine, leucine and isoleucine biosynthesis. The Lipid metabolism mainly pointed to alpha linolenic acid metabolism and steroid biosynthesis.

The experiment were expected to get the related genes of jujube easy cracking fruit or resistance cracking fruit by the KEGG enrichment analysis of DEGs, the α-linolenic acid is the basic substances of cell membrane and enzymes (SanGiovanni et al., 2005). The result of down-regulated differences in gene enrichment of metabolic pathway suggested that α-linolenic acid metabolism in cracking fruit types were weaker than in resistance cracking fruit types; therefore, α-linolenic acid metabolic pathway can be one of the research targets for cracking fruit study.

The number Ko (ko00592) in the α-linolenic acid metabolic pathway corresponding to KEGG pathway can analyse the relationship of genes and KEGG, and map the information into the pathway. The α-linolenic acid metabolism process with KEGG pathway was shown in Figure 5.

Table 3 Enrichment analysis of differentially expressed genes by KEGG

#	KEGG name term	Classification	P-value	Q-value	Number
1	Cysteine and methionine metabolism [PATH:ko00270]	Amino acid metabolism	2.89E-06	0.000890265	11
2	Flavonoid biosynthesis [PATH:ko00941]	Biosynthesis of other secondary metabolites	4.59E-11	1.41E-08	11
3	Alpha linolenic acid metabolism [PATH:ko00592]	Lipid metabolism	8.11E-07	0.000249766	8
4	Glutathione metabolism [PATH:ko00480]	Metabolism of other amino acids	0.00014206	0.043753378	8
5	Circadian rhythm - plant [PATH:ko04712]	Environmental adaptation	5.29E-06	0.001628153	4
6	ko01220	-	2.84E-05	0.008752995	4
7	Stilbenoid, diarylheptanoid and gingerol biosynthesis [PATH:ko00945]	Biosynthesis of other secondary metabolites	6.33E-05	0.019485288	4
8	Steroid biosynthesis [PATH:ko00100]	Lipid metabolism	0.00015861	0.048852201	4
9	Valine, leucine and isoleucine biosynthesis [PATH:ko00290]	Amino acid metabolism	7.25E-05	0.022328	4
10	Valine, leucine and isoleucine biosynthesis [PATH:ko00290]	Carbohydrate metabolism	0.00014	0.043056	2

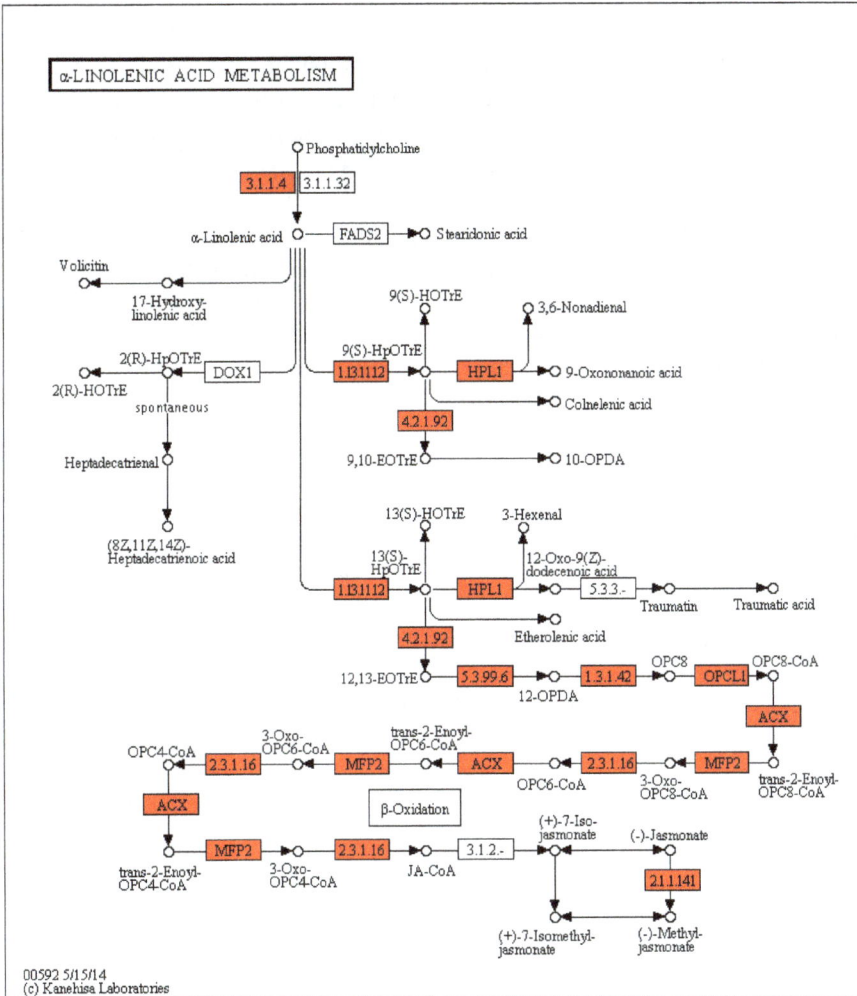

Figure 5. Alpha linolenic acid metabolism with KEGG pathway

Functional annotation of differentially expressed genes

The main gene annotations of DEGs are given in Table 4. The screened 391 items DEGs were functionally annotated, 92 items genes annotated, 45 items genes defined, and 299 items genes still not annotated. The annotated genes related to jujube cracking fruit were Aquaporin PIP, Tubulin, Calreticulin, and Calmodulin; there were 5, 6, 2 and 2 in DEGs, respectively.

Aquaporin PIP is a protein channel in the plasma membrane that facilitates water movement across the membrane, regulating the water balance inside and outside the cell, which are the down-regulated DEGs, suggesting the expression of Aquaporin PIP in resistance to cracking fruit types are higher than that in easy to cracking fruit types. Previous studies showed that

With the development of the fruit, the water in fruit was reduced, and water absorbing capacity was increased. Then when there was water in the fruit surface and dived in water potential gradient, the water would go into the fruits, thus when the turgor pressure within the fruit exceed the skin strength, it would be the cracking fruits (Wang et al., 2013).

And in the resistance to cracking fruit types, the Aquaporin PIP could promote the water transportation inside and outside cells, which reduced the turgor pressure within the fruit to a certain extent, thus reduce the risk of cracking fruits.

Tubulin is the important component of the cytoskeleton, it acts to keep the cells' shape, involved in cell division, cell movement, intracellular material transportation, and assists all kinds of organelles to complete their respective functions (Rao and Zhang, 2013). In our study, we annotated α-tubulin and β-tubulin to down-regulated DEGs, which could promote the high expression of microtubule protein gene, thus reducing the happening of cracking fruit types.

Calreticulin is one of the calcium binding proteins on the endoplasmic reticulum, with the biological function of chaperone, adjusting the steady state of Ca2+, cell adhesion and gene expression regulation (Liu, 2013), Calmodulin is one of the conservative strong regulatory proteins in the structure of biological cell by adjusting a series of enzymes to affect the plant growth, development and stress (Hoeflich and Ikura, 2002). Li (2014) screened 13 genes associated with calcium transportation from the candidate genes related to cracking fruit by DEGs. In our study, the Calreticulin and Calmodulin showed high expression in resistance to cracking fruit types, suggesting the cracking fruit had important relationship with Ca transportation and regulation.

Table 4 The main gene annotation of DEGs

Unigene	Name	Definition	Length	up/down	P-value	Q-value
comp54052_c0_seq1	PIP	AQUAPORIN PIP	1416	down	1.79E-12	2.21E-09
comp40997_c0_seq1	PIP	AQUAPORIN PIP	1508	down	6.84E-13	9.69E-10
comp16213_c0_seq1	PIP	AQUAPORIN PIP	315	down	3.99E-05	1.20E-02
comp16225_c0_seq1	PIP	AQUAPORIN PIP	315	down	3.99E-05	1.20E-02
comp45258_c1_seq1	PIP	AQUAPORIN PIP	1231	up	0.00012	0.03147
comp22094_c0_seq1	TUBA	TUBULIN ALPHA	984	down	0.00011	0.02806
comp10051_c0_seq1	TUBA	TUBULIN ALPHA	704	down	2.65E-06	1.16E-03
comp15927_c0_seq1	TUBA	TUBULIN ALPHA	984	down	4.83E-05	1.41E-02
comp49817_c0_seq1	TUBA	TUBULIN ALPHA	2035	down	4.10E-11	4.33E-08
comp49059_c0_seq1	TUBA	TUBULIN ALPHA	1978	down	2.23E-11	2.43E-08
comp36379_c0_seq1	TUBB	TUBULIN BETA	853	down	1.60E-06	7.43E-04
comp45120_c0_seq1	CALM	CALMODULIN	1210	down	0.000171	0.04298
comp45967_c0_seq1	CALM	CALMODULIN	1150	down	0.000169	0.04249
comp41077_c0_seq1	CALR	CALRETICULIN	1584	down	1.23E-06	0.00059
comp42453_c0_seq1	CALR	CALRETICULIN	1687	down	2.65E-06	0.00116

CONCLUSIONS

In summary, 5 gene fragments annotated Aquaporin PIP, 6 gene fragments annotated Tubulin, 2 gene fragments annotated Calreticulin and Calmodulin have been obtained by the differential gene analysis of two near-isogenic gene pools. All the gene fragments above were down-gene of differential genes. The result suggested that the express level of Aquaporin PIP, Tubulin, Calreticulin and Calmodulin in cracking-resistant fruit were higher than that in sensitive cracking fruit.

It was found that α-linolenic acid metabolism was related to fruit cracking of Chinese jujube according to the GO functional enrichment and KEGG enrichment analysis by building two near-isogenic pools with the same genetic background and different cracking fruit traits. It was concluded that cracking fruit process of Chinese jujube is closely linked to the cellular structure because α-linolenic acid is the basis substance of cell membrane and enzyme (SanGiovanni et al., 2005).The previous researches obtained similar results (Zhou et al., 1999; Xin et al., 2006; Liu et al., 2015).

ACKNOWLEDGEMENTS

This work was supported by the Project (14226809D, 16226313-7) from the Science and Technology Department of Hebei Province, the National Science and Technology Support Project (2013BAD14B03) and the project (2005DKA21003) from National Forest Genetic Resources Platform of China.

REFERENCES

Ashburner M., Ball C.A., Blake J. A., et al., 2000. Gene ontology: tool for the unification of biology. Nature Genetics, 25(1):25-29.

Audic S., Claverie J.M., 1997. The significance of digital gene expression profiles. Genome Research, 7(10):986-995.

Grabherr M.G., Haas B.J., Yassour M., et al., 2011. Full-length transcriptome assembly from RNA-Seq data without a reference genome. Nature Biotechnology, 29(7):644-652.

Hoeflich K.P., Ikura M., 2002. Calmodulin in action: diversity in target recognition and activation mechanisms. Cell, 108(6):739-742(4).

Kanehisa M., 1997. A database for post-genome analysis. Trends in Genetics Tig, 13(9):375-376.

Kanehisa M., Goto, S., 1999. Kegg: kyoto encyclopedia of genes and genomes. Nucleic Acids Research, 27(1):29-34(6).

Kasai S., Hayama H., Kashimura Y., et al., 2008. Relationship between fruit cracking and expression of the expansin gene MdEXPA3 in 'Fuji' apples (Malusdomestica Borkh.). Scientia Horticulturae, 116(2):194-198.

Li W.C., Wu J.Y., Zhang H.N., et al., 2014. De novo assembly and characterization of pericarp transcriptome and identification of candidate genes mediating fruit cracking in Litchi chinensis Sonn. International Journal of Molecular Sciences, 15(10):17667-17685.

Liu X.M., 2013. The structure, function and evolution of the Calreticulin. Science Technology Information (16):226-226.

Liu Z.G., Lu Y.Q., Zhao J., et al., 2015. The effects of water absorbing dynamics and pericarp microstructure on fruit cracking in Chinese jujube. Journal of Plant Genetic Resources, 16(1):192-198(7).

Mao Y.M., Shen L.Y., Bi P., et al., 1998. Comparative study on cracking resistance of different Chinese jujube cultivars. The First Academic Symposium Proceedings of Dried Fruit Production and Scientific Development, 145-151.

Moctezuma E., Smith D.L., Gross K.C., 2003. Antisense suppression of a β-galactosidase gene (TB G6) in tomato increases fruit cracking. Journal of Experimental Botany, 54(390):2025-2033.

Mortazavi A., Williams B.A., Mccue K., et al., 2008. Mapping and quantifying mammalian transcriptomes by RNA-Seq. Nature Methods, 5(7):621-628.

Rao G.D., Zhang J.G., 2013. Advances of studies on plant tubulin gene. World Forestry Research. 26(3):17-20

Reiner A., Yekutieli D., Benjamini Y., 2003. Identifying differentially expressed genes using false discovery rate controlling procedures. Bioinformatics, 19(3):368-375.

Sangiovanni J.P., Chew E.Y., 2005. The role of omega-3 long-chain polyunsaturated fatty acids in health and disease of the retina. Progress in Retinal and Eye Research, 24(1):87–138.

Simonsen, K.L., Mcintyre, L.M., 2005. Implementing false discovery rate control: increasing your power. Oikos, 108(3):643–647.

Wakasa Y., Hatsuyama Y., Takahashi A., et al. 2003. Divergent expression of six expansin genes during apple fruit ontogeny. European Journal of Horticultural Science, 68(6):253-259.

Wang Y., Lu W., Li J., et al., 2006. Differential expression of two expansin genes in developing fruit of cracking-susceptible and -resistant litchi cultivars. Journal of the American Society for Horticultural Science American Society for Horticultural Science, 131(1):118-121.

Wang Z.L., Tian Y.P., Liu M.J., et al., 2011. Crack resistance and its mechanism of different Chinese jujube cultivars. Nonwood Forest Research, 29(3):74-77

Wang, B.M., Ding, G.X., Wang, X.Y., et al., 2013. Changes of histological structure and water potential of huping jujube fruit cracking. Scientia Agricultura Sinica, 46(21):4558-4568.

Xin Y.W., Ji X., Liu H., 2006. Observation and studies on peel and pulp growing characters of different crack in jujube fruit varieties. Chinese Agricultural Science Bulletin, 22(11):253-257.

Yuan Z., Lu Y.Q., Zhao J., et al., 2013. Screening and evaluation of germplasms with high resistance to fruit cracking in Ziziphus jujuba Mill. China Agriculture Science, 46(23):4968-4976.

Zhou J.Y., Mao Y.M., Shen L.Y., et al., 1998. Study on the relationship between microstructure and fruit cracking of Chinese jujube. The Academic Symposium Proceedings of Dried Fruit Production and Scientific Development, 266-267.

16

ISOLATION AND BIOINFORMATICS ANALYSIS OF GLUTAMYL-TRNA REDUCTASE IN CHINESE JUJUBE

ZG LIU, J ZHAO, MJ LIU
Hebei Agricultural University, 071001, Baoding, Hebei, China
Corresponding authors email: lmj1234567@aliyun.com; zhaojinbd@126.com

Abstract

Tetrapyrroles, such as chlorophyll and heme, are integral to the metabolism of all living organisms. Glutamyl-tRNA reductase (GluTR) is the first unique enzyme in the tetrapyrrole biosynthetic pathway in plants. We firstly cloned the GluTR complete sequences (1733 bp) from Chinese jujube (Ziziphus jujuba Mill., belonging to the family Rhamnaceae), named as ZjGluTR (GenBank accession no. KF530842). ZjGluTR shares high similarity (90%) to those of Malus domestica and Prunus mume through BLASTX analysis. It contains a 1659-bp ORF and encodes a predicted polypeptide of 552 amino acids, with an estimated molecular mass of 60.0 kDa and a theoretical pI of 8.55. The speculated formula of ZjGluTR is $C_{2615}H_{4318}N_{756}O_{804}S_{25}$. ZjGluTR has a chloroplast transit peptide which contains 61 amino acids in the N terminal. The subcellular localization result showed that ZjGluTR protein exists in chloroplasts. ZjGluTR has three typical functional domains, i.e. GluTR N-terminal domain (101-259 aa), NAD(P)-binding Rossmann-fold domain (266-422 aa) and GluTR dimerization domain (425-534 aa). The molecular phylogenetic tree of GluTR indicated that the family Rhamnaceae has a close genetic relationship with the family Rosaceae. Our studies on the GluTR using molecular biology and bioinformatics approaches would play an important role in chlorophyll metabolic research of Chinese jujube.

Key words: Chinese jujube, GluTR, isolation, bioinformatics analysis.

INTRODUCTION

Tetrapyrroles, such as chlorophyll and heme, are integral to the metabolism of all living organisms. The first common precursor molecule of all tetrapyrroles is derived from 5-aminolevulinic acid (ALA). It is synthesized through two distinct biosynthetic routes: Firstly, in humans, animals, fungi and the α-group of the proteobacteria, the condensation of succinyl coenzyme A and glycine with the release of CO_2 is catalyzed by ALA-synthase (Kikuchi et al., 1958; Neuberger, 1968; Avissar et al., 1989; Ferreira, 1995). Secondly, the older pathway, utilizing the C_5-skeleton of glutamate, was first discovered in plants (Beale and Castelfranco, 1973). Subsequently, the C_5-pathway was found to be common to plants, green algae, archaea and most bacteria (Schön et al., 1986; Jahn et al., 1992). In the first dedicated step, the NADPH-dependent reduction of glutamyl-tRNA to glutamate-1-semialdehyde (GSA) is catalyzed by glutamyl-tRNA reductase (GluTR) (Mayer et al., 1987;

Chen et al., 1990; Jahn et al., 1992, Verkamp et al., 1992; Vothknecht et al., 1996, 1998). In the subsequent reaction, GSA is transaminated by the pyridoxal/pyridoxamine 5'-phosphate-dependent glutamate-1-seminaldehyde-2, 1-aminomutase (GSAM) to form ALA (Grimm, 1990; Jahn et al., 1991; Smith et al., 1992; Ilag and Jahn, 1992). In plants, all of the components required for such a conversion are located in the chloroplasts.

Other pathways of ALA formation in plants have been reported (Ramaswamy and Nair, 1973; Meller and Gassman, 1982), but efforts to thoroughly characterize these 'alternative ALA-forming pathways' have been unsuccessful. The existence of the C_5 pathway is widely accepted. Furthermore, lethal effect produced in some lines by antisense HEMA1 (Kumar et al., 2000) and antisense GSA (Höfgen et al., 1994) favors the C_5 pathway as a sole source for ALA biosynthesis.

Currently, studies on ALA biosynthesis in Chinese jujube (*Ziziphus jujuba* Mill.)-have not been reported. We carried out relevant research on the *GluTR* using molecular biology and bioinformatics approaches, which would provide a foundation for chlorophyll metabolic studies of Chinese jujube.

MATERIALS AND METHODS

Materials

Ziziphus jujuba Mill. 'Xingguang' was used as material to isolate total RNA. The fresh young leaves were collected, then frozen with liquid nitrogen rapidly and kept at -80 °C.

Methods

Total RNA isolation

Isolation of total RNA was carried out according to the instructions of improved CTAB method (Zhao et al., 2009). DNase I treatment was applied to remove contaminating genomic DNA. First-strand cDNA was synthesized as described by TaKaRa RNA PCR Kit (AMV) Ver.3.0 (TaKaRa).

Amplification of the full length cDNA of GluTR

Homology cloning method was used to obtain full-length cDNA of *GluTR* in Chinese jujube, the pair primers of 5' end-primer (5'-ATGGCCGTGTCGACCAGT T-3') and 3' end-primer (5'-GAGGATGTTGCCTCTTATTC-3') were used in this study. PCR was performed in a volume of 25 μL containing 15.5 μL of ddH$_2$O, 2 μL of the first strand cDNA, 2.5 μL of ExTaq DNA polymerase buffer, 2 μL of MgCl$_2$, 0.5 μL of dNTPs, 0.5 μL of ExTaq DNA polymerase, and 1 μL of each primer. PCR were optimized to consist with the following parameters, i.e. denaturation of 5 min at 95 °C; 35 cycles of 30 s at 94 °C, 60 s at 55 °C and 90 s at 72 °C; extension at 72 °C for 10 min. PCR products were separated on 1% agarose gels. Amplified fragment was cloned into pMD-19T vector and sequenced by Beijing ZhongKe XiLin Biotechnology CO., Ltd.

Analysis of cDNA and protein sequences

cDNA sequence was analyzed using BLAST (http://www. http://blast.ncbi.nlm. nih.gov/)

and the ORF FINDER (http://www.ncbi.nlm.nih.gov/ gorf/gorf.html). The protein sequence was analyzed by Protparam (http://expasy.org/ tools/protparam.html), WoLF PSORT (http://wolfpsort.seq.cbrc.jp), ChloroP Server 1.1 (http://www.cbs.dtu.dk/services/ ChloroP/), Pfam (http://pfam.sanger.ac.uk) and SWISS-MODEL Workspace (http://www.expasy.ch/swissmod/ SWISS-MODEL.html). Homology tree was educed according to MEGA 6 using neighbor-joining method.

RESULTS AND DISCUSSIONS

Cloning of the full-length cDNA of *ZjGluTR*

Approximately 1750-bp fragment was identified by homology cloning method (Figure 1). The fragment was cloned into a cloning vector, and sequenced subsequently. A 1733-bp expressed sequence tags (EST) was obtained after removing vector and adapter sequences. We named the sequence as *ZjGluTR* (GenBank accession no. KF530842). BLASTX analysis showed that the 1733-bp sequence shared high similarity (90%) to those of *Malus domestica* (XP008378624.1) and *Prunus mume* (XP008219178.1).

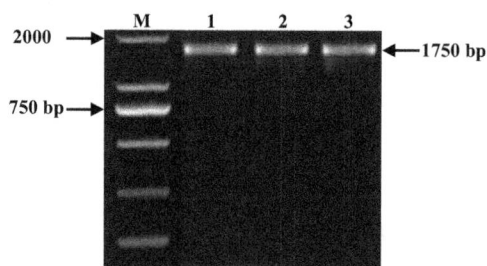

Figure 1. The PCR result of *ZjGluTR* by homologous cloning. Note: line 1, 2 and 3 were three replication

Protein sequence analysis and characterization of ZjGluTR

The full-length cDNA sequence of *ZjGluTR*, containing a 1659-bp ORF, encoded a predicted polypeptide of 552 amino acids, with an estimated molecular mass of 60.0 kDa and a theoretical pI of 8.55. The speculated formula of protein was $C_{2615}H_{4318}N_{756}O_{804}S_{25}$.

The results predicted from Search Pfam showed that ZjGluTR contained three functional domains, including GluTR N-terminal domain, Shikimate/quinate 5-dehydrogenase domain and GluTR dimerisation domain. GluTR N-terminal domain and GluTR dimerisation domain were typical characteristics of GluTR.

Relevant literature reported that the precursor protein of Glutamyl-tRNA reductase has a transit peptide in the N terminal. Therefore, we carried out transit peptide prediction for the ZjGluTR. The results showed that the ZjGluTR has a chloroplast transit peptide which contains 61 amino acids in the N terminal. At the same time, WoLF PSORT was used to predict the ZjGluTR subcellular localization. The result showed that ZjGluTR protein exists in plant

chloroplasts with the identity of 79%. This result was consistent with the ZjGluTR chloroplast transit peptide prediction.

Alignment result by homology modeling showed that ZjGluTR has three typical functional domains (Figure 2), such as GluTR N-terminal domain (101-259 aa), NAD(P)-binding Rossmann-fold domains (266-422 aa) and GluTR dimerization domain (425-534 aa). The results were consistent with Schubert's reports that each GluTR monomer consists of three distinct domains (Schubert et al., 2002). In addition, Shikimate/quinate 5-dehydrogenase domain (262-414 aa) was also predicted, which was not the typical characteristic of GluTR. Meanwhile, the tertiary structure of ZjGluTR was constructed by SWISS-MODEL Workspace (Figure 3b), which has the same tertiary structure as *Methanopyrus kandleri* GluTR (Figure 3a) (Schubert et al., 2002).

```
IPR000343: Glutamyl-tRNA reductase, Family
   TIGR01035: 104 - 533  hemA: glutamyl-tRNA reductase

IPR000343: Glutamyl-tRNA reductase, Family
   SSF69742: 101 - 259  Glutamyl tRNA-reductase catalytic, N-terminal domain

IPR006151: Shikimate/quinate 5-dehydrogenase, Domain
   PF01488: 262 - 414  Shikimate_DH

noIPR: unintegrated, unintegrated
   SSF51735: 266 - 422  NAD(P)-binding Rossmann-fold domains

noIPR: unintegrated, unintegrated
   SSF69075: 425 - 534  Glutamyl tRNA-reductase dimerization domain
```

Figure 2. Alignment results of *ZjGluTR* homology modeling

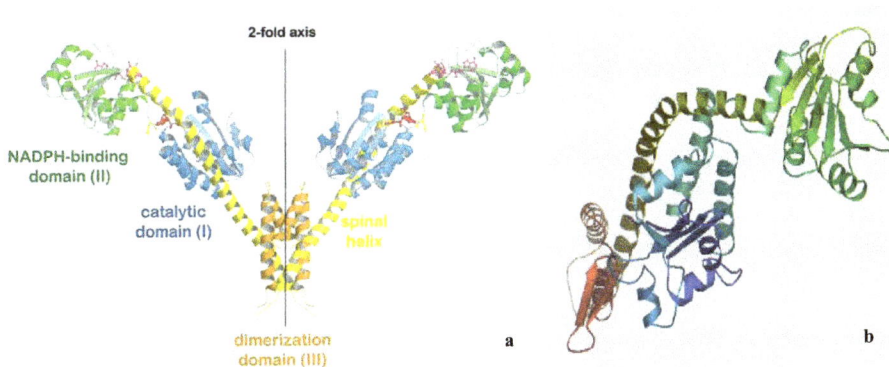

Figure 3. A schematic diagram of the *Methanopyrus kandleri GluTR* dimer viewed perpendicular (a) (Schubert et al., 2002) and the tertiary structure of *ZjGluTR* (b)

The phylogenetic tree of GluTR

The molecular phylogenetic tree of ZjGluTR and other 11 plant species was constructed by MEGA 6 (Figure 4). This tree showed that GluTR of *Z. jujuba* (belonging to the family Rhmnaceae) was firstly clustered with that of *Malus domestica* and *Prunus mume* (the family Rosaceae) which indicated that ZjGluTR should have a closely genetic relationship with that of Rosaceae.

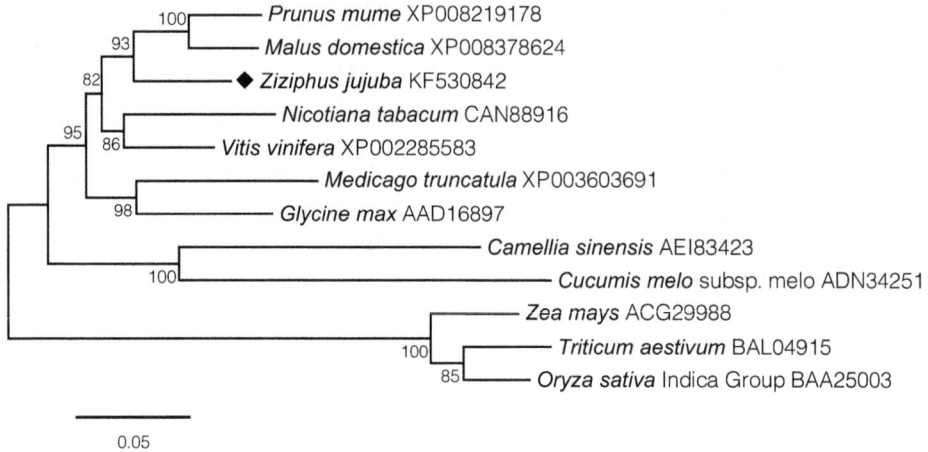

Figure 4. The phylogenetic tree of *ZjGluTR* and other 11 plant species

CONCLUSIONS

GluTR is the first committed enzyme in plant tetrapyrrole biosynthesis and is likely to be involved in the control of this metabolic pathway. The tetrapyrrole biosynthetic pathway provides the vital cofactors and pigments for photoautotrophic growth (chlorophyll) (Czarnecki et al., 2011). Some of the transgenic studies showed that plant chlorophyll deficiency, ranging from patchy yellow to total yellow. Moreover, the plants that completely lacked chlorophyll failed to survive under the growth conditions (Kumar et al., 2000; Höfgen et al., 1994). These observations suggest that suppression of the enzymes of the C_5 pathway affect the growth of the plant. Therefore, studies on the regulated synthesis of tetrapyrroles, including heme and chlorophyll, are important. Thus, cloning and bioinformatics analysis of *ZjGluTR* is very helpful for molecular biology research of chlorophyll synthesis in Chinese jujube.

REFERENCES

Vissar Y J, Ormerod J G, Beale S I, 1989. Distribution of δ-aminolevulinic acid biosynthetic pathways among phototrophic bacterial groups. Archives of microbiology, 151(6): 513-519.

Beale S I, Castelfranco P A, 1973. ^{14}C incorporation from exogenous compounds into δ-aminolevulinic acid by greening cucumber cotyledons. Biochemical and biophysical research communications, 52(1): 143-149.

Chen M W, Jahn D, O'Neill G P et al., 1990. Purification of the glutamyl-tRNA reductase from Chlamydomonas reinhardtii involved in delta-aminolevulinic acid formation during chlorophyll biosynthesis. Journal of Biological Chemistry, 265(7): 4058-4063.

Czarnecki O, Hedtke B, Melzer M et al., 2011. An Arabidopsis GluTR binding protein mediates spatial separation of 5-aminolevulinic acid synthesis in chloroplasts. The Plant Cell Online, 23(12): 4476-4491.

Ferreira G C, 1995. Heme biosynthesis: biochemistry, molecular biology, and relationship to disease. Journal of bioenergetics and biomembranes, 27(2): 147-150.

Grimm B, 1990. Primary structure of a key enzyme in plant tetrapyrrole synthesis: glutamate 1-semialdehyde aminotransferase. Proceedings of the National Academy of Sciences, 87(11): 4169-4173.

Höfgen R, Axelsen K B, Kannangara C G et al., 1994. A visible marker for antisense mRNA expression in plants: inhibition of chlorophyll synthesis with a glutamate-1-semialdehyde aminotransferase antisense gene. Proceedings of the National Academy of Sciences, 91(5): 1726-1730.

Ilag L L, Jahn D, 1992. Activity and spectroscopic properties of the Escherichia coli glutamate 1-semialdehyde aminotransferase and the putative active site mutant K265R. Biochemistry, 31(31): 7143-7151.

Jahn D, Chen M W, Söll D, 1991. Purification and functional characterization of glutamate-1-semialdehyde aminotransferase from Chlamydomonas reinhardtii. Journal of Biological Chemistry, 266(1): 161-167.

Jahn D, Verkamp E, 1992. Glutamyl-transfer RNA: a precursor of heme and chlorophyll biosynthesis. Trends in biochemical sciences, 17(6): 215-218.

Kikuchi G, Kumar A, Talmage P et al. The enzymatic synthesis of δ-aminolevulinic acid. Journal of Biological Chemistry, 1958, 233(5): 1214-1219.

Kumar A M, Söll D, 2000. Antisense HEMA1 RNA expression inhibits heme and chlorophyll biosynthesis in Arabidopsis. Plant physiology, 122(1): 49-56.

Mayer S M, Beale S I, Weinstein J D, 1987. Enzymatic conversion of glutamate to delta-aminolevulinic acid in soluble extracts of Euglena gracilis. Journal of Biological Chemistry, 262(26): 12541-12549.

Meller E, Gassman M L, 1982. Biosynthesis of 5-aminolevulinic acid: two pathways in higher plants. Plant Science Letters, 26(1): 23-29.

Neuberger A, 1968. Biochemical basis and regulatory mechanisms of porphyrin synthesis. Proceedings of the Royal Society of Medicine, 61(2): 191-193.

Ramaswamy N K, Nair P M, 1973. δ-Aminolevulinic acid synthetase from cold-stored potatoes. Biochimica et Biophysica Acta (BBA)-Enzymology, 293(1): 269-277.

Schön A, Krupp G, Gough S, et al., 1986. The RNA required in the first step of chlorophyll biosynthesis is a chloroplast glutamate tRNA. Nature, 322: 281-284.

Schubert W D, Moser J, Schauer S et al., 2002. Structure and function of glutamyl-tRNA reductase, the first enzyme of tetrapyrrole biosynthesis in plants and prokaryotes. Photosynthesis research, 74(2): 205-215.

Smith M A, Kannangara C G, Grimm B, 1992. Glutamate 1-semialdehyde aminotransferase: anomalous enantiomeric reaction and enzyme mechanism. Biochemistry, 31(45): 11249-11254.

Verkamp E, Jahn M, Jahn D, et al., 1992. Glutamyl-tRNA reductase from Escherichia coli and Synechocystis 6803. Gene structure and expression. Journal of Biological Chemistry, 267(12): 8275-8280.

Vothknecht U C, Kannangara C G, Von Wettstein D, 1996. Expression of catalytically active barley glutamyl tRNAGlu reductase in Escherichia coli as a fusion protein with glutathione S-transferase. Proceedings of the National Academy of Sciences, 93(17): 9287-9291.

Vothknecht U C, Kannangara C G, von Wettstein D, 1998. Barley glutamyl tRNA^{Glu} reductase: Mutations affecting haem inhibition and enzyme activity. Phytochemistry, 47(4): 513-519.

Zhao J, Liu Z C, Dai L et al., 2009. Isolation of Total RNA for Different Organs and Tissues of Ziziphus jujuba Mill. Journal of Plant Genetic Resources, 111-117.

17

EFFECT OF CITRIC ACID AND TREATMENTS ON PRESERVATION OF ASCORBIC ACID IN PROCESSING OF CHINESE JUJUBE JUICE

Zhihui ZHAO, Sujuan GONG, Lili WANG, Mengjun LIU

Jujube Research Center, Hebei Agricultural University,
289 Lingyusi Street, Baoding, China
Corresponding author email: lmj1234567@aliyun.com

Abstract

Effects of citric acid and treatments on preservation of ascorbic acid in processing of Chinese jujube juice were investigated. In our research, we chose for testing Ziziphus jujuba Mill. 'Zanhuangdazao'. Before crushing the jujube fruits, we tried three trials including 1% salt solution treating the jujube fruit about 0.5 h, 5% sugar solution treating the jujube fruit less 1h and putting the jujube fruit in 60°C water about 5 minutes, and the preserving rate of ascorbic acid were 68.2%, 68.9% and 121.9% respectively. During the crushing, we put the 0.10% citric acid into the mix of water and jujube pulps, this method could raise the preserving rate of ascorbic acid to 79% compared to the contrast (29.9%). The study of pasteurization, boiling water sterilization and autoclaving indicated that pasteurization was the best sterilization mean with the preserving rate of ascorbic acid to 44.2%, boiling water sterilization was the second with the preserving rate of ascorbic acid to 44.1%, and the autoclaving was far worse than the others with the preserving rate of ascorbic acid only to 19.4%.

Key words: chinese jujube juice, citric acid, ascorbic acid, preserving rate, high performance liquid chromatography.

INTRODUCTION

Chinese jujube (*Ziziphus jujuba* Mill.), a native plant of China, belongs to the genus *Ziziphus* (Rhamnaceae) and is a medicinal plant of China (Liu et al., 2009), is used as a significantly traditional Chinese medicine and invigorant(Li et al., 2009). It has crucial activities just as nourishing blood, fitting brain, activities of sedation, calm the nerves, antitumor effect, anti-aging and boost immunity(Sweetman, 2005; Gioia et al.,2008; Vidovic et al.,2008;Yu,2008;Yang et al.,2008.) . The contents of cAMP, flavones, vitamins, dietary fiber, polysaccharide, triterpene acid are remarkable (Guil-Guerrero et al., 2004; San et al., 2010; Zhao et al., 2010) in the jujube fruits, particularly the ascorbic acid. Its average content reaches to 300-600 mg/100 g based on fresh weight, even some can reach to 800-900 mg/100 g. Especially the white maturing fruit involves considerable amounts of vitamins (Gao et al., 2011). The content of ascorbic acid is higher than that in the other fruits, which known as their highly ascorbic acid, such as kiwi fruit, orange, lemon, and so on. Nowadays, there are many kinds of jujube products, such as dried jujube, honeyed jujube,

jujube wine, jujube vinegar, jujube tea, jujube juice, etc.. Ascorbic acid is generally recognized as safe (GRAS) when used in accordance with Good Manufacturing Practices (21CFR182. 2000), is relatively inexpensive, and is widely recognized by consumers as a beneficial nutrient (vitamin C) (Kokkinidou et al., 2014). Vitamin C is a very important vitamin for human nutrition that is supplied by fruits and vegetables. Ascorbic acid is the main biologically active form of vitamin C. As a potent antioxidant, it has the capacity to eliminate several different free radicals (Scherera et al., 2012). However, many nutrients will be destroyed in the processing of products, especially ascorbic acid (Sheetal et al., 2013; Mapson et al., 1958; Maria et al., 2013). The ascorbic acid is so active and easily be broken by external condition just like heat, oxygen, light ray and pH. So, the content of ascorbic acid is low in the current jujube products. Therefore, it is very necessary to research a new method to protect ascorbic acid. In our research, we chose *Ziziphus jujuba* Mill. 'Zanhuangdazao' as the material and investigated the effect of citric acid and treatments on preservation of ascorbic acid.

MATERIALS AND METHODS

Ziziphus jujuba Mill. 'Zanhuangdazao' gained from Zanhuang, Hebei province was selected. The jujube fruits were washed under running water to remove the adhering mud particles followed by double distilled water and drained completely. Then we divided the fruits into three groups. Immerged the first group fruits into 1%, 2% and 3% salt solution for 0.5 h, 1.5 h, 2.5 h, 3.5 h, 4.5 h. Put the second group into 5%, 10% and 15% sugar solution for 0.5 h, 1.5 h, 2.5 h, 3.5 h, 4.5 h. At last treated the others by 60°C water for 5 min, 10 min, 15 min, 20 min, 25 min. The contents of ascorbic acid in jujube fruits were detected by high performance liquid chromatographic methods. To prevent the loss of vitamins, all operations were performed in the absence of direct sun light, using amber glassware. All tests were performed in triplicate.

During the crushing, we put the 0.01%, 0.05% and 0.10% citric acid into the mix of water and jujube pulps, then determined the content of ascorbic acid in the juice and metered the preservation rate of ascorbic acid. All tests were performed in triplicate.

To protect ascorbic acid in the jujube juice, we investigated three kinds of sterilization methods, just like pasteurization (65°C, 30 min), boiling water sterilization (100°C, 15 min) and autoclaving (121°C, 10 min). The results were compared by one-way analysis of variance (ANOVA) and Duncan's test was carried out to identify significant differences between the mean values. All analyses were performed using the software SPSS 17.0 and differences at $P < 0.05$ were considered statistically significant.

RESULTS AND DISCUSSIONS

Effect of preprocessing on preserve rate of ascorbic acid

As can be observed in Figure 1, after 1% salt solution treating the jujube fruit about 0.5 h, the preserve rate of ascorbic acid was 68.2% which higher than the contrast (66.8%).

As Figure 2 showed, 5% sugar solution treating the jujube fruit less 1 h, the preserve rate of ascorbic acid was 68.9%, it was also better than the contrast (66.8%).

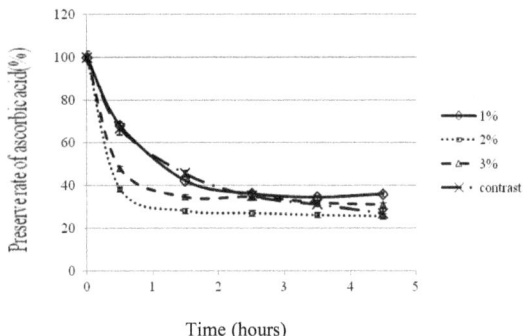

Figure 1. Influence of salt solution to the preserving rate of ascorbic acid

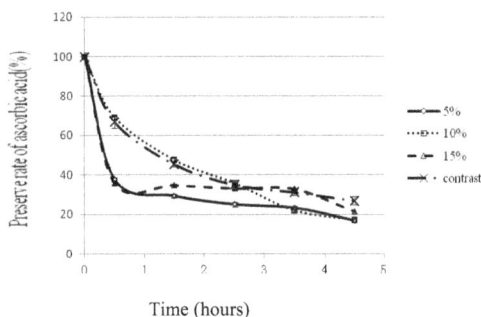

Figure 2. Influence of sugar solution to the preserving rate of ascorbic acid

Figure 3 indicated that putting the jujube fruit in 60°C water about 5 minutes, the preserving rate of ascorbic acid reached to 121.9% which obviously higher than the contrast (66.8%), this will have practical value to the processing industries of jujube fruits.

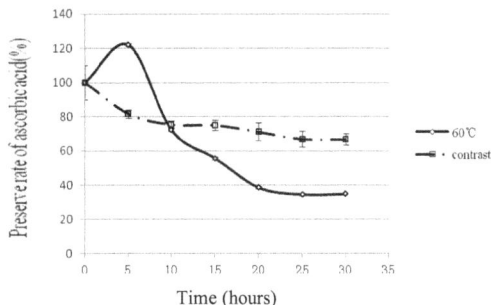

Figure 3. Influence of softening treatment to the preserving rate of ascorbic acid

Effect of citric acid on preservation of ascorbic acid

As can be observed in Figure 4, during the crushing, putting into 0.01% and 0.05% citric acid could rise the preserve rate of ascorbic acid to 43% and 46% respectively, compared by the contrast (29%), especially after the 0.10% citric acid treatment, the preserve rate of ascorbic acid reach to 79%, much higher than he contrast (29%).

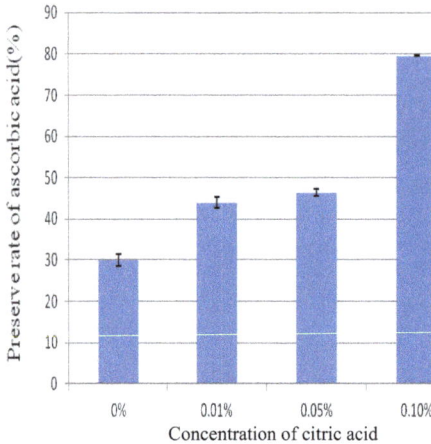

Figure. 4 Influence of citric acid on the preserving rate of ascorbic acid

Influence of the sterilization conditions on preservation of ascorbic acid

The results of the sterilization conditions (Table 1) indicated that the result of pasteurization and boiling water sterilization were ideal with the content of ascorbic acid were 226.5 mg/100 g and 225.9 mg/100 g, the preserve rate of ascorbic acid reached to 44.2% and 44.1%. However the result of autoclaving was less than the others.

Table 1 comparison of sterilization of jujube juice

Sterilization conditions	Content of ascorbic acid (mg/100g)
Pasteurization (65°C, 30 min)	226.5±3.8a
boiling water sterilization (100°C, 15 min)	225.9±1.9a
Autoclaving (121°C, 10 min)	99.4±3.6b

Each value is expressed as means ± standard deviation of n=3 determinations;
Different letters within a column (a-d) denote significant differences (p<0.05) between cultivars.

CONCLUSIONS

Putting the jujube fruit in 60°C water about 5 minutes before crushing fruits could keep the preserving rate of ascorbic acid to 121.9%.

During the crushing, putting the 0.10% citric acid into the mix of water and jujube pulps could raise the preserving rate of ascorbic acid to 79% compared to the contrast (29.9%). These methods were proved to be useful technique to increase stability of ascorbic acid.

ACKNOWLEDGEMENTS

This research work was carried out with the support of Hebei Provincial Department of Education (QN2015232), Top-notch Talent Support Project of Hebei province and Young Academic Leader of Hebei Agricultural University.

REFERENCES

Liu MJ, Wang M., 2009. Germplasm Resources of Chinese Jujube, Forestry Publishing. Beijing, China.

Li JW, Ding SD, Ding XL., 2005. Comparison of antioxidant capacities of extracts from five cultivars of Chinese jujube, Process Biochemistry, 40, 3607-3613.

Gao YR, Zhao ZH and Liu MJ., 2011. Determination of ascorbic acid and dehydroascorbic acid in Chinese jujube by high performance liquid chromatography, Asian Journal of Chemistry, 23, 3989.

Gioia M.G., Andreatta P., Boschetti S., Gatti R., 2008. Development and validation of a liquid chromatographic method for the determination of ascorbic acid, dehydroascorbic acid and acetaminophen in pharmaceuticals. Journal of Pharmaceutical and Biomedical Analysis, 48, 331-339.

Guil-Guerrero J.L, Delgado A.D, Matallana M.C, GONZÁLEZ, Isasa M.E. Torija., 2004. Fatty acids and carotenes in some Ber (Ziziphus jujuba Mill.) varieties, Plant Foods for Human Nutrition (Formerly Qualitas Plantarum), 59, 23-27.

Kokkinidou S. Floros J. D. LaBorde L. F., 2014. Kinetics of the thermal degradation of patulin in the presence of ascorbic acid. J Food Sci, 79(1): 108-114.

Mapson L.W., Tomalin A. W., 1958. Preservation of ascorbic acid in rose hips during storage. J. Sci. Food Agric, 9: 424–430.

Maria D.C. Antunes, Daniela Rodrigues, Vasilios Pantazis, et al., 2013. Nutritional Quality Changes of Fresh-cut Tomato during Shelf Life. Food Sci. Biotechnol, 22(5): 1229-1236.

San B, Yildirim A N., 2010. Phenolic, alpha-tocopherol, beta-carotene and fatty acid composition of four promising jujube (Ziziphus jujuba Mill.) selections. Journal of Food Composition and Analysis, 23, 706-710.

Scherera R., Rybkaa A. C. P., Ballus C.A, et al., 2012. Validation of a HPLC method for simultaneous determination of main organic acids in fruits and juices. Food Chemistry, 135: 150–154.

Sheetal Gupta, Gowri B. S., Jyothi Laksh mi A., Prakash J., 2013. Retention of nutrients in green leafy vegetables on dehydration. J Food Sci Technol, 50(5):918–925.

Sweetman SC, 2005. Dose adjustment in renal impairment: Response from Martindale: the Complete Drug Reference, 331-292.

Vidovic S, Stojanovic B, Veljkovic J, Prazic-Arsic L, Roglic G, Manojlovic D., 2008. Simultaneous determination of some water-soluble vitamins and preservatives in multivitamin syrup by validated stability-indicating high-performance liquid chromatography method. Journal of Chromatography, 1202(2), 155-162.

Yang SM, Yang XM, 2008. Research trends and treatment prospects of ascorbic acid. China Medical Herald, 28(5): 11- 13.

Yu CL, 2008. Research and new influence of ascorbic acid. Jilin Medical Journal, 29(21): 145- 147.

Zhao ZH, Liu MJ, 2010. Study of properties of crude water soluble polysaccharides from *Ziziphus jujuba* Mill. "Jinsixiaozao", Journal of Agricultural University of Hebei, 33, 58-04.

18

THE PROPAGATION OF TWO RED AND BLACK CURRANT VARIETIES BY HARDWOOD CUTTINGS COMBINING SUBSTRATE AND ROOTING STIMULATORS

Adrian ASĂNICĂ[1], Valerica TUDOR[1], Dorin SUMEDREA[2], Răzvan Ionuț TEODORESCU[1], Adrian Peticilă[1], Alexandru IACOB[1]

[1]University of Agronomic Sciences and Veterinary Medicine of Bucharest, 59 Marasti Blvd., District 1, Bucharest, Romania
[2]Research Institute for Fruit Growing Pitesti, 402 Mărului Street, Mărăcineni 117450, Romania

Corresponding author email: adrian.asanica@horticultura-bucuresti.ro

Abstract

The planting material production is a particularly attractive segment of fruit growing field, both through the possibility of maximizing profits on small lands and because of the increased interest manifested by growers in the horticultural sector, especially in berries. The currant is specie that manages to propagate more easily than other berry shrubs, but rooting capacity vary from species to species and depending on different climatic and technological factors. The present paper supports the nursery activity by developing a complex experimental module that identifies the best solutions for black and red currant cuttings as a way of vegetative propagation. The thickness of the cuttings, rooting substrates and rooting hormonal stimulants (obtained in laboratory and commercial products) are taken into account. Two varieties of currants were used: 'Tinker' cultivar for black currant and 'Elite' cultivar for red currant. The rooting percentage of cuttings was very similar for the black and currant varieties (around 77%). For both currants, the thinner cuttings gave the best results in terms of rooted cuttings percentage. Perlite + sand substrate achieved the highest rooting percentage of the cuttings but with the lowest vegetative growths. Among the stimulators used for rooting, the IBA hormone (regardless of the concentration used) achieved the best percentages of cuttings rooted in both currant varieties. IBA 1000 ppm gave the best results on the length of the shoots on both currant varieties. Peat + sand positively influenced the root volume in the red currant and perlite + peat mixture for black currant. The volume of the root system in the red currant was approximately 2 times larger than the black currant. IBA 1000 and IBA 1500 conducted to the most extensive radicular system to both currant varieties.

Key words: rooting hormons, cuttings thickness, 'Elite', 'Tinker', root system volume.

INTRODUCTION

In horticulture, vegetative propagation is widely used to multiply elite plants obtained in breeding or selected from natural populations (Hartman et al., 1990). Propagation of currant's varieties is still one of the most cost-effective methods for vegetative regeneration as any other woody species (Koyuncu and Senel, 2003; De Klerk et al., 1999). It reclaims cuttings from the previous season's growth (minimum one year old) harvested during the dormant stage from the mother plant (Sandor, 2007).

Induction of roots is a process regulated by timing (Szecsko et al., 2002), environmental and endogenous factors including hormones. Auxin is one of the major endogenous hormones involved in the adventitious rooting process (Wiesman et al, 1988; Pop et al., 2011).

Well known sources of growth hormones for rooting of cuttings are the IBA (indole-3-butyric acid), IAA (indole-3-acetic acid) and NAA (α-naphthalene acetic acid) in different concentrations (Khudhur et al., 2015; Uniyal et al., 1993), together with a lot of commercial products nowadays used as root promoters (solutions or powders).

In a previous research, Pandey et al. (2011) concluded that IBA treatment had shown not only an improvement of the percent rooting at *Ginkgo biloba* L. but also a positive effect on the subsequent growth and survival rate of the plantlets.

IBA has recorded the highest recognition among all auxins used for rooting.

Treatments of the currants hardwood cuttings with furostanol glycosides (Caulet et al., 2012), for one hour, leaded to an increasing of roots number and contributed to obtain more qualitative black currant cuttings of Deea, Abanos and Ronix varieties in comparison with Radistim 2% (commercial rooting stimulator product).

Combined and softwood cuttings have been experienced for several currant varieties by Siksnianas and Sasnauskas (2006) using IAA, NAA and ascorbic acid as rooting growth stimulators. Substrate used was 2 parts of peat and 1 part of sand and artificial mist was provided. Results showed that for cultivars of black currant, rooting of combined cuttings was higher and the NAA 25 g/l-1 offered the highest quality of rooting.

There is always a need for improvement and in this regard, the current study was conducted to determine the most suitable way to propagate two of the most cultivated varieties of black and red currant by hardwood cuttings. Thickness of the cuttings, rooting substrate and stimulator concentrations influence has been took into consideration.

MATERIALS AND METHODS

The experiment was established in the Vegetation House of the University of Agronomic Sciences and Veterinary Medicine of Bucharest and was carried out between February and June of 2016.

The biological material consists of two varieties of currant: 'Elite' (red currant) and 'Tinker' (black currant).

The cuttings were made using annual dormand branches harvested from mother plants located in the experimental field of the Faculty of Horticulture, Bucharest.

Two categories of cuttings have been executed (figure 1):
- for red currant: thin cuttings (0.26-0.35 mm) and thick cuttings (0.36-0.78 mm)
- for black currant: thin cuttings (0.3-0.48 mm) and thick cuttings (0.49-1.89 mm)

The length of the cuttings was of 15 cm with minimum 3 nodes.

Cuttings were afterwards treated with different rooting hormones and concentrations as follows:
- IBA 500 ppm (solution)
- IBA 1000 pm (solution)
- IBA 1500 ppm (solution)

- Razormin (solution of aminiacids, macroelements and microelelments)
- Rhizopon AA1% (powder, IAA 1%)

Figure 1. Cuttings of red (left) and black currant (right) made with different thickness

Except the control, all the cuttings were submerged in the hormonal solutions for 5 minutes with the basal part or covered with powder in the situation of Rhizopon product.

Rooting substrate was represented by equal share of the following parts (figure 2).

Figure 2. Rooting substrate variants:
top Peat + Perlite = 1:1
middle Peat + Sand = 1:1
bottom Perlite + Sand = 1:1

At the end, 720 cuttings resulted: 360 of red currant and 360 of black currant. Half of them (360 cuttings) were in the first thickness category and the other half in the second one. In every pot it were placed 10 cuttings of currant according to the experimental model A number of 72 pots with the capacity of 3 liter each were customized (figure 3).

Figure 3. Experimental module of currant cuttings grouped by substrate, thickness and hormonal treatment (4.04.2016)

On 7th of June, 2016, all the cuttings were removed from the pots and several determinations and observations were performed: rooting percentage, vegetative growth of the rooted cuttings, root system volume.

All data were processed by analyze of variance and Duncan's multiple range test at confidence level of 95% (P≤0.05) using XLSTAT software.

RESULTS AND DISCUSSIONS

Five months after the experiment start, the entire plot was disassembled, taking each variant one by one in careful observation and gathering data related to the rooting success and overall cuttings quality.

In this respect, we remarked very small differences concerning the rooting ratio of the two currant species. As an overall percentage, 'Elite' variety recorded 77.22% rooting percentage; very close to the black currants 'Tinker' ones of 77.36% (table 1).

Depending on the cuttings thickness, the rooting percentages were quite variable; significantly differences have been registered at red currant thinner cuttings for control and variants treated with Razormin.

At black currant, the cuttings thickness did not influence much the rooting percentage, only 10.84% difference was noticed between thicker and thinner cuttings.

Comparing the two species on this issue, only few rooting percentages (3.33%) more have been registered between thicker black currant cuttings than red currants cuttings.

Same observation was made in the case of thinner cuttings (3.61%) but in favour of red currants.

The rooting substrate played a very important role in the rooting percentage of the cuttings, at the end of the experiment, the highest values been revealed for the perlite with sand mixture (more than 80%).

We find a relatively close hierarchy between substrate types, the second place being peat + perlite which averaged 77.08%. Peat + sand had a slightly weaker effect, with lower black currant (62.92%) and almost 10% lower than 'Elite' red currant variety.

Table 1. Rooting percentage of currant cuttings depending on cultivar, thickness of cutting, substrate and hormones

Red currant	%	Hormone/Product						
Cutting type	Substrate	IBA 500	IBA 1000	IBA 1500	Razormin	Rhizopon	Control	Average
Thin	Peat + Perlite	100	90	60	0	80	80	**68.33a**
	Peat + Sand	100	60	60	60	60	0	**56.67a**
	Perlite + Sand	90	100	100	20	90	80	**80.00a**
	Average	**96.67a**	**83.33a**	**73.33ab**	**26.67b**	**76.67ab**	**53.33ab**	**68.33a**
Thick	Peat + Perlite	70	100	90	30	100	80	**78.33a**
	Peat + Sand	80	100	80	100	70	90	**86.67a**
	Perlite + Sand	100	90	100	80	90	100	**93.33a**
	Average	**83.33a**	**96.67a**	**90.00a**	**70.00a**	**86.67a**	**90.00a**	**86.11a**
	Overall Average (%)	*90.00a*	*90.00a*	*81.67a*	*48.33a*	*81.67a*	*71.67a*	*77.22a*
Black currant	%	Hormone/Product						
Cutting type	Substrate	IBA 500	IBA 1000	IBA 1500	Razormin	Rhizopon	Control	Average
Thin	Peat + Perlite	80	80	100	80	60	20	**70.00ab**
	Peat + Sand	60	60	60	55	60	50	**57.50b**
	Perlite + Sand	100	100	90	60	80	100	**88.33a**
	Average	**80.00a**	**80.00a**	**83.33a**	**65.00a**	**66.67a**	**56.67a**	**71.94ab**
Thick	Peat + Perlite	80	90	100	90	90	100	**91.67a**
	Peat + Sand	90	90	60	10	70	90	**68.33b**
	Perlite + Sand	70	90	90	90	90	100	**88.33ab**
	Average	**80.00a**	**90.00a**	**83.33a**	**63.33a**	**83.33a**	**96.67a**	**82.78ab**
	Overall Average (%)	*80.00a*	*85.00a*	*83.33a*	*64.17a*	*75.00a*	*76.67a*	*77.36a*

*Duncan's multiple range test (P≤0.05)

Regarding the effect of rooting stimulants on the percentage of root formation, we noted a fairly good influence of the IBA hormone regardless of the concentration in which it was used. A negative response was also noticed at Razormin, which practically had an antagonistic effect, the rooting percentages being lower for both species but more evident in the red bark (48.33%). By stimulating with the IBA 500 and IBA 1000, the red currant has reached a very good rooting percentage, 90% of the cuttings have been rooted.

Observations and calculations were made on the type of the vegetative growths resulting from the evolution of the apical buds of the cutting. Thus, more shoots could be counted than leaf rosettes in the case of thicker cuttings; the thinner ones, even if they have recorded the best rooting percentage, have led to the formation of more rosettes of leaves (table 2).

The rooting hormones showed a more pronounced effect on the black currant related to the percentages of cutting's shoots, where the IBA 1500 and descending to the IBA 500 induced the occurrence of larger shoots. For red currant, the differences between the compounds used to stimulate the rooting of the cuttings were less evident; the control formed the highest number of shoots compared to the other cuttings.

Also, the highest number of rosettes was observed in the perlite + sand substrate (figure 4).

Table 2. Share of shoots and rosettes / currants cuttings (5 months from the begining)

Red currant	Substrate	IBA 500		IBA 1000		IBA 1500		Razormin		Rhizopon		Control		Average	
		S %	R %	S %	R %	S %	R %	S %	R %	S %	R %	S %	R %	S* %	R* %
Thick cutting	Peat + Perlite	7	3	88.8	11.1	83.3	16.6			87.5	12.5	10		85.94a	14.06a
	Peat + Sand	7	3	83.3	16.6	66.6	33.3	5	5	83.3	16.6			70.66a	29.34a
	Perlite + Sand	77.7	22.2	9	1	7	3	10		77.7	22.2	10		85.92a	14.08a
	Average	72.59	27.41a	87.40	12.59a	73.33	26.66	75.00	25.00a	82.87	17.13a	100	0.00	80.84a	19.16a
Thin cutting	Peat + Perlite	10		8	2	9	1	66.6	33.3	8	2	7	2	81.94a	18.06b
	Peat + Sand	7	2	7	3	5	5	9	1	57.1	42.8	66.6	33.3	68.13ab	31.87b
	Perlite + Sand	5	5	33.3	77.7	7	3	12.	87.	33.3	77.7	9	1	44.49b	55.51a
	Average	75.00	25.00	57.41	42.59a	70.00	30.00	56.39	43.61a	53.12	46.88	77.22	22.78	64.86a	35.14a
	Overall mean	73.80a	26.20a	72.41a	27.59a	71.67a	28.33a	65.69a	34.31a	67.99a	32.00a	88.61a	11.39a	72.85a	27.15a

Black currant	Substrate	IBA 500		IBA 1000		IBA 1500		Razormin		Rhizopon		Control		Average	
		S %	R %	S %	R %	S %	R %	S %	R %	S %	R %	S %	R %	S* %	R* %
Thick cutting	Peat + Perlite	75	25	75	25	90	10	62.5	37.5	83.33	16.66	50	50	72.64b	27.36b
	Peat + Sand	100	0	100	0	100	0	100	0	100	0	100	0	100.00a	0.00c
	Perlite + Sand	30	70	50	50	44.44	55.55		100		100	20	100	24.07c	79.26a
	Average	68.33a	31.67a	75.00a	25.00a	78.15a	21.85a	54.17a	45.83a	61.11a	38.89a	56.67a	50.00a	65.57b	35.54a
Thin cutting	Peat + Perlite	62.5	37.5	88.88	11.11	100	0	55.55	44.44	66.66	33.33	100	0	78.93ab	21.06c
	Peat + Sand	77.77	22.22	66.66	33.33	100	0	100	0	71.42	28.57	100	0	85.98a	14.02c
	Perlite + Sand	42.85	57.14	20	80	0	100		100		100		100	10.48c	89.52a
	Average	61.04a	38.95a	58.51a	41.48a	66.67a	33.33a	51.85a	48.15a	46.03a	53.97a	66.67a	33.33a	58.46b	41.54b
	Overall mean	64.69a	35.31a	66.76a	33.24a	72.41a	27.59a	53.01a	46.99a	53.57a	46.43a	61.67a	41.67a	62.02a	38.54a

*S-shoot; R-rosette of leaves
*Duncan's multiple range test (P≤0.05)

Figure 4. The bigger share of rosettes formed by red currants cuttings in perlite + sand substrate (first two rows in the front) - 21.05.2016

The vigour of the rooted cuttings was highlighted by the length of the total growth of the cutting (figure 5).

The length of the shoots was influenced by all the experimental factors analysed and are presented in table 3 for each type of currant.

Regarding the type of cutting, we can say that this factor did not directly influence the length of the shoot/s, the differences being very small in the red currant and somewhat larger in the black currant (only 1.18 cm).

The same thing cannot be said about the type of substrate, which had a decisive role in plant height. Therefore, for the black currant, the most effective was the peat + sand substrate, which achieved an average of 16.79 cm. At the opposite side, there is the perlite with sand, the substrate which generated the lowest length of the shoot, and the biggest share of the rosettes/cutting. For the red currant, the best substrate proved to be peat + perlite.

Table 3. The length of the shoots/cutting (cm)

Red currant	Substrate	IBA 500	IBA 1000	IBA 1500	Razormin	Rhizopon	Control	Average
Thick cutting	Peat + Perlite	11.07	12.50	14.20	0.00	20.14	12.00	**11.65a**
	Peat + Sand	13.40	9.40	11.00	8.33	13.00	0.00	**9.19a**
	Perlite + Sand	8.57	7.88	5.67	5.50	7.00	6.25	**6.81a**
	Average	**11.01a**	**9.93a**	**10.29a**	**4.61a**	**13.38a**	**6.08a**	**9.22a**
Thin cutting	Peat + Perlite	13.14	10.50	12.89	8.00	11.25	11.33	**11.19a**
	Peat + Sand	10.50	9.07	11.25	11.22	10.63	11.08	**10.63ab**
	Perlite + Sand	7.20	4.00	4.57	6.00	6.00	5.67	**5.57c**
	Average	**10.28a**	**7.86a**	**9.57a**	**8.41a**	**9.29a**	**9.36a**	**9.13b**
	Overall mean	**10.65ab**	**8.89ab**	**9.93ab**	**6.51b**	**11.34a**	**7.72ab**	**9.17ab**
Black currant	Substrate	IBA 500	IBA 1000	IBA 1500	Razormin	Rhizopon	Control	Average
Thick cutting	Peat + Perlite	7.17	8.91	8.05	4.8	12.4	2	**7.22b**
	Peat + Sand	17	22.17	19.83	11.75	17.16	15.75	**17.28a**
	Perlite + Sand	1.75	2.7	2.5	0	0	2	**1.49c**
	Average	**8.64a**	**11.26a**	**10.13a**	**5.52a**	**9.85a**	**6.58a**	**8.66ab**
Thin cutting	Peat + Perlite	12.2	13.31	12.8	12.8	15	7.05	**12.19b**
	Peat + Sand	16.71	18.67	22	14	13.2	13.22	**16.30a**
	Perlite + Sand	3.67	2.5	0	0	0	0	**1.03c**
	Average	**10.86a**	**11.49a**	**11.60a**	**8.93a**	**9.40a**	**6.76**	**9.84b**
	Overall mean	**9.75ab**	**11.38a**	**10.86a**	**7.23bc**	**9.63abc**	**6.67c**	**9.25abc**

*Duncan's multiple range test (P≤0.05)

Figure 5. The stimulating effect of IBA hormone upon the length of the thicker black currants cuttings rooted in the sand + perlite substrate (7.06.2016)

To illustrate the quality of the cuttings, we proceed to assess the root volume of the black and red currant cuttings.

As can be seen from table 4, the black currant developed a bigger root system, which also explains the length of the shoots on same substrates.

In the red currant, both for the thicker and the thinner cuttings, the substrate that favoured the formation of a larger root volume was peat + sand (figure 6).

In the black currant, the differences were somewhat smaller, highlighting peat mixed with pearl or sand.

Table 4. The root volume of the currants cuttings (mm^3)

Red currant	Substrate	IBA 500	IBA 1000	IBA 1500	Razormin	Rhizopon	Control	Average
Thick cutting	Peat + Perlite	6.10	5.22	7.00	1.00	4.80	6.00	**5.02a**
	Peat + Sand	6.17	6.50	7.43	4.67	7.00	6.00	**6.29a**
	Perlite + Sand	3.89	4.80	3.40	7.00	4.67	3.75	**4.58a**
	Average	**5.39a**	**5.51a**	**5.94a**	**4.22a**	**5.49a**	**5.25a**	**5.30a**
Thin cutting	Peat + Perlite	4.29	2.50	3.56	2.67	3.90	4.20	**3.52bc**
	Peat + Sand	3.33	4.22	6.67	5.00	4.57	4.89	**4.78a**
	Perlite + Sand	3.14	2.60	2.60	2.44	2.60	2.80	**2.70c**
	Average	**3.59a**	**3.11a**	**4.27a**	**3.37a**	**3.69a**	**3.96a**	**3.67b**
	Overall mean	**4.49a**	**4.31a**	**5.11a**	**3.80a**	**4.59a**	**4.61a**	**4.48a**
Black currant	Substrate	IBA 500	IBA 1000	IBA 1500	Razormin	Rhizopon	Control	Average
Thick cutting	Peat + Perlite	2.60	3.11	2.00	1.75	3.71	2.50	**2.61abc**
	Peat + Sand	2.80	4.33	3.25	3.50	2.25	2.57	**3.12a**
	Perlite + Sand	2.20	2.00	2.22	2.80	2.88	2.10	**2.37bc**
	Average	**2.53a**	**3.15a**	**2.49a**	**2.68a**	**2.95a**	**2.39a**	**2.70ab**
Thin cutting	Peat + Perlite	2.57	2.67	2.60	2.22	2.22	2.40	**2.45a**
	Peat + Sand	1.60	2.25	2.50	2.40	2.25	2.50	**2.25a**
	Perlite + Sand	2.22	1.80	1.33	1.56	1.67	0.80	**1.56c**
	Average	**2.13a**	**2.24a**	**2.14a**	**2.06a**	**2.05a**	**1.90a**	**2.09ac**
	Overall mean	**2.33a**	**2.69a**	**2.32a**	**2.37a**	**2.50a**	**2.15a**	**2.39a**

*Duncan's multiple range test (P≤0.05)

IBA 1000 and IBA 1500 presented the highest efficiency in providing a broader radicular system, managing the highest values in both species of currant. But the differences were not significantly evident from the other variants.

Figure 6. Detail of the root system generated by red currants thick cuttings stimulated with IBA 1000 ppm in peat + sand substrate (left) versus perlite + sand (right)

Regardless of the variants analysed, we can emphasize the clear difference between the ability to form more roots / cutting in the case of the black currant compared to the red currant. Thus, about 65% of the total roots belong to the black currant cuttings while only 35% of the red currant.

CONCLUSIONS

The rooting percentage of cuttings was of 77.22% for the black currant variety 'Tinker' and 77.36% for the 'Elite' red currant variety.

For both varieties, the thinner cuttings, with a diameter between 0.3-0.48 mm (black currant) and 0.26-0.35 mm (red currant) gave the best results in terms of rooted cuttings percentage.

The perlite + sand substrate achieved the highest rooting percentage of the cuttings but with the lowest vegetative growths, also with higher share of rosettes vs shoots.

Among the stimulators used for rooting, the IBA hormone (regardless of the concentration used) achieved the best percentages of cuttings rooted in both currant varieties.

The mixture of perlite + peat for 'Tinker' variety, has contributed to the appearance of a larger number of shoots than the other tested substrates.

For the black currant, the sandy peat substrate favoured the development of the most vigorous shoots, its average length being 16.79 cm with a maximum of 22.17 cm.

The IBA 1000 ppm gave the best results on the length of the shoots on both currant varieties.

The root volume in the red currant was positively influenced by the peat + sand mixture regardless to the thickness of the cuttings.

For black currant, better results of root volume were recorded when the perlite + peat mixture was used as substrate for rooting.

The volume of the root system in the red currant was approximately 2 times larger than the black currant.

IBA 1000 and IBA 1500 gave the most extensive radicular system to both currant varieties.

ACKNOWLEDGEMENTS

This work was supported by a grant of the Romanian National Authority for Scientific Research and Innovation, CNCS-UEFISCDI, project number PN-II-RU-TE-2014-4-0749.

REFERENCES

Caulet R.P., Onofrei O., Morariu A., Iurea D., Gradinaru G., 2012. Effect of furostanol glycoside treatments in plant material production in currants (*Ribes* sp.), Lucrari Stiintifice–Seria Horticultura, Ed. "Ion Ionescu de la Brad", Iasi, 54 (2): 231- 237.

De Klerk G.J., Van der Krieken W., De Jong J., 1999. Review the formation of adventitious roots: new concepts, new possibilities. *In Vitro* Cellular & Developmental Biology - Plant, 35(3): 189-199.

Gehlot A., Gupta R. K., Tripathi A., Arya I. D., 2014. Vegetative propagation of *Azadirachta indica*: effect of auxin and rooting media on adventitious root induction in mini-cuttings. Adv. For. Sci., 1(1): 1-9.

Hartmann H.T., Kester D.E., Jr. Davies F.T., 1990. Plant Propagation: Principles and Practices. 5th Edn., Prentice-Hall Inc., Englewood, Cliffs, New Jersey, USA.

Khudhur S.A., Omer T.J., 2015. Effect of NAA and IAA on Stem Cuttings of Dalbergia Sissoo (Roxb). Journal of Biology and Life Science, 6(2): 208-220

Koyuncu F., Şenel E., 2003. Rooting of Black Mulberry (*Morus nigra* L.) Hardwood Cuttings. Journal of Fruit and Ornamental Plant Research, 11: 53-57.

Kroin J., 2016. Hortus Plant Propagation from Cuttings. A guide to using plant rooting hormones by foliar and basal methods. Third Edition. Hortus USA Corp, PO Box 1956, New York NY 10113. support@hortus.com

Pandey A., Tamta S., Giri D., 2011. Role of auxin on adventitious root formation and subsequent growth of cutting raised plantlets of *Ginkgo biloba* L. International Journal of Biodiversity and Conservation, 3(4): 142-146.

Pop T.I., Pamfil D., 2011. Auxin Control in the Formation of Adventitious Roots. Not Bot Hort Agrobot Cluj, 39 (1): 307-316

Sandor F., 2007. Vegetative propagation techniques. Perrenial crop support series. Jalalabad, Afganistan, 2007-003-AFG, Roof of Peace, 13-15.

Siksnianas T., Sasnauskas A., Šterne D., 2006. The propagation of currants and gooseberries by softwood and combined cuttings. Agronomijas Vēstis 9: 135-139.

Szecsko V., Csikos A., Hrotko K., 2002. Timing of hardwood cuttings in the propagation of plum rootstocks. Acta Hortic. 577: 115-119

Szecsko V., Csikos A., Hrotko K., 2002. Timing of hardwood cuttings in the propagation of plum rootstocks. Acta Hortic. 577: 115-119.

Uniyal R.C., Prasad P., Nautiyal A.R., 1993. Vegetative propagation in *Dalbergia sericea*: Influence of growth hormones on rooting behaviour of stem cuttings. J. Trop. For. Sci., 6: 21-25.

Wiesman Z., Riov J., Epstein E., 1988. Comparison of movement and metabolism of indole-3-acetic acid and indole-3-butyric acid in mung bean cuttings. Physiologia Plantarum, 74(3): 556-560.

19

ENGINEERING PROPERTIES OF THE ŞIRE GRAPE
(*VITIS VINIFERA* L. CV.)

Reşat ESGİCİ[1], Gültekin ÖZDEMİR[2], Göksel PEKİTKAN[3], Konuralp ELİÇİN[3],
Ferhat ÖZTÜRK[4], Abdullah SESSİZ[3]

[1]Dicle University, Bismil Vacational High School, Diyarbakır, Turkey
[2]Dicle University, Faculty of Agriculture, Department of Horticulture, Diyarbakır, Turkey
[3]Dicle University, Faculty of Agriculture, Department of Agricultural Machinery and Technologies
Engineering, Diyarbakır, Turkey
[4]Dicle University, Faculty of Agriculture, Department of Field Crops, Diyarbakır, Turkey
Corresponding author: asessiz@dicle.edu.tr

Abstract

Turkey will continue to acting an important role in grape production and raisin exportation in the world because of its large number of grape varieties, favorable ecological conditions and large amount of production areas. Turkey is the one of the gene center of grapevines, for this reason it possesses over 1600 grape varieties. Grapevine varieties are generally harvested by hand; however, the feasibility of using a mechanical harvester is some engineering properties such as physical and mechanical properties must be consideration. In this study, some physical and mechanical properties of grape berries and canes of local variety Şire (Vitis vinifera L. cv.) were determined depend on phenological stages. This research was performed at commercial vineyard in Dicle, the town of Diyarbakır, which is located in the southeastern part of Turkey. Cutting properties were measured by The Lloyd LRX plus materials testing machine. Grape berries length, width, thickness, arithmetic and geometric mean diameter, sphericity, roundness, detachment force (FDF), weight (W), the ratio of FDF/W, skin firmness, total soluble solids content, pH, total acidity and cane of grapevine shearing force, shearing strength, upper yield, shearing energy were determined. The test results indicated that very significant correlations were found between axial dimensions of grape berries, and physical dimensions, mechanical and pomological properties. The ratio of FDF/W decreased depending on phenological stages. Berry weight was lowest at the Veraison (1.60 g). The grape berry skin firmness decreased from 1.174 N to 0.766 N with phenological stages. TSSC values varied from 20.40 to 16.20 %, pH of grape (3.39-3.65) values increased with phenological stages, whereas the total acids were slight changed and reduced from 0.876 to 0.669 %. Cutting properties of Şire grapevine cane has been changed with phenological stages. Shearing force and energy requirement increased with increase internode diameter of canes. Shearing force values changed between 472.38 N and 119.57 N.

Key words: Şire, grape berry, grape cane, physical properties, mechanical properties, engineering properties

INTRODUCTION

Grape is an important product for the economy of Turkey. Turkey is sixth largest producer of worldwide with an estimated production of 4 million tons in 550,000 ha production area in 2016. It is the biggest exporter of raisin grapes. Each year over 200,000 tons golden coloured raisins is exported all over the world. The grape export is 170,000 tons valued at 133 million \$ (Anonymous, 2016). To maintain this values and to be leading of the world's, production costs must be reduced, especially pruning, harvesting and transporting. One way of reducing production cost is use of mechanical harvesting. Mechanical harvesting of viticulture for juice is used many developed countries such

as USA, France and Italia and such as country, there is valuable effort for developed and improved mechanical practices. But mechanical harvesting is not common in Turkey because grape juice sector not has improved and grape price is low, vineyards are not suitable for mechanization applications, especially in southeastern part of Turkey. Grape harvesting is made by hand. Hand harvesting is labor intensive. In fact, mechanical harvesting has not been improved and damaged product is very high. So, use of mechanization application should be increased.

Percentage presence of undamaged berries and axial dimensions are an important quality criteria both table grape and juice industry. Therefore, the economic value of grape mostly depends on the presence of undamaged grape berries.

Mechanization of agriculture particularly harvest and after harvest has been produced big demand on the knowledge of physical and mechanical properties of products. Mechanical and physical properties of plants are important criteria in the design of machines.

The importance of understanding the physical properties of fruits is to design of machines and processes for harvesting, handling and storage of agricultural materials and for converting these materials into foods. Some of these properties include the dimensional size, shape, sphericity, bulk density, true density, porosity, geometric mean diameter, projected area, surface area, mass, volume, etc.

The knowledge related to shape and physical dimensions, is useful in sorting and sizing of fruits and determining how many fruits can place in shipping containers. These properties depend on the species, variety, diameter, maturity, moisture content and cellular structure (Mohsenin, 1986; Persson, 1987; Altuntaş and Yıldız, 2007; Nazari Galedar et al., 2008; Skubisz, 2001).

Also, the variation in the physical properties of plant branches and the resistance of cutting equipment have to be known in order to understand the behavior of material with respect to different operation of conditions.

Knowing those properties will be useful industry, academia, research Institutes, consumers, manufacturer of machines and producers of food processing equipment.

Especially, information on plant properties and the power or energy requirement of equipment has been very valuable for selecting design and operational parameters (Persson, 1987; Georget et al. 2001; Emadi et al., 2004; Voicu et al., 2011; Ghahraei et al., 2011; Hoseinzadeh and Shirneshan, 2012). Perhaps, the stem of plants cutting energy is one of the main parameters for optimizing design of cutting elements in harvesting and pruning machines (Alizadeh et al., 2011).

Therefore, comparative performance of cutting elements applied in harvester and pruning machine design can be judge by their cutting energy requirements, cutting force and stress applied (Chakraverty et al., 2003; Alizadeh et al., 2011; Sessiz et al., 2013).

Cutting strength and cutting energy are related to the stem mechanical and physical properties. Therefore, such information is very important for the suitable design of grape pruning knife and pruning machine and harvesters for efficient use of energy (Sessiz et al., 2015). With the increasing scarcity of manual labor for vineyard pruning and harvesting operations, mechanized vine pruning and harvesting has received much attention.

Mechanically harvested grapes could have as good as and sometimes better quality than hand-harvested grapes when the grapes are harvested cool and delivered promptly to the processing unit (Morris, 2000). A review of the literature revealed little information on direct cutting properties of cutting grape canes. Romano et al. (2010) determined cutting force for certain vine branches such as Cabernet, Sauvignon and Chardonnay in different regions in Italy. The tests were conducted in the laboratory and the results were processed to show if the manual forces dispensed during cutting were a function of diameters and cultivated varieties. Sessiz et al. (2015) determined cutting properties of some grape varieties in Turkey.

Studies showed that the cutting properties are valuable information for suitable design of grape pruning knifes, pruning machines and harvesters for efficient energy use. Data on physical properties of agro-food materials are valuable because they are needed as input to models predicting the quality and behavior of produce in pre-harvest, harvest situations and they aid the understanding food processing (Nesvadba et al., 2004). Therefore, to successful mechanization of grapes, we must know exactly physical and mechanical properties of grapes.

The specific objectives of this study were to: (1) determine the relationship between the basic berry and cane physical and mechanical properties of Şire grape variety in different phenological stages, (2) determine the relationship between grapevine internodes of cane's cutting properties and berry detachment force (FDF) and berry shell rupture force, (3) development of empirical model between berry axial dimensions and arithmetic, geometric mean diameter, sphericity, roundness properties, (4) to determine relationship between phenological stages and pomological properties of grape, (5) determine the relationship between berry detachment force from cluster (FDF) and berry shell rupture force under compressive load and pomological properties of grape.

MATERIALS AND METHODS

Sample preparation and measuring
This study was performed with Şire (*Vitis vinifera* L.) local grape variety (Figure 1).
The samples were obtained from a commercial vineyard (Figure 1.) in Diyarbakır province, which is located in the southeastern part of Turkey.

Figure 1. Research vineyard area and Şire grape variety.

The grape berry and cane cutting tests were carried out during the different phenological stages of the veraison (30 August), 15 days after veraison (15 September) and harvesting time (30 Semptember) in 2016.
Grape canes which have five internode and different diameter were randomly harvested by hand from vineyard.
Harvested and collected canes which have different internode and clusters were transported to laboratory of Department of Agricultural Machinery and Technologies Engineering, University of Dicle and preservation in a refrigerator at 5 ^{0}C until the time of the cutting tests.
This study was conducted in two phases.
The first phase consist of the determination of ripening grape berries length, width, thickness, arithmetic and geometric mean diameter, sphericity, roundness, force detachment (FDF), weight (W), FDF/W, skin firmness, total soluble solids content, pH, total acidity were measured.
In the second phase, grapevine cane cutting shearing force, shearing strength, upper yield, shearing energy, specific shearing energy were determined.

Measurement of Grape Berry Axial Dimensions and Other Physical Properties
The physical properties were measured at three different phenological stages during the harvest season.
In all experiment, in order to determine the initial moisture content of grape canes, three samples of 30 g were weighed and dried in an oven of 105 ^{0}C for 24 hours (ASABE, 2006; Sessiz et al., 2007), after oven drying, samples were removed from oven and kept for 15 minutes in a desiccator for moisture equilibrium. Then samples reweighed to obtain the final moisture content using the gravimetrical method.
The weights were measured using electronic scales with a capacity of 1.2 kg and with a precision of 0.01 g. The moisture content levels of internode of cane were determined at 38.64 %, 48.00 % and, 51.76% w.b. The results were evaluated according to these moisture content values.
To determine the dimensional size of grape, 25 berrries randomly taken from four grape clusters at each phenological stage and the three linear dimensions namely, length, width and thickness were measured by using an electronic micrometer with a reading accuracy within 0.01 mm. These geometric dimensions were determined at the sample place in the middle of the berry.
The, geometric mean diameter, sphericity, roundness, and surface area of individual berries were calculated using the following equations (Mohsenin 1986; Deshpande et al., 1993; Baryeh, 2002; Aydin, 2002; Zare et al., 2012; Sessiz et al., 2013).

$$Da = \frac{(L + W + T)}{3}$$
$$Dg = (LWT)^{1/3}$$

$$\emptyset = \frac{(LWT)^{1/3}}{L} = \frac{Dg}{L}$$

$$Ro = \frac{W}{L} \times 100$$

$$S = \pi D_g^2$$

Where L is the length (mm), W is the width (mm), T is the thickness (mm), D_a is arithmetic mean diameter (mm), D_g is geometric mean diameter (mm), $\emptyset$ is sphericity (%) , Ro is roundness (%), and S is surface area (mm^2).

Measurement of Grape Berry Mechanical and Pomological Properties

Grape berry detachment force, skin firmness, and the FDF/W ratio are important mechanical properties for fruit harvesting (Sessiz and Özcan, 2007; Putri et al., 2015), and firmenss is the resistance of the individual fruit to deformation under applied forces (Renny et al., 2015). These parameters were measured at three different phenological stages during the harvest period. Thickness, width and of berries were measured with a micrometer to within 0.01 mm. Grape berries were weighed by means of a digital balance with 0.01g. FDF and skin firmness were measured by using a pull digital force gauge (Model FG-20, Lutron Instrument) with stainless steel cone head adapter which apical angle 86^0. The maximum value recorded for each test, 25 grape berries were randomly selected from grape cluster and measured. The maximum skin firmness and FDF values were recorded by the force gauge while probe passing inside in grape fruit in Newton (N) (Jha et al., 2006). The digital force gauge is shown in Figure 2.

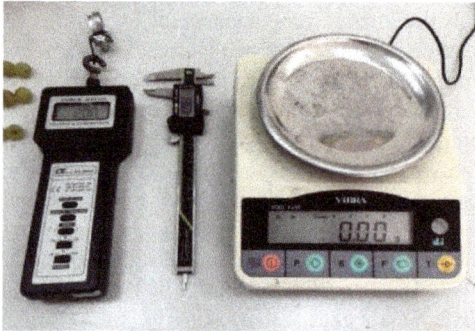

Figure 2. Force gauge

The grape berry phomological properties were measured at three different phenological stages during the harvest season. Total Soluble Solids Content (TSSC) (by refractometer), pH (by pH meter) and total acidity (bye Digital Burette) values were measured (Ozdemir et al., 2016).

Measurement of Grapevine Cane Cutting Properties

The mechanical properties the shearing force, shearing strength and shearing energy were determined along the canes from first internode to fifth internode in three phenlogical stages.

Prior to the tests, the grapevine canes were cutted into five different groups (Figure 3). Five internodes of grape canes were named first to fifth from the top toward the bottom. Internode cutting locations were marked on randomly selected canes. Internodes (between two nodes were considered a internode) based on their internode of cane mean diameter ranging from 6.5 mm to 10.5 mm (6.5, 7.5, 8.5, 9.5, 10.5 mm). The cane cutting diameters were measured before the test using a caliper. The ranges of internode diameter of cane (mm) values were converted to cross-section area in mm^2. Testing was completed as rapidly as possible in order to reduce the effects of drying. All the cutting measurements were performed on the same day of harvesting.

Figure 3. Grapevine canes

The cutting tests were conducted by Lloyd LRX Plus Materials Testing Machine (Figure 4), which allows determination of the relationship between cutting strength and deformation. It has a single column with a crosshead travel range of 735 mm. In the compression tests, the test samples were placed on the machine loading table in its flat position. Loading was applied vertical direction. The cutting knife was steel, 50 mm width, 6 mm thickness and the blade angle of 17∘???. Cutting measurement were performed at 100 mm/min fixed loading speed for all tests.

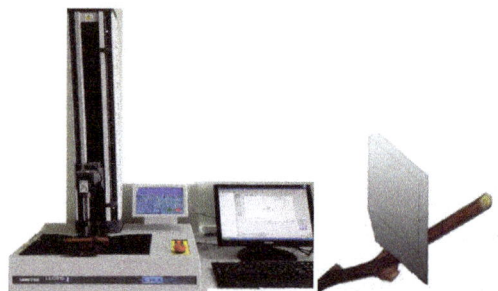

Figure 4. The Lloyd LRX Plus Materials Testing Machine and cutting blade

The peak shearing strength, obtained from the shearing force findings, was determined by the following equation (Mohsenin, 1986; Beyhan, 1996; Amer Eissa et al., 2008; Zareiforoush et al., 2010; Tavakoli M., 2011; Sessiz et al., 2013):

$$\sigma s = \frac{F}{A}$$

Where: σs is the maximum shearing strength in (MPa), Fmax is the maximum shearing force in (N) and A is the cross-sectional area in (mm^2).

The cutting energy was calculated by measuring the surface area under the force-deformation curve (Yore et al., 2002; Chen, et al., 2004; Nazari Galedar, et al., 2008; Ekinci et al., 2010; Zareiforoush, et al., 2010; Heidar and Chegini, 2011; Sessiz et al., 2015; Nowakowski, 2016). The cutting energy and displacement was calculated by material testing machine. A computer data acquisition system recorded all the force-displacement curves during the cutting process. A typical force-deformation curve for grapevine cane under compression is shown in Figure 5.

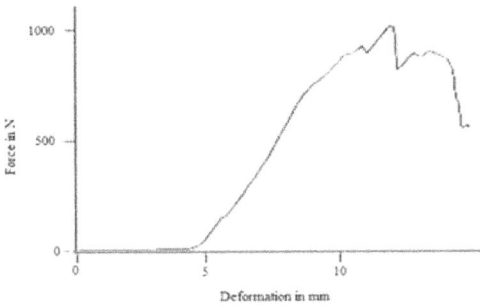

Figure 5. Typical force-deformation curve

The first peak corresponds to the yield point at which cane damage was initiated. The second peak corresponds to maximum compressive force. This bio yield point is characterized by the fact that any further compression yields no increase in applied load (Mohsenin, 1986; Lu and Siebenmorgen, 1995).

Data Analysis

The experiment was planned as a completed randomized plot design, and, data were analyzed using The General Linear Model (GLM). Mean separations were made for significant effects with LSD and the means were compared at the 1% and 5% levels of significance using the Tukey multiple range tests in JMP software, version 11.

RESULTS AND DISCUSSION

Grape Berry Properties Axial Dimensions and Physical Properties

The physical properties of grape berries at different phenological stages are presented in Table 1. The comparison of means indicated that there was a significant difference (p<0.05) among phenological stages values. The results indicated that the data obtained from the measured values gave the following significant correlations between axial dimensions and other physical properties such as arithmetic mean diameter, geometric mean diameter, surface area, sphericity and roundness. The axial dimensions of grape berries increased significantly (p<0.05) over the period of maturity time (the harvest season being usually prolonged for some 20-30 days). Dimensions of Şire grape fruits varied from 14.03 to 16.16 mm in length, 12.94 to 15.51 mm in width, and 12.78 to 15.51 mm in thickness, with average values of 15.32, 14.28, and 14.36 mm, respectively. These dimensions can be used in design of sorting and separating machine and food industry for grape. Similar results were reported by Khodaei ve Akhijahani (2012) for Rasa grape variety. Also, results show that arithmetic mean diameter, geometric mean diameter, surface area, sphericity and roundness of grape berries increases with along phenological stages (P<0.05).

Table 1. Axial dimensions and physical properties of Şire grape berries at different phenological stages

Phenological Stages	Dimensions (mm)			Arithmetic Mean Diameter (mm)	Geometr Mean Diameter (mm)	Area (mm²)	Sphericity (%)	Roundness (%)
	Length (mm)	Width (mm)	Thickness (mm)					
Veraison	14.03b[1]	12.94c	12.78c	13.24c	12.22c	551.4c	0.943c	0.921b
15 days after Veraison	15.78a	14.49b	14.80b	15.09b	15.08b	716.7b	0.950b	0.930b
Harvest	16.16a	15.43a	15.51a	15.70a	15.69a	776.2a	0.971a	0.955a
Mean	15.32	14.28	14.36	14.67	14.33	681.43	0.954	0.935
LSD	0.385	0.378	0.414	0.362	0.362	33.91	0.01	0.01

[1]means followed by the same letter in each column are not significantly different by Tukey's multiple range test at the 5% level

The relationship between axial dimensions and the other physical properties can be calculated by the regression equation shown in Table 2. The high correlation was found between axial dimensions and the other physical properties values. So, the equations can be used to predict the arithmetic mean diameter, geometric mean diameter, surface area, sphericity and roundness of Şire grape variety as a function of axial dimensions and maturity time.

Table 2 . Regression equations of cutting properties as a function of three axial dimensions[1]

Parameters	Regression equation	R^2
Arithmetic mean diameter (mm)	$Y = -1.7 \times 10^{-14} + 0.333L + 0.333\,W + 0.333\,T$	1.00
Geometric mean diameter (mm)	$Y = -0.00204 + 0.3174L + 0.339\,W + 0.343\,T$	0.999
Surface area (mm2)	$Y = -660 + 29.06\,L + 30.98\,W + 31.43T$	0.999
Sphericity (%)	$Y = 0.9548 - 0.04187\,L + 0.02237\,W + 0.0224\,T$	0.994
Roundness (%)	$Y = 0.934 - 0.06127\,L + 0.0654\,W + 0.000171T$	0.995

[1]L: Length (mm), W: Width (mm), T: Thickness

Mechanical and Pomological Properties

The values of obtained from the test results the berry detachment force from grape cluster, berry weights, FDF/W ratio and, berry shell firmness are shown in Table 3. As shown in Table 3 very high correlation was observed between the ratio of FDF/W and phenological stages. The ratio of FDF/W decreased depending on phenological stages and maturity time. Fruit weight was lowest at the beginning of the harvest (1.60 g). The fruit detachment force from the grape cluster stalk was reduced from 1.57 N to 1.27 N. This value is valuable for maturity criteria of grape fruit because FDF/W ratio is an important parameter of mechanical harvesting. Similar results were observed between skin firmness and phenological stages. The grape berry skin firmness decreased from 1.17 N to 0.766 N with harvesting period. The reason for this trend is that the water content of fruit increased with maturity. According to these results we can express that there is a high correlation between FDF/W ratio and fruit skin firmness. The maximum firmness was observed as 1.174 N when the TSSC was 16.20 at the first date of harvest (Table 3 and Table 4). Total Soluble Solids Content (16.20 - 20.40 %) and pH (3.39-3.65) values increased with harvesting date, whereas the total acidity were slight changed and reduced from 0.876 to 0.669 %. Similar results were reported by Ozdemir et al. (2016) for different wine grape variety.

Table 3. Some mechanic properties of Şire grape berries at different phenological stages

Phenological Stages	Properties			
	Detachment Force (FDF), N	Weight (W), g	(FDF/W), N/g	Skin firmness(N)
Veraison	1.57	1.60	0.98	1.174
15 days after Veraison	1.51	2.13	0.84	0.859
Harvest	1.27	2.49	0.51	0.766

Table 4. Some pomological properties of Şire grape berries at different phenological stages

Phenological Stages	Properties			
	TSSC (%)	pH	Acidity (%)	Maturit Index
Veraison	16.20	3.39	0.876	18.43
15 days after Veraison	18.23	3.50	0.775	23.52
Harvest	20.40	3.65	0.669	30.49

Grapevine Cane Cutting Properties

The test results of the cutting properties are shown in Table 5. As shown in the table, the phenological stages has significant effect on the cutting properties of grapevine canes (P <0.01). It can be seen from table 5 that the shearing force, shearing strength, upper yield, energy requirement and specific cutting energy has increased depending on phenological stages. There was no significant difference between the mean shearing strength for phenological stages. However, there was a significant difference between strength for internodes' diameter of cane (Table 5). It was observed that the minimum values cutting force, cutting strength, upper yield, energy requirement and specific cutting energy were obtained at a date of 30.08.2016 as 649.78 N, 13.29 MPa, 588.8, 5.02 J and 0.0944 J.mm^{-2}, while maximum values of were obtained at date of 23.09.2013 as 823.16 N, 16.11 MPa, 723.4 N, 6.25 J, and 0.1049 J mm^{-2}, respectively.

Table 5. The average cutting properties and phenological stages

Phenological Stages	Shearing force (N)	Shearing strength (Nmm^{-2})	Upper Yield (N)	Shearing energy (Joule)
Veraison	649.78[b][1]	13.29	588.8[b]	5.02[b]
15 days after Veraison	819.48[a]	15.87	680.2[ab]	6.69[a]
Harvest	823.16[a]	16.11	723.4[a]	6.25[a]
Mean	764.14	14.95	664	5.99
LSD	78.59	ns	0.098	1.13
R^2	0.941	0.575	0.886	0.872
Upper value	853.75	16.23	744	6.86
Lower value	674.53	13.67	584	5.12
Std dev	298.28	4.262	4.0	2.90
Std err mean	44.46	0.635	26.8	0.43

[1]means followed by the same letter in each column are not significantly different by Tukey's multiple range test at the 5 % level.

The results of the cutting properties depending on diameter of internodes of grapevine cane are shown in Table 6. The results shown in Table 6 indicate that the shearing force, shearing strength, upper yield, shearing energy and specific shearing energy requirement increased with increase internode diameter of canes. The significant differences were found between all of internodes of diameter at a 5 % probability level. The maximum values of shearing force shearing, upper yield and energy were obtained at 86.54 mm^2 cross-sectional area as 1.197 N, 1.060 N, and 10.16 J, respectively, while the maximum shearing strength was obtained at 70.84 mm^2 cross-section area (9.5 mm^2 diameter) as 19.675 MPa. Especially, the cross-sectional area has a significant influence on cutting properties and energy. The shearing energy values varied from 3.509 J to 10.16 J depend on diameter. The effect of stem diameter on the maximum cutting force and cutting energy is consistent with Chen et al. (2004), who reported that both the cutting energy and maximum cutting force are directly proportional to the cross-sectional area of hemp stalk. Similar results were found by Sessiz at al. (2013) for the olive sucker and Sessiz et al (2015) for grape sucker.

Table 6. The relationship between average cutting properties and cross-sectional area

Cross-sectional area (mm^2)	Shearing force (N)	Shearing strength (Nmm^{-2})	Upper Yield (N)	Shearing energy (Joule)
33.16 (6.5)^2	472.38d[1]	14,245b	432.0 c	3,509c
44.15 (7.5)	626.45c	14,189b	540.4bc	5,004bc
56.71 (8.5)	726.89bc	12,817b	601.7b	5,092bc
70.84 (9.5)	798.38b	19,675a	686.2b	6,192b
86.54 (10.5)	1196.57a	13.82b	1,060a	10.16a
Mean	764.14	14.95	664.06	5.99
LSD	119.38	Ns	14.97	1,719

[1]means followed by the same letter in each column are not significantly different by Tukey's multiple range test at the 5 % level.
[2] diameter of internode (mm)

CONCLUSIONS

The tests results indicated that the data obtained from the measured values gave the significant correlations between axial dimensions and other physical properties such as arithmetic mean diameter, geometric mean diameter, surface area, sphericity and roundness. Dimensions of Şire grape berries varied from 14.03 to 16.16 mm in length, 12.94 to 15.51 mm in width, and 12.78 to 15.51 mm in thickness, with average values of 15.32, 14.28, and 14.36 mm, respectively. These dimensions can be used in design of sorting and separating machine and food industry for grape.

The results show that arithmetic mean diameter, geometric mean diameter, surface area, sphericity and roundness of grape fruits increases with along phenological stages (P<0.05). Very high correlation was observed between the ratio of FDF/W and phenological stages. The ratio of FDF/W decreased depending on phenological stages and maturity time. Fruit weight was lowest at the beginning of the harvest (1.60 g).

The grape berry skin firmness decreased from 1.17 N to 0.766 N with harvesting period. TSSC values varied from 20.40 to 16.20.

Total soluble solids content (16.20-20.40 %) and pH (3.39-3.65) values increased with harvesting date, whereas the total acids were slight changed and reduced from 0.876 to 0.669 %. Cutting properties of sire grape cane has been changed with harvesting time. Shearing force and energy requirement increased with increase internode diameter of canes. Shearing force values varied between 472.38 N and 1,196.52 N.

The maximum shearing force and energy requirement were determined the last harvesting time.

REFERENCES

Anonymous, 2016.Turkish Statistical Institute Agriculture Databases. http://www.turkstat.gov.tr/PreTabloArama.do?metod=search&araType=vt

ASABE Standards, 2006. S358.2: 1:1 Measurement – Forages. 52nd edn. American Society of Agricultural Engineers, St Joseph MI.

Altuntas E., Yildiz M., 2007. Effect of moisture content on some physical and mechanical properties of faba bean (Vicia faba L.) grains, J. Food Eng., vol. 78, pp.174-183, 2007.

Alizadeh M.R., F.R.Ajdadi, Dabbaghi A, 2011. Cutting energy of rice stem as influenced by internode position and dimensional characteristics of different varieties. AJCS 5(6):681-687, ISSN:1835-2707.

Amer Eissa A.H., Gomaa A.H,. Baiomay M.H., Ibrahim A.A, 2008. Physical and mechanical charactesristics for some agricultural residues. Journal of Agricultural Engineering, 25(1),121-146. ASAE Standards, 2006. S358.2: 1:1 Measurement –Forages. 52nd ed. American Society of Agricultural Engineers, St Joseph MI

Aydin C., 2002. Phsical properties of hazel nuts. Biosystems engineering. 82(3), 297-303

Baryeh E.A., 2002. Physical properties of millet .Journal of Food Engineering 51, 39-46.

Beyhan M.A., 1996. Determination of shear strength of hazelnut sucker. Journal of Agriculture Faculty OMU,11(3),167-181.

Chakraverty A., P.R Singh, Raghavan G.S.V., Ramaswamy H.S., 2003. Handbook of postharvest technology. 1st edn, Marcel Dekker Inc, New York.

Chen Y., Gratton J.L., Liu J. 2004. Power requirements of hemp cutting and conditioning. Biosystems Engineering, 87(4), 417–424.

Deshpande S.D, Bal S., Ojha T.P., 1993. Physical properties of soybean. Journal of Agricultural Engineering Research, 56, 89-98

Georget D.MR., Smith A.C., Waldron K.W.,2001. Effect of ripening on the mechanical properties of Portuguese and Spanish varieties of olive (*Olea europaea* L) J Sci Food Agric 81:448-454.

Ekinci K., Yilmaz D., Ertekin C., 2010. Effects of moisture content and compression positions on mechanical properties of carob pod (*Ceratonia siliqua* L.). In African Journal of Agricultural Research, vol. 5, 2010, no. 10, pp. 1015–1021.

Emadi B., Kosse V., Yarlagadda P., 2004. Relationship between mechanical properties of pumpkin and skin thickness. International Journal of Food Properties, 8(2), 277-287.

Ghahraei O., Ahmad D., Khalina A., Suryanto H.,Othman J. 2011. Cutting tests of kenaf stems. Transactions of the ASABE, 54(1), 51-56.

Heidari A., Chegini G.R.,2011. Determining the shear strength and picking force of rose flower. Agricultural Engineering. Ejpau 14(2), 13.

Hoseinzadeh B., Shirneshan A. 2012. Bending and shearing characteristics of canola stem. American-Eurasian J. Agric. & Environ.Sci., 12 (3), 275-281.

Jha S.N. , Kingsly A.R.P., Sangeeta C,. 2006. Physical and mechanical properties of mango during growth and storage for determination of maturity. Journal of Food Engineering,73–76.

Khodaei J., Akhijahani H.S.,2012. Some physical properties of Rasa grape (*Vitis vinifera* L.).World Applied Sciences Journal 18 (6): 818-825

Lu R., Siebenmorgen T. J.,1995. Correlation of head rice yield to selected physical and mechanical properties of rice kernels. Transactions of the ASAE .VOL. 38(3):889-894.

Mohsenin N.N. ,1986. Physical properties of plant and animals materials. 2nd edition. New York, NY: Gordon and Breach Science Publishers.

Morris J.R. ,2000. Past, Present, and future of vineyard mechanization. Proceeding ASEV 50 th Anniv. Ann. Mtg. Seatle, WA, Vol. 51, 155-164.

Nazari G.M., Tabatabaeefar A., Jafari A., Sharifi A., Rafiee S., 2008. Bending and shearing characteristics of alfalfa stems. Agricultural Engineering International the CIGR Ejournal. Manuscript FP 08 001. Vol. X. May.

Nesvadba N., Houska M., Wolf W., Gekas V., Jarvis D., Sadd P.A., johns A.I., 2004. Database of physical properties of agro-food materials. Journal of Food Engineering 61, 497-503.

Nowakowski T. ,2016. Empirical model of unit energy requirements for cutting giant miscanthus stalks depending on grinding process parameters. Annals of Warsaw University of Life Sciences – SGGW, Agriculture No 67 (Agricultural and Forest Engineering) 2016: 63–70.

Ozdemir G., Sogut A.B., Pirinccioglu M., Kizil G., Kizil M.,2016. Changes in the phytochemical components in wine grape varieties during the ripening period. Scientific Papers B, Horticulture Vol. LX, 85-93.

Persson S., 1987. Mechanics of cutting plant material. ASAE Publications, St Joseph, MI, USA

Putri R., Yahya E., Adam A., Aziz N.M., Samsuzana, A.A.,2015. Correlation of moisture content to selected mechanical Properties of Rice Grain

samples. International Journal on Advanced Science Engineering Information Technology. Vol.5, ISSN: 2088-5334.

Romano E., Bonsignore R., Camillieri D., Caruso L., Conti A., Schillaci G.,2010. Evaluation of hand forces during manual vine branches cutting. International Conference Ragusa SHWA, September 16-18, 2010 Ragusa Ibla Campus- Italy. Work Safety and Risk Prevention in Agro-food and Forest Systems.

Renny E.P., Yahya A., Adam N.M., Aziz, Samsuzana, A.A.,2015. Correlation of moisture content to selected mechanical properties of rice grain samples. (2015). International Journal on Advanced Science Engineering Information Technology. Vol.5, ISSN: 2088-5334.

Sessiz A., Özcan M.T., 2006. Olive removal with Pneumatic branch shaker and abscission chemical .Journal of Food Engineering, Volume: 76 Issue: 2 Pages: 148-153.Elsevier, London.

Sessiz A., Esgici R., Kızıl S., 2007. Moisture-dependent physical properties of caper (*Capparis* ssp.) Fruit. Journal Of Food Engineering", 79, 1426-1431. Elsevier, London.

Sessiz A., Elicin A.K., Esgici R., Ozdemır G., Nozdrovický L., 2013. Cutting properties of olive sucker. Acta Technologica Agriculturae. The Scientific Journal for Agricultural Engineering, The Journal of Slovak University of Agriculture in Nitra. Vol: 16(3), 80–84.

Sessiz A., Esgici R., Özdemir G., Eliçin A.K., Pekitkan F.G., 2015. Cutting properties of different grape varieties, Agriculture & Forestry, Vol. 61. Issue 1: 211-216.

Skubisz G., 2001. Development of studies on the mechanical properties of winter rape stems. International Agrophysics, 15,197-200. Taghijarah H., Ahmadi H., Ghahderijani M.

Tavakoli M., 2011. Shearing characteristics of sugar cane (*Saccharum officinarum* L.) 157 stalks as a function of the rate of the applied force. AJCS 5(6), 630-634.

Voicu G., Moiceanu E., Sandu M., Poenaru I.C., Voicu P., 2011. Experiments regarding mechanical behaviour of energetic plant miscanthus to crushing and shear stress. *Engineering For Rural Development Jelgava*, 26.- 27.05.2011.

Yore M.W., Jenkins B.M, Summers M.D., 2002. Cutting properties of rice straw. *Paper Number: 026154.*ASAE Annual International Meeting / CIGR XVth World Congress

Zare D., Salmanizade F., Safiyari H.,2012. Some Phyical and mechanical properties of Russian olive fruit. World Academiy of Science, Engineering and Technology. International Journal and biological, Biomolecular, Agricultural, Food and Biotechnologicval Enginerring. Vol:6(9), 668-671.

Zareiforoush H., Mohtasebi S.S., Tavakoli H., Alizadeh M.R., 2010. Effect of loading rate on mechanical properties of rice (*Oryza sativa* L.) straw. Australian Journal of Crop Science, 4(3), 190–195.

20

PRECISION VITICULTURE TOOLS TO PRODUCTION OF HIGH QUALITY GRAPES

Gultekin OZDEMIR[1], Abdullah SESSIZ[2], Fatih Goksel PEKITKAN[2]

[1]Dicle University, Faculty of Agriculture, Department of Horticulture, Diyarbakir, Turkey
[2]Dicle University, Faculty of Agriculture, Department of Agricultural Machinery and Technologies
Engineering, Diyarbakir, Turkey
Corresponding author email: gozdemir@dicle.edu.tr

Abstract

Grapes are the most widely grown commercial fruit crop in the world, and also one of the most popular fruit crops for horticultural production. Grape growers constantly search the ways in order to maximize their profits all over the world. It becomes to be important to use new information technologies to increase to overall returns. Precision Viticulture (PV) refers to the application of new and emerging information technologies to the production of grapes to improve the efficacy of production, maximize the quality of production, minimize the environmental footprint of production and minimize the risk associated with production for the grower and processer. Precision viticulture depends on new and emerging technologies such as global positioning systems (GPS), meteorological and other environmental sensors, satellite and airborne remote sensing, and geographic information systems (GIS) to assess and respond to variability. It can be possible that take under control such as soil fertility, fertilizer application norm, disease, water, weed, harvesting, and environmental management by precision viticulture systems in vineyard. So, to reduce inputs such as fertilizer, water, pesticides and to increase yield and quality of grape berries, we must to increase precision technologies in our vineyards. In this review, Precision Viticulture tools will be demonstrated to producing of high quality grapes. Finally, this study will also help grape growers and government agencies that provide new information and technologies such as Remote Sensing to growers in order to detect some factors affecting to maximize grape production.

Key words: Geographical Information Systems, Grape, Precision Viticulture, Remote Sensing, Vineyard Management.

INTRODUCTION

Precision viticulture is precision farming applied to optimize vineyard performance, maximizing grape yield and quality while minimizing environmental impacts and risk (Proffitt et al., 2006; Urretavizcaya et al., 2017). This is accomplished by measuring local variation in factors that influence grape yield and quality (soil, topography, microclimate, vine health, etc.) and applying appropriate viticulture management practices (trellis design, pruning, fertilizer application, irrigation, timing of harvest) (Bramley and Hamilton, 2004; Bramley, 2005).

Among the benefits of precision viticulture reduction of fertilizer costs, reduction of pesticide application costs, minimization of environmental pollution, increase of product yield, more accurate information management due to more efficient information production, operating records required for sales and after sales production periods.

Precision viticulture is based on the premise that high in-field variability for factors that affect vine growth and grape ripening warrants intensive management customized according to local conditions. Precision viticulture depends on new and emerging technologies such as global positioning systems (GPS), meteorological and other environmental sensors, satellite and airborne remote sensing, and geographic information systems (GIS) to assess and respond to variability (Matese and Di Gennaro, 2015).

Several authors have studied precision viticulture in different countries (Bramley et al., 2000; Bramley et al., 2003; Bramley, 2001; Bramley and Williams, 2001; Bramley and Lamb, 2003; Bramley and Hamilton, 2004, 2007; Taylor, 2004; Tisseyre et al., 2001; Arno et al., 2005; Arno, 2008; Penn, 1999; Carothers, 2000; Aho, 2002; Matese and Di Gennaro, 2015).

Vineyards are characterized by a high heterogeneity due to structural factors such as the morphological characteristics, and other dynamics such as cropping practices and seasonal weather (Bramley, 2003).

This variability causes different vine physiological response, with direct consequences on grape quality (Smart, 1985). Vineyards therefore require a specific agronomic management to satisfy the real needs of the crop, in relation to the spatial variability within the vineyard (Proffit et al., 2006). The introduction of new technologies for supporting vineyard management allows the efficiency and quality of production to be improved and, at the same time, reduces the environmental impact.

This paper presents a review of applications used in precision viticulture to production of high quality grapes.

Precision Viticulture Applications

Precision viticulture is still relatively new in that yield monitoring technology for wine grapes has only been commercially available in Australia since the 1000 vintage, and there is still only one brand of grape yield monitor on the market (although at least three others are currently under development). Nevertheless, this technology, along with other tools such as different global positioning systems (dGPS) and geographical information systems (GIS), promotes the capacity for grape and wine producers to acquire detailed geo-referenced information about vineyard performance and to start using this to tailor production of both grapes and wine according to expectations of vineyard performance, and desired goals in terms of both yield, quality and the environment (Figure 2) (Bramley and Proffitt, 1999 and 2000).

Viticulture precision process (Figure 1) begins with yield mapping and the acquisition of complementary information followed by interpretation and evaluation of the information leading to implementation of targeted management. This is followed by further observation. The process of data acquisition and use is therefore continuous, and improvements to management, incremental. Over time, data collected during the observation stage take on a predictive value (Bramley, 2001; Arno et al., 2017).

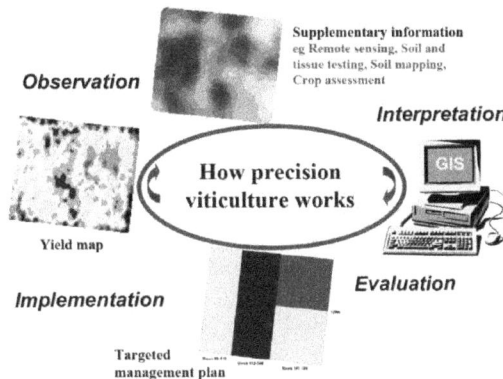

Figure 1. The process of precision viticulture (Bramley, 2001)

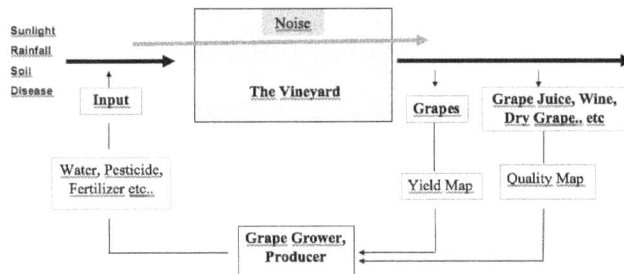

Figure 2. Viticulture input and output process

Nowadays different precision viticulture applications have helped grape growers to produce high quality grapes (Goldammer, 2015).

Terroir Management

Precision agriculture suitability to improve vineyard terroir management (Bouma, 2015). The tools of Precision Viticulture enable both growers, winemakers and researchers to see that terroir may vary within vineyards. Indeed, vineyards producing wines that are deemed characteristic of a region, may in fact be capable of producing contrasting wines from different areas within the same management units (Bramley and Hamilton, 2007).

According to Urretavizcaya et al. (2017) the early definition of within vineyard zones combining NDVI, ECa and BN data was successful, since the zones delineated allowed a differentiation of grape batches with different characteristics at harvest. Interestingly, the inclusion of a variable related to sink size (in this case the number of bunches per plant) provided the most efficient classification, which makes its consideration highly advisable for any PV work aimed at zone delineation for grape quality estimation.

Canopy Management

Canopy and vigor monitoring is the area of greatest adoption by the growers and the wineries for several reasons. It is possible to get timely, high-resolution information during the growing period, which may be relevant for canopy management, fertilization, and irrigation.

Arno et al. (2017) studied mapping the leaf area index (LAI) by using mobile terrestrial laser scanners (MTLS) is of significance for viticulture. Three different row length sections of 0.5, 1, and 2 m have been tested. Data analysis has shown that models required to estimate LAI differ significantly depending on the scanned length of the row; the model required to estimate LAI for short sections (0.5 m) is different from that required for longer sections (1 and 2 m).

According to Luo et al. (2016), grapes are likely to have collisions and be damaged by manipulations when harvesting grape clusters. For this reason, to conduct an undamaged robotic harvesting, they attempted locating the spatial coordinates of the cutting points on a peduncle of grape clusters for the end-effector and determining the bounding volume of the grape clusters for the motion planner of the manipulator using binocular stereo vision. As a result of the study, they found that cutting point detection success rate was approximately 87% and this method that it could be used on harvesting robots.

According to Berenstein et al. (2010) while much of modern agriculture is based on mass mechanized production, advances in sensing and manipulation technologies may facilitate precision autonomous operations that could improve crop yield and quality while saving energy, reducing manpower, and being environmentally friendly. They focused on autonomous spraying in vineyards and presented four machine vision algorithms that facilitate selective spraying. Researchers tested all image-processing algorithms on data from movies acquired in vineyards during the growing season. Results showed that 90% accuracy of grape cluster detection leading to 30% reduction in the use of pesticides (Gatti, et al., 2009; Goldammer, 2015).

Tang et al. (2016) conducted a study on non-productive vine canopy estimation through proximal and remote sensing. They asserted that non-productive canopy detection in a vinicultural block is a key factor in reducing the drain on infrastructure and improving management practices and current methods are significant in cost, biased, and do not provide information on location of non-productive canopy. Researchers announced that results indicate the success of semi-supervised method in providing a useful measure of non-productive canopy at the phenological stage of veraison; laying the groundwork for improved methods in this area. They also stated that these methods provide practical outputs that lay the foundations for improving management decisions in an automatic and low-cost manner at different times in the season.

Reis et al. (2012) states that one of the most demanding tasks in wine making is harvesting, even for humans, the environment makes grape detection difficult, especially when the grapes and leaves have a similar color, which is generally the case for white grapes. In this reason, they proposed a system for the detection and location, in the natural environment, of bunches of grapes in color images. In this study, they stated that system is able to distinguish between white and red grapes, and at the same time, it calculates the location of the bunch stem. They also reported that system achieved 97% and 91% correct classifications for red and white grapes, respectively.

Escola et al. (2013) developed and tested an orchard sprayer prototype that running a variable-rate algorithm to adapt the volume application rate to the canopy volume in orchards on a real-time. They divided prototype was into three parts: the canopy characterization system (using a LiDAR sensor), the controller executing a variable-rate algorithm, and the actuators. As a result, they observed a strong relationship between the intended and the sprayed flow rates ($R^2 = 0.935$) and between the canopy cross-sectional areas and the sprayed flow rates ($R^2 = 0.926$). In addition, they state that when spraying in variable-rate mode, the prototype achieved significantly closer application coefficient values to the objective than those obtained in conventional spraying application mode.

Gil et al. (2013) announced that the structural characteristics of the canopy are a key consideration for improving the efficiency of the spray application process for tree crops. However, they state that obtaining accurate data in an easy, practical, and efficient way is an important problem to be solved. Researchers developed and tested a sprayer prototype for the suspension plant for this purpose. They electronically measured variations in canopy width along the row crop using several ultrasonic sensors placed on the sprayer and used to modify the emitted flow rate from the nozzles in real time; the objective during this process is to maintain the sprayed volume per unit canopy volume. As a result, they estimated that 21.9% less pesticides could be used compared to traditional pesticide applications. In addition, they announced that this result is in accordance with the results of similar research on automated spraying systems.

Llorens et al. (2010) compared two different spray application methods during different crop stages of three vine varieties. A conventional spray application with a constant volume rate per unit ground area was compared with a variable rate application method designed to compensate electronically for measured variations in canopy dimensions. An air-blast sprayer with individual multi-nozzle spouts was fitted with three ultrasonic sensors and three electro valves on one side, in order to modify the emitted flow rate of the nozzles according to the variability of canopy dimensions in real time. As a result, they obtained the better leaf deposits and 58% saving in application volume with variable rate method.

Crop Load Monitoring

Crop load management in vineyards is important for the consistent production of both quality fruit and mature wood. "Crop load" is the ratio of exposed leaf area to fresh fruit weight. Too much leaf area promotes shading and reduces fruit quality and sometimes bud fruitfulness. Too little leaf area per unit of fruit delays ripening and reduces vine size. Measures of crop load are useful to growers in evaluating success of vineyard management practices. The Ravaz index which uses the ratio of yield to pruning weight to estimate crop load is one common metric (Figure 3).

Research into PV is still in its infancy, and to date relatively little has been published in this field. Current and future research into PA (PV) have many of different priorities: environmental economics, production quality assessment methods and new technologies for crop monitoring (Arno et al., 2009).

In the context of precision viticulture, remote sensing in the optical domain offers a potential way to map crop structure characteristics, such as vegetation cover fraction, row orientation or leaf area index, that are later used in decision support tools. Weiss and Baret (2017) studied to Using 3D Point Clouds Derived from UAV RGB Imagery to Describe Vineyard 3D Macro-Structure.

Figure 3. (A) Spectron. (B) Multiplex hand device
sensors for grape quality proximal monitoring, which
allows quality maps to be realized
(Matese and Di Gennaro, 2015)

Berry Quality Management

The NDVI image is an excellent tool to design
quality, sampling zones based upon the NDVI
classifications.

Source to sink size ratio, i.e.: the relative
abundance of photosynthetically active organs
(leaves) with regards to photosynthate
demanding organs (mainly bunches), is widely
known to be one of the main drivers of grape
oenological quality. However, due to the
difficulty of remote sink size estimation,
Precision Viticulture (PV) has been mainly
based on within-field zone delineation using
vegetation indices. This approach has given
only moderately satisfactory results for
discriminating zones with differential quality.
Urretavizcaya et al. (2017) investigate an
approach to delineate within-vineyard quality
zones that includes an estimator of sink size in
the data-set. Zone delineation was performed
using Normalized Difference Vegetation Index
(NDVI), soil apparent electrical conductivity
(ECa) and bunch number (BN) data.

Irrespective of the seasonal factors which affect
the mean concentration of berry rotundone,
variation in the land (soil, topography)
underlying the vineyard is a consistent driver of
within-vineyard variation in this important
grape-derived flavour and aroma compound
(Bramley et al., 2017).

Harvest Management

The proper ripening of grapes is the key to
obtain a high-quality wine and another grape
product. Ripening is a temporal process that is
influenced, in addition to uncontrollable
climate factors, by the spatial distribution of the
vineyard and planted variety. It is a complex
process that cannot be characterized by a single
parameter; rather, it is a modification of the

profile of the compounds of the grape.
Melendez et al. (2015) analyzed the joint
evolution of twelve physicochemical
parameters determined in red grapes from four
different varieties, in sixteen representatives (in
both geographical and edaphic point of view)
plots belonging to the Qualified Designation of
Origin (DOC) Rioja. Samples were collected in
September 2009 during four consecutive weeks
prior to harvest.

Disease Management

Disease from insects, pathogens, and other
infectious organisms can become a serious
problem. In some cases, disease development
on grapevines occur rapidly and results in
entire vineyards incurring injury to various
degrees. For example, grapevines are
susceptible to powdery mildew infection early
in the growing season. Patricia et al. (2009)
studied to field monitoring for grapevine
leafroll virus and mealybug in pacific
northwest vineyards.

Oberti et al. (2014) conducted a study on the
automatic detection of powdery mildew on
grapevine leaves by image analysis. They
announced that powdery mildew is a major
fungal disease for grapevine (*Vitis vinifera* L.)
as well as for other important specialty crops,
causing severe damage, including yield loss
and depreciation of wine or produce quality.
According to researchers proximal optical
sensing is a major candidate for becoming the
preferred technique for identification of foci for
powdery mildew in grapevine and other
specialty crops, but detection sensitivity of
symptoms in the early-middle stage can yield
largely limited results due to the combination
of small dimensions, low density, and spatial
arrangement of thin fungal structures. They
processed multi-spectral images from different
angles of vine leaves under laboratory
conditions. As a result, researchers found that
detection sensitivity generally increases as the
view angle is increased, with a peak value
obtained for images acquired at 60°.

Oberti et al. (2016) developed an agricultural
robot equipped with a new precision-spraying
end-effector with an integrated disease-sensing
system based on R-G-NIR multispectral
imaging. Researchers tested the robotic system

on four different replicates of grapevine canopy plots prepared in a greenhouse setup by aligning potted plants exhibiting different levels of disease. They announced that the results indicated that the robot could automatically detect and spray from 85% to 100% of the diseased area within the canopy and to reduce the pesticide use from 65% to 85% when compared to a conventional homogeneous spraying of the canopy.

Water Management

With water becoming a more scarce and managed commodity, better management is required. Most vineyard blocks do not have the same water requirements due to differences in soil type and topography within the same vineyard. Irrigation systems have been developed that can apply the correct amount of water where it is needed.

Vineyard water status is a key aspect to reach a control about yield and quality parameters and is linked to irrigation system management. Stem and leaf water potential, in several day times, was used for monitoring, controlling and managing irrigation with good correlations with soil and plant water status and with the vegetation index (Cancela et al., 2017).

Thermal imaging can become a readily usable tool for crop agricultural water management, since it allows a quick determination of canopy surface temperature that, as linked to transpiration, can give an idea of crop water status. In the last years, the resolution of thermal imaging systems has increased and its weight decreased, fostering their implementation on Unmanned Aerial Vehicles (UAV) for civil and agricultural engineering purposes. This approach would overcome most of the limitations of on site thermal imaging, allowing mapping plant water status at either field or farm scale, taking thus into account the naturally existing or artificially induced variability at those scales. Santesteban et al. (2017) studied to evaluate to which extent high-resolution thermal imaging allows evaluating the instantaneous and seasonal variability of water status within a vineyard. The information provided by thermal images proved to be relevant at a seasonal scale as well, although it did not match seasonal trends in water status

but mimicked other physiological processes occurring during ripening. Therefore, if a picture of variations in water status is required, it would be necessary to acquire thermal images at several dates along the summer.

Cancela et al. (2017) studied to test the discrimination of homogenized areas in traditional Galician vineyards of *Vitis vinifera* (L) cv. Albarifio, using a vegetation index and soil electrical conductivity and their relations with plant and soil measures (stem water potential and soil water content) and productivity and quality parameters.

Environmental Monitoring

Manually monitoring environmental parameters (e.g., humidity, temperature, soil moisture etc.) in the vineyard is not only time consuming but difficult to respond to in a timely manner when conditions change rapidly over space and time. Wireless sensor networks (WSNs), have been found to be suitable for collecting real time data for different parameters pertaining to weather, crop, and soil in developing solutions for vinicultural processes related to growing grapes. The development of wireless sensor applications in viticulture has made it possible to increase efficiency, productivity, and profitability of vineyard operations (Goldammer, 2015).

Wireless Sensor Networks

Wireless Sensor Networks (WSNs) have existed for many years and had assimilated many interesting innovations. Advances in electronics, radio transceivers, processes of IC manufacturing and development of algorithms for operation of such networks now enable creating energy-efficient devices that provide practical levels of performance and a sufficient number of features (Dziadak et al., 2016).

Wireless sensor networks deployed in vineyards is used for monitoring site conditions such as temperature, wind speed, wind direction, rainfall, solar-radiation, relative humidity, soil-moisture, soil-temperature, sap flow, and leaf wetness, for management decision making purposes. For example, Wireless sensor networks is used in the following applications:

A wireless sensor network for precision viticulture (Figure 4): The NAV (Network Avanzato per il Vigneto – Advanced Vineyard Network) system is a wireless sensor network designed and developed with the aim of remote real-time monitoring and collecting of micrometeorological parameters in a vineyard. The system includes a base agrometeorological station (Master Unit) and a series of peripheral wireless nodes (Slave Units) located in the vineyard. The Master Unit is a typical single point monitoring station placed outside the vineyard in a representative site to collect agrometeorological data. It utilizes a wireless technology for data communication and transmission with the Slave Units and remote central server (Matese et al., 2009).

Figure 4. Some sensors employed in wireless sensor networks for proximal sensing in vineyards. (A) Soil moisture (Spectrum Technologies Aurora, IL, USA). (B) Leaf wetness (Decagon Devices Inc. Pullman, WA, USA). (C) Grape temperature. (D) Dendrometer (GMR Strumenti SAS Scandicci, Italy). (E) Sap Flow (Fruition Sciences Inc., Montpellier, France) (Matese and Di Gennaro, 2015).

Normalized Difference Vegetation Index

Agricultural remote sensing products are frequently based on so-called spectral vegetation indices (SVIs), formed as various combinations of visible and near-infrared (NIR) spectral channels of digital imagery. SVIs are radiometric variables that are useful for mapping relative variations in canopy density. One common SVI is the normalised difference vegetation index (NDVI), formulated as (NIR-red)/(NIR+red). Many commercial wine grape growers in coastal California are now using NDVI imagery, generally acquired at maximum foliar expansion, to delineate management zones, identify problems, and re-develop properties. Agricultural remote sensing

products are frequently based on so-called spectral vegetation indices (SVIs), formed as various combinations of visible and near-infrared (NIR) spectral channels of digital imagery. SVIs are radiometric variables that are useful for mapping relative variations in canopy density. One common SVI is the normalised difference vegetation index (NDVI), formulated as (NIR-red)/ (NIR+red). Many commercial wine grape growers in coastal California are now using NDVI imagery, generally acquired at maximum foliar expansion, to delineate management zones, identify problems, and re-develop properties (Johnson, 2003).

By measuring the health and vigor of vegetation, NDVI can help vineyard managers fine-tune irrigation patterns. NDVI is directly related to the amount of photosynthetically active radiation that a plant may absorb (Kavak et al., 2014).

Soil Mapping

Soil electrical conductivity (EC) has been widely used to interpret soil spatial variability. Initially used to assess soil salinity, the use of EC in soil studies has expanded to include: mapping soil types; characterizing soil water content and flow patterns; assessing variations in soil texture, compaction, organic matter content, and pH; and determining the depth to subsurface horizons, stratigraphic layers or bedrock, among other uses. Variation of conductivity across soil types is the one of the main advantages of using this technology.

Weed Control

Typical vineyards may be infested by up to 20 weed species, of which three or four are dominant in terms of number of plants and land area covered. The distribution of weed species across a vineyard is "patchy" in nature. Some areas will be densely populated by weed while others will have few or no weeds. Densely populated patches often occur along vineyard edges, but may be found anywhere in the vineyard where the environment and management have favored the establishment and survival of weeds. The composition of weed species varies across a vineyard, and different patches may be dominated by different species. In addition to weed density varying spatially in a vineyard, it also may vary temporally and can be strongly influenced by weather (Goldammer, 2015).

Yield Monitoring

Grape yield maps are of fundamental importance for the development of PV (Arno et al., 2009). Yield monitoring refers to the "on-the-go" collection of both yield and positional data by the yield monitor and DGPS as the harvester travels along. The output in the form of yield maps allows growers and wine producers the ability to identify areas of different crop yield, and in some cases, different fruit quality attributes, within individual vineyard blocks. Yield maps do not require ground trothing since they represent actual as opposed to surrogate measures. Ground trothing is the process of gathering data in the vineyard that either complements or disputes remote sensing data collected by aerial photography or satellite (Gatti et al., 2009).

Kicherer et al.(2017) studied automatic image-based determination of pruning mass as a determinant for yield potential in grapevine management and breeding. Researchers calculated the mass of dormant pruning wood with the assistance of an automated image-based method for estimating the pixel area of dormant pruning wood. The evaluation of digital images in combination with depth map calculation and image segmentation is a new and non-invasive tool for objective data acquisition.

According to Aquino et al. (2015) one of the main challenges being faced by the scientific community in viticulture is early yield prediction. They have announced that flowering as well as fruit set assessment is of special interest since these two physiological processes highly influence grapevine yield. In addition, reported that an accurate fruit set evaluation can only be performed by means of flower counting. For this purpose, they presented a new methodology for segmenting inflorescence grapevine flowers in digital images. Thus, they found that values for Precision and Recall were 83.38% and 85.01%, respectively.

CONCLUSIONS

Precision viticulture is very new technology in Turkey. Hoverer, recently, precision viticulture has been received much attention in vineyard in the developed country. Different precision viticulture applications have been using and helped grape growers to produce high quality grapes. Precision viticulture depends on new and emerging technologies such as global positioning systems (GPS), meteorological and other environmental sensors, satellite and airborne remote sensing, and geographic information systems (GIS) to assess and respond to variability. It can be possible that take under control such as soil fertility, fertilizer application norm, disease, water, weed, harvesting, and environmental management by precision viticulture systems in vineyard. So, to reduce inputs such as fertilizer, water, pesticides and to increase yield and quality of grape berries, we must to increase precision technologies in our vineyards.

REFERENCES

Aho J.E., 2002. NASA providing new perspectives on vineyard management. Vineyard and Winery Management. 28(4):74–77.

Aquino A., Millan, B., Gutiérrez, S., Tardáguila, J., 2015. Grapevine flower estimation by applying artificial vision techniques on images with uncontrolled scene and multi-model analysis. Computers and Electronics in Agriculture, Vol. 119, 92-104.

Arnó J., 2008. Variabilidad Intraparcelaria en Viña y uso de Sensores Láser en Viticultura de Precisión [doctoral thesis]. [Laser sensor in Precision Viticulture to describe intra-field variability in the vineyard] Lleida: University of Lleida, Spanish.

Arnó J., Bordes X., Ribes-Dasi M., Blanco R., Rosell J.R., 2005. Esteve J. Obtaining grape yield maps and analysis of within-field variability in Raimat (Spain). Proceedings of the 5th European Conference on Precision Agriculture June 8–11, Uppsala, Sweden. 899–906.

Arno J., Escola A., Rosell-Polo J.R., 2017. Setting the optimal length to be scanned in rows of vines by using mobile terrestrial laser scanners.

Arnó J., Martinez Casasnovas J.A., Ribes Dasi M., Rosell J.R., 2009. Precision Viticulture. Research topics, challenges and opportunities in site-specific vineyard management. Spanish Journal of Agricultural Research 7(4):779-790.

Berenstein R., Ben Shahar O., Shapiro A., Edan Y., 2010. Grape clusters and foliage detection algorithms for autonomous selective vineyard sprayer. Intelligent Service Robotics, 3:233–243.

Bouma I., 2015. Interactive comment on Precision agriculture suitability to improve vineyard terroir management by J.M. Terron et al. Soil Discuss, 1, C447-C451.

Bramley R, Pearse B, Chamberlain P., 2003. Being profitable precisely – a case study of precision viticulture from Margaret River. Australian and New Zealand Grapegrower and Winemaker, 473a:84–87.

Bramley R., 2003. Smarter thinking on soil survey. Australian and New Zealand Wine Industry Journal. 18(3):88–94.

Bramley R.G.V., 2001. Progress in the development of precision viticulture – variation in yield, quality and soil properties in contrasting Australian vineyards. In: Currie LD, Loganathan P, editors. Precision Tools for Improving Land Management: Proceedings of the Workshop held by the Fertilizer and Lime Research Centre in Conjunction with the NZ Centre for Precision Agriculture at Massey University, Palmerston North, New Zealand, 14–15 February, 25–43.

Bramley R.G.V., 2005. Understanding variability in winegrape production systems. 2. Within vineyard variation in quality over several vintages. Australian Journal of Grape and Wine Research 11: 33-42.

Bramley R.G.V., Hamilton R.P., 2004. Understanding variability in winegrape production systems. 1. Within vineyard variation in yield over several vintages. Australian Journal of Grape and Wine Research 10: 32-45.

Bramley R.G.V., Hamilton R.P., 2007. Terroir and Precision Viticulture: are they compatible. J. Int. Sci. Vigne Vin, 41(1):1-8.

Bramley R.G.V., Lamb D.W., 2003. Making sense of vineyard variability in Australia. Proceedings IX Congreso Latinoamericano de Viticultura y Enología. November 24–28, Santiago, Chile. 35–54.

Bramley R.G.V., Proffitt A.P.B., 1999. Managing variability in viticultural production. The Australian Grape grower and winemaker, 427.

Bramley R.G.V., Proffitt A.P.B., Corner R.J., Evans T.D., 2000. Variation in grape yield and soil depth in two contrasting Australian vineyards. Australian and New Zealand Second Joint Soils Conference; December 3–8, Lincoln, New Zealand. 29–30.

Bramley R.G.V., Siebert T.E., Herderich M.J., Krstic M.P., 2017. Patterns of within-vineyard spatial variation in the 'pepper' compound rotundone are temporally stable from year to year. Australian Journal of Grape and Wine Research, 23(1):42-47.

Bramley R.G.V., Williams S.K., 2001. A protocol for winegrape yield maps. Proceedings of the 3rd European Conference on Precision Agriculture. June 18–21, Montpellier, France, 773–778.

Cancela, J.J., Fandino M., Rey B.J., Dafonte J., Gonzalez X.P., 2017. Discrimination of irrigation water management effects in pergola trellis system vineyards using a vegetation and soil index. Agricultural Water Management, 183:70-77.

Carothers J., 2000. Imagery technology meets vineyard management. Practical Winery and Vineyard, 21(1):54–62.

Dziadak B., Makowski L., Michalski A., 2016. Survey of Energy Harvesting systems for Wireless sensor networks in Environmental Monitoring. Metrology and Measurement Systems. 23(4):495-512.

Escola A., Rosell-Polo J.R., Planas S., Gil E., Pomar J., Camp F., Llorens J., Solanelles F., 2013, Variable Rate Sprayer. Part 1 – Orchard Prototype: Design, Implementation and Validation. Computers and Electronics in Agriculture 95, 122-135.

Gatti M., Dosso P, Maurino M., Merli M.C., Bernizzoni F., Pirez F.J. Plate B., Bertuzzi G.C., Poni S., 2009. MECS-VINE (R): A New Proximal Sensor for Segmented Mapping of Vigor and Yield Parameters on Vineyard Rows. Sensors, 16(12):1-21.

Gil E., Llrens J., Liop J., Fabregas X., Escola A., Rosell-Pol J.R., 2013. Variable Rate Sprayer. Part 2 – Vineyard Prototype: Design, Implementation and Validation. Computers and Electronics in Agriculture 95:136-150.

Goldammer T., 2015. Grape Grower's Handbook. A Guide to Viticulture for Wine Production. Apex Publishers. 728.

Johnson L.F., 2003. Temporal stability of an NDVI-LAI relationship in a Napa Valley vineyard. Australian Journal of Grape and Wine Research. 9(2):96-101.

Kavak M.T., Karadogan S., Ozdemir G., 2014. Investigating Vineyard Areas of Egil County of Diyarbakır Using Remote Sensing and GIS Techniques. International Mesopotamia Agriculture Congress, 22-25 September, 307-316, Diyarbakır, Turkey.

Kicherer A., Klodt M., Sharifzadeh S., Cremers D., Toepfer R., Herzog K., 2017. Automatic image-based determination of pruning mass as a determinant for yield potential in grapevine management and breeding. Australian Journal of Grape and Wine Research. 23(1):120-124.

Llorens J., Gil E., Llop J., Escola A., 2010. Variable rate dosing in precision viticulture: Use of electronic devices to improve application efficiency. Crop Protection, 29, 239-248.

Luo L., Tang Y., Zou X., Ye M., Li G., 2016. Vision-based extraction of spatial information in grape clusters for harvesting robots. Biosystems Engineering, Vol. 151, 90-104.

Matese A., Di Gennaro S.F., 2015. Technology in precision viticulture: a state of the art review. International Journal of Wine Research. 7:69-81.

Matese A., Di Gennaro S.F., Zaldei A., Genesio L., Vaccari F.P., 2009. A wireless sensor network for precision viticulture: The NAV system. Computers and Electronics in Agriculture 69:51-58.

Melendez E., Sarabia L.A., Ortiz M.C., 2015. Parallel factor analysis for monitoring data from a grape harvest in Qualified Designation of Origin Rioja including spatial and temporal variability. Chemometrics and Intelligent Laboratory Systems. 146:347-353.

Oberti R., Marchi M., Tirelli P., Calcante A., Iriti M., Borghese A.N., 2014. Automatic detection of powdery mildew on grapevine leaves by image analysis: Optimal view-angle range to increase the sensitivity. Computers and Electronics in Agriculture, Vol. 104, 1-8.

Oberti R., Marchi M., Tirelli P., Calcante, A., Iriti, M., Tona, E., Hočevar, M., Baur, J., Pfaff, J., Schütz, C., Ulbrich, H., 2016. Selective spraying of grapevines for disease control using a modular agricultural robot. Biosystems Engineering, Vol. 146, 203-215.

Penn C., 1999. Grape growers gravitating toward space age technologies. In Wine Business, Monthly, Wine Communications Group, Sonoma, CA, USA.

Proffit T, Bramley R, Lamb D, Winter E., 2006. Precision Viticulture – A New Era in Vineyard Management and Wine Production. Winetitles Pty Ltd., Ashford, South Australia; 1–90.

Proffitt, T., Bramley R., Lamb D., Winter, E., 2006. Precision Viticulture: A New Era in Vineyard Management and Wine Production. WineTitles, Adelaide. ISBN 978-0-9756850-4-4.

Reis M.J.C.S., Morais R., Peres E., Pereira C., Contente O., Soares S., Valente A., Baptista J., Ferreira, P.J.S.G., Bulas Cruz J., 2012. Automatic detection of bunches of grapes in natural environment from color images. Journal of Applied Logic, Vol. 10(4):285-290.

Santesteban L.G., Di Gennaro S.F., Herrero-Langreo A., Miranda C., Royo J.B., Matese A., 2017. High-resolution UAV-based thermal imaging to estimate the instantaneous and seasonal variability of plant water status within a vineyard. Agricultural Water Management. 183:49-59.

Skinkis P.A., Dreves A.J., Walton V.M., Martin R.R., 2009. Field Monitoring for Grapevine Leafroll Virus and Mealybug in Pacific Northwest Vineyards. Oregon State University ExtensionService. EM 8985.

Smart R.E., 1985. Principles of grapevine canopy management microclimate manipulation with implications for yield and quality. American Journal of Enology and Viticulture. 1985;36(3):230–239.

Tang J., Woods M., Cossell S., Liu S., Whitty M., 2016. Non-Productive Vine Canopy Estimation through Proximal and Remote Sensing IFAC-PapersOnLine, Vol.49(16):398-403.

Taylor J.A., 2004. Digital Terroirs and Precision Viticulture: Investigations into the Application of Information Technology in Australian Vineyards [doctoral thesis]. Sydney: University of Sydney.

Tisseyre B., Mazzoni C., Ardoin N., Clipet C., 2001. Yield and harvest quality measurement in precision viticulture – application for a selective vintage. Proceedings of the 3rd European Conference on Precision Agriculture. June 18–21, Montpellier, France. 133–138.

Urretavizcaya I, Royo J.B., Miranda C., Tisseyre B., Guillaume S., Santesteban L.G., 2017. Precision Agriculture. 18(2):133-144.

Weiss M., Baret F., 2017. Using 3D Point Clouds Derived from UAV RGB Imagery to Describe Vineyard 3D Macro - Structure. Remote Sensing, 9(2): 111.

21

DETERMINATION OF TOTAL PHENOLIC AND FLAVONOID CONTENT OF BERRY SKIN, PULP AND SEED FRACTIONS OF ÖKÜZGÖZÜ AND BOĞAZKERE GRAPE CULTIVARS

Gültekin ÖZDEMIR[1], Mihdiye PİRİNÇÇİOĞLU[2], Göksel KIZIL[2], Murat KIZIL[2]

[1]Dicle University, Faculty of Agriculture, Department of Horticulture, Diyarbakir, Turkey
[2]Dicle University, Faculty of Science, Department of Chemistry, Diyarbakir, Turkey

Corresponding author email: gozdemir@dicle.edu.tr

Abstract

Grape cultivars (Vitis vinifera L.) are believed to have health benefits due to their antioxidant activityand phenolic content. Thus, scientists have conducted research to explore their positive effects on many human diseases. The aim of this study was to determinetotal phenolic and flavonoid contents of berry pulp, seed and skin of Öküzgözü and Boğazkere red wine grape cultivars grown in Turkey. In conclusion, it was found that total phenolic (µg GAE/mg) and flavonoid content in Öküzgözü and Boğazkere grape cultivars showed importantdifferences according to the berry skin, pulp, seed and research years. The highest phenolic content was found in Öküzgözü berry pulp 803.00 µg GAE/mg in 2012 year. When the flavonoid amounts are compared, it has been determined that the total flavonoid amount varied from 5.08 µg QUE/mg to 111.55 µg QUE/mg. The highest flavonoid content was found in the Öküzgözü grape berry skin in 2011 year (111.55 µg QUE/mg). This study showed that these grapes are a potential source of phenolic and flavonoid compounds. It can be concluded that selected grape varieties and their parts can be considered a good source of phenolics.

Key words: Grape, Öküzgözü, Boğazkere, Berry, Phenolic, Flavonoid.

INTRODUCTION

Turkey is one of the top producers of grape. It has 467,093 ha of vineyards and a production of 4.1 million tons. Over 74 million tons of grapes are grown worldwide on more than 7.1 million ha. Turkey ranks fifth in terms of growing area, after Spain, France, China, and Italy, and ranks sixth in production after China, Italy, USA, Spain and France (Anonymous, 2014).

In Turkey, grapes have been mainly grown as table grapes (52%), for raisins (38%), and for fruit juice and wine (10%), with around 80 standard cultivars grafted onto mainly six standard rootstocks, in nine viticultural regions. Turkey has about 7% of the world's area of vineyards, and produces 6.4% of the world's grape production. In addition, productivity in Turkey has improved by about 40% in the last 15 years, from 6654 kg ha^{-1} in 1998 to 9249 kg ha^{-1} in 2012 (Soylemeoglu et al., 2016).

Turkey is the one of the gene center of grapevines, for this reason it possesses over 1200 grape varieties. Nearly all grape varieties grown in Turkey are european-type grapes (*Vitis vinifera* L.).

The types and concentration of the phenolic compounds depend on the grape variety, ripening, climatic conditions, wine making practices (the use of enzymes, maceration conditions, and fermentation temperature), and ageing (Kelebek et al., 2007).

Among the types of grapes, especially red grapes and grape juice, the major phenolic compounds found in red wine are called as flavonoids, anthocyanins and flavonols (Rice-Evans et al., 1996; Singleton, 1982; Palomino et al., 2000). It is reported that these substances that are important in terms of human health and found in grape(Morris and Cawthon, 1982; Bravdo et al., 1985; Matthews and Anderson, 1988; Iland, 1989; Nadal and Arola, 1995; De La Hera Orts et al., 2005; Pirinccioglu et al., 2012; Ozdemir et al., 2016) vary according to the varieties of the grapes (Landrault et al., 2001), climate and soil conditions of the place where it grows (Spayd et al., 2002; Mateus et al., 2001).

Among the climate features of the vineyard areas, the place and vector issues especially the temperature, humidity and sunlight are encountered as the important factors affecting the synthesis the of phenolic compounds.

Some authors have studied the total phenolic and flavonoid contents of different grape cultivars and ecological regions in Turkey (Deryaoglu and Canbas, 2003; Aras, 2006; Orak, 2007; Babalik et al., 2009; Baydar et al., 2009; Uluocak, 2010; Kelebek, 2009; Bayir, 2011; Cangi et al., 2011; Toprak, 2011; Kaplama, 2012; Pehlivan et al., 2015).

Öküzgözü and Boğazkere are red grape cultivars of *Vitis vinifera* L. grown in eastern Turkey, especially Elazig, Malatya, and Diyarbakir provinces. It is an important red grape variety for Turkey, which produces well-balanced and characteristics wines, with fruity notes such as strawberry, cherry, and blackberry-like odours (Cabaroglu et al., 2002; Kelebek et al., 2007)

The aim of the present study was to determine the total phenolic and flavonoid content of berry skin, pulp and seed of Öküzgözü and Boğazkere grape cultivars.

MATERIALS AND METHODS

Plant Material

This research was carried out in the Dicle University Department of Horticulture and Chemistry in 2011, 2012 and 2013 years. In the research, Öküzgözü and Boğazkere (*Vitis vinifera* L.) Turkish wine grape cultivars were used as biological material (Figures 1 and 2).

Grape varieties are grown in Elazığ (Sün Village) province.

Fresh grapes from Boğazkere and Öküzgözü cultivars were manually harvested at optimum maturity in the 2011, 2012 and 2013 vintage in Elazig province and transported to the Plant Physiology Laboratory at the Department of Horticulture, University of Dicle, located in Diyarbakir, Turkey.

Determination of total phenolic and flavonoid contents

Total phenolic and flavonoid content of the grapes obtained with different part of berries (skin, pulp, seed) (Figure 1, Figure 2) from Boğazkere and Öküzgözü grape cultivars.

Total phenolic content was determined according to Le et al. (2007); gallic acid (10–180 µg/mL) was used as standard. Samples of 40 µL of extract solution (1 mg/mL) were mixed with 200 µL Folin–Ciocalteau's phenol reagent 10% in water. After 4 min of incubation, 0.4 mL of 20% Na_2CO_3 was added. The reaction tubes were further incubated for 2 h at room temperature and the absorbance was measured at 760 nm. The concentration of total phenolic compounds in the extract was determined as µg of gallic acid equivalents per mg of extract (µg GAE/mg) (Kada et al., 2016; Ozdemir et al., 2016).

Total flavonoid content was quantified according to Bahorun et al. (1996) using quercetin (2–20 µg/mL) as standard. Briefly, samples of 1 mL of extract solution (1 mg/mL) were incubated in the presence of 1 mL of $AlCl_3$ (2%) for 10 min at room temperature. The absorbance was measured at 430 nm. Total flavonoid content was expressed as µg quercetin equivalent per mg of extract (µg QUE/mg) (Kada et al., 2016; Ozdemir et al., 2016).

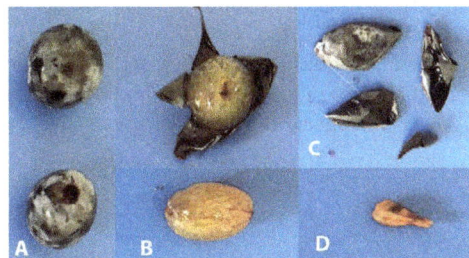

Figure 1. Öküzgözü (*Vitis vinifera* L.cv) (A) berry (B) skin, (C) pulp and (D) seed

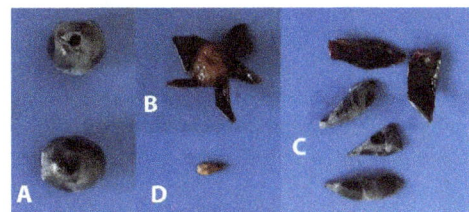

Figure 2. Boğazkere (*Vitis vinifera* L. cv) (A) berry (B) skin, (C) pulp and (D) seed

RESULTS AND DISCUSSIONS

As a result of the study, total phenolic (µg GAE/mg) content of the grape cultivars showed

differences according to the berry fractions (skin, pulp and seed) and years.
Total phenolic content varied from 85.45 µg GAE/mg to 126.70 µg GAE/mg in Öküzgözü and Boğazkere grape berry skin.
The highest values in Öküzgözü berry skin were found in 2013 year (average 89.58 µg GAE/mg).
The maximum amount of phenolic content was found in Boğazkere variety in 2012 (126.70 µg GAE/mg) (Table 1).

Table 1. Total phenolic content (µg GAE/mg) in grape berry skin

Cultivars	Total phenolic content			
	2011	2012	2013	Average
Öküzgözü	85.45	81.25	102.05	89.58
Boğazkere	100.55	126.70	107.00	111.42

Total phenolic content varied from 493.70 µg GAE/mg to 766.40 µg GAE/mg in Öküzgözü and Boğazkere grape berry pulp.
The maximum amount of phenolic content was found in Öküzgözü variety in 2013 (766.40 µg GAE/mg).
The least amount was found in the Boğazkere grape variety in 2011 (493.70 µg GAE/mg).
The average values in berry skin were found in 89.58 µg GAE/mg in Öküzgözü and 523.43 µg GAE/mg in Boğazkere variety (Table 2).

Table 2. Total phenolic content (µg GAE/mg) in grape berry pulp

Cultivars	Total phenolic content			
	2011	2012	2013	Average
Öküzgözü	704.40	803.00	766.40	757.93
Boğazkere	493.70	546.60	530.00	523.43

Total phenolic content varied from 157.60 µg GAE/mg to 340.40 µg GAE/mg in grapes berry seed.
The average values in berry seed were found in 182.75 µg GAE/mg in Öküzgözü and 329.45 µg GAE/mg in Boğazkere variety.
The highest amount of phenolic content was found in Boğazkere variety in 2012 (340.40 µg GAE/mg) (Table 3).

Table 3. Total phenolic content (µg GAE/mg) in grape berry seed

Cultivars	Total phenolic content			
	2011	2012	2013	Average
Öküzgözü	157.60	183.30	207.35	182.75
Boğazkere	327.70	340.40	320.25	329.45

The total flavonoid content found in the berry skin was detected to vary from 36.16 to 111.55 µg QUE/mg. It has been detected that the flavonoid content in the skin was found in Öküzgözü grape variety in 2011 year (111.55 µg QUE/mg) and the lowest one was found in Boğazkere variety in 2013 year (36.16 mg QUE/mg). The average values in berry skin were found to be: 108.44 µg QUE/mg in Öküzgözü and 48.35 µg QUE/mg in Boğazkere variety (Table 4).

Table 4. Total flavonoid content (µg QUE/mg) in grape berry skin

Cultivars	Total flavonoid content			
	2011	2012	2013	Average
Öküzgözü	111.55	107.01	106.77	108.44
Boğazkere	54.11	54.79	36.16	48.35

Total flavonoid content varied from 17.20 µg QUE/mg to 39.66 µg QUE/mg in Öküzgözü and Boğazkere grape berry pulp. The highest flavonoid content in the pulp was found to be 39.66 µg QUE/mg in Boğazkere grape variety in 2013 year. The lowest one was found to be 17.20 µg QUE/mg in Öküzgözü variety in 2011 year. The average flavonoid values in berry pulp were found to be: 17.32 µg QUE/mg in Öküzgözü and 29.65 µg QUE/mg in Boğazkere variety (Table 5).

Table 5. Total flavonoid content (µg QUE/mg) in grape berry pulp

Cultivars	Total flavonoid content			
	2011	2012	2013	Average
Öküzgözü	17.20	17.40	17.37	17.32
Boğazkere	24.54	24.76	39.66	29.65

Total flavonoid content varied from 5.08 µg QUE/mg to 11.23 µg QUE/mg in grape berry seed.

The highest flavonoid content in the seed was found to be 11.23 µg QUE/mg in Boğazkere grape variety in 2013 year.

The lowest one was found to be 5.08 µg QUE/mg in Öküzgözü variety in 2012 year. The average flavonoid values in berry seed were found to be: 5.14 µg QUE/mg in Öküzgözü and 9.88 µg QUE/mg in Boğazkere variety (Table 6).

Table 6. Total flavonoid content (µg QUE/mg) in grape berry seed

Cultivars	Total flavonoid content			
	2011	2012	2013	Average
Öküzgözü	5.20	5.08	5.16	5.14
Boğazkere	8.08	10.34	11.23	9.88

It has been identified in different research that total phenolic amounts vary according to the variety and year and decreased during maturation period (Doshi et al., 2006; Navarro et al., 2008; Jin et al., 2009). Saidani Tounsia et al., (2009) examined the total phenolic amount in the methanol extract of three types of red grape seeds and found the equivalent of respectively 427.00 mg/100g, 218.00 mg/100g and 112.81 mg/100g of gallic acid for dry weights of Muscat, Shiraz and Carignan varieties. A similar study was carried out by Hogan et al. (2009) and it was examined the total phenolic content of the three types of Virginia black wine grapes in various regions of northern France and Cabernet in Virginia black wine grapes in northern France as Cabernet Franc 1, Cabernet Franc 2 and Cabernet Franc 3 and, as a result, they were identified to be equivalent of respectively 1.82 ± 0.07 mg/g, 1.47 ± 0.05 mg/g, 0.63 ± 0.02 mg/g of gallic acid.

As a result of their study, Ozden and Vardin (2009) have found that the total phenolic compound concentration of some grape varieties grown in Sanliurfa conditions such as Merlot, Chardonnay, Cabernet Sauvignon and Shiraz (V. vinifera L.) grape varieties vary from 1805 mg/kg to 3170 mg/kg in terms of total phytochemical properties. While the highest concentration of phenolic compounds was found in Chardonnay variety, the lowest concentration was found in Shiraz variety.

In their study, Gokturk Baydar et al., (2011) have identified grape seeds and skin extracts belonging to Cabernet Sauvignon, Kalecik Karasi and Narince grape varieties, antioxidant properties of wine and the content of phenolic compounds. Total phenolic content was determined to vary from 522.49 to 546.50 mg GAE g^{-1} in seed extracts and from 22.73 to 43.75 mg GAE g^{-1} in skin extracts and from 217.06 to 1336.21 mg L^{-1} in wine. The radical scavenging effects of the samples and reducing capacities varied depending on grape varieties, the parts of the grape and the wine type.

Kanner et al. (1994) analyzed total phenolic compound amounts by harvesting the grapes in optimal harvest ripeness and the study was conducted with seven different table (Miabell Concord, Flame Seedless, Emperor, Thomson Seedless, Red Globe and Red Malaga) and seven different wine (Calzin Petite Shiraz, Merlot, Cabernet Sauvignon, Cabernet Franc, Sauvignon Blanc and Chardonnay) grapes. They reported that phenolic compounds in wine grapes vary from 230 to 1236 mg/l and Calzin and Petit Shiraz grape varieties have the highest phenolic content.

As a result of the analysis made in grape varieties examined in Elazig conditions, the total amount of phenolic and flavonoid content in the pulp, skin and seed of the berry were found to vary greatly among varieties and research years (Tables 1, 2, 3, 5 and 6).

CONCLUSIONS

It was found that total phenolic (µg GAE/mg) and flavonoid content in Öküzgözü and Boğazkere grape cultivars showed important differences according to the berry skin, pulp, seed and years.

In this research, the highest values in terms of the amount of phenolic content have been determined in the case of the pulp 803.00 µg GAE/mg in Öküzgözügrape variety.

The highest phenolic content was noticed in Öküzgözü berry seed being207.35 µg GAE/mg in 2013 year. In the skin of in Öküzgözü variety, in 2013 phenolic content was found to be 102.05 µg GAE/mg.

When the flavonoid amounts are compared, it has been determined that the total flavonoid amount varied from 5.08 μg QUE/mg to 111.55 μg QUE/mg. The highest flavonoid content was found in the Öküzgözü grape berry skin in 2011 year (111.55 μg QUE/mg).

Among plant-derived foods, fruits and vegetables are natural sources that are rich in phenolic substances. Today, it is clear that the increase in escaping from the artificial substances will increase the significance of the natural phenolic substances. Besides the use opportunities in the fields of food, ladder and pharmacology, it is seen that understanding the mechanism of action of phenolic substances with significant effects on human health and it is important to investigate the paths to be able to use technologically (Ozdemir et al., 2016).

ACKNOWLEDGEMENTS

The authors thank Dicle University Scientific Research Project Coordinatory for its funding of this research (Project Number: 15-ZF-14).

REFERENCES

Anonymous, 2014. Food and Agriculture Organization of United Nations Crops (FAO) Database. http://www.fao.org/faostat/

Aras O., 2006. Determination of the Total Carbohydrate, Protein, Mineral Substances and Phenolic Compounds of Grape and Grape Products. Suleyman Demirel University, Graduate School of Natural and Applied Sciences M.Sc. Thesis, 59p.

Babalik Z. Cetin S. Hallac Turk F., Gokturk Baydar N., 2009. Determination of Phenolic Compounds of Cavus Grape Cultivars under Different Training Systems. 7th Symposium on Viticulture and Technologies, 5-9 October, Turkey.

Bahorun T, Gressier B, Trotin F, Brunet C, Dine T, Luyckx M, Vasseur J, Cazin M, Cazin JC, Pinkas M. 1996. Oxygen species scavenging activity of phenolic extracts from hawthorn fresh plant organs and pharmaceutical preparations. Arzneimittelforschung. 46:1086–1089.

Baydar G. N., Ozkan G., Yasar S., 2007. Evaluation of the Antiradical and Antioxidant Potential of Grape Extracts. Food Control, 18, 1131-1136.

Bayir A., 2011. Investigation of Phenoloic Content and Antiradical Activity of Grape, Mulberry and Myrtle. Akdeniz University, Graduate School of Natural and Applied Sciences PhD Thesis, 147p.

Bravdo B.A., Hepner Y., Loigner C., Cohen S., Tabacman H., 1985. Effect of irrigation and crop level on growth, yield, and wine quality of Cabernet

Sauvignon. American Journal of Enology and Viticulture. 36: 132-139.

Cabaroglu T., Canbas A., Lepoutre J.P., Gunata Z., 2002. Free and bound volatile composition of red wines Vitis vinifere L. cv. Öküzgözü and Boğazkere grown in Turkey. America Journal of Enology and Viticulture 53:64-68.

Cangi R., Saracoglu O., Uluocak E., Kilic, D., Sen A., 2011. The Chemical Changes of Some Wine Grape Varieties During Ripening Period in Kazova (Tokat) Ecology. Igdir Univ. J. Inst. Sci. & Tech. 1(3): 9-14

De La Hera Orts M.L., Martinez-Cutillas A., Opez-Roca J.M., Gomez-Plaza E., 2005. Effect of moderate irrigation on grape composition during riperning. Spanish Journal of Agricultural Research. 3: 352-361.

Deryaoglu A., Canbas A., 2003. Physical and Chemical Changes Occured During Maturation of Okuzgozu Grape Variety Grown in Elazig Region. The Journal of Food. 28(2):131-140.

Doshi P., Adsule P., Banerjee K., 2006. Phenolic Composition and Antioxidant Activity In Grapevine Parts and Berries (Vitis vinifera L.) cv. Kishmish Chornyi (Sharad Seedless) During Maturation. International Journal of Food Science and Technology, 41 (Supplement 1), 1–9.

Gokturk Baydar N., Babalik Z., Hallac Turk F., Cetin E.S., 2011. Phenolic Composition and Antioxidant Activities of Wines and Extracts of Some Grape Varieties Grown in Turkey. Journal of Agricultural Sciences, 17(2011), 67-76.

Hogan S., Zhang L., Li J., Zoecklein B., Zhou K., 2009. Antioxidant properties and bioactive components of Norton (Vitis aestivalis). Food Science And Technology, 42, 1269-1274.

Iland P., 1989. Grape berry composition-the influence of environmental and viticultural factors, Australian Grape grower & Winemaker. 302: 13-15.

Jin Z.M., He J.J., Bi H.Q., Cui X.Y., Duan C.Q., 2009. Phenolic Compound Profiles in Berry Skins From Nine Red Wine Grape Cultivars in Northwest China. Molecules, 14(12), 4922-4935.

Kada S., Bouriche H., Senator A., Demirtaş İ., Özen T., Toptanci B.Ç., Kızıl G., Kızıl M., 2016. Protective activity of Hertia cheirifolia extracts against DNA damage, lipid peroxidation and protein oxidation. Pharmaceutical Biology, 55(1):330-337.

Kanner J., Frankel E., Granit R., German B., Kinsella J.E., 1994. Natural Antioxidants in Grape and Wines. Ibid. 42, 64-69.

Kaplama P., 2012. Antioxidant Activities Anthocyanin Profiles and some Physical and Chemical Properties of Grape Cultivars Grown in Erzincan. Ataturk University, Graduate School of Natural and Applied Sciences M.Sc. Thesis, 87p.

Kelebek H., 2009. Researches on the Phenolic Compounds Profile of Okuzgozu, Bogazkere and Kalecik Karasi Cultivars Grown in Different Regions and Their Wines. Cukurova University, Graduate School of Natural and Applied Sciences PhD Thesis, 251p.

Kelebek H.K., Canbas A., Cabaroglu T., Selli S., 2007. Improvement of antocyanin content in the cv

Öküzgözü wines by using pectolytic enzymes. Food Chemistry 105(1):334-339

Landrault N, Poucheret P, Ravel P, Gasc, Cros G, Teissedre PL., 2001. Antioxidant capacities and phenolic l evels of French wines from different varieties and vintages. J. Agr. Food Chem, 49: 3341-3348.

Le K, Chiu F, Ng K., 2007. Identification and quantification of antioxidants in Fructus lycii. Food Chem. 105:353–363.

Mateus N., Proença S., Ribeiro P., Machado J.M., De Freitas V., 2001. Grape and wine polyphenolic composition of red Vitis vinifera varieties concerning vineyard altitude, Cienc. Technol., Aliment., 3(2): pp.102-110.

Matthews M., Anderson, M., 1988. Fruit ripening in Vitis vinifera L responses to seasonal water deficits, American Journal of Enology and Viticulture, 39:313-320.

Morris J.R., Cawthon D.L., 1982. Effect of irrigation, fruit load, and potassium fertilization on yield, quality, and petiole analysis of concord (Vitis vinifera L) grapes, American Journal of Enology and Viticulture, 33: 145-148.

Nadal M., Arola L., 1995. Effects of limited irrigation on the composition of must and wine of Cabernet Sauvignon under semi-arid conditions, Vitis, 34: 151-154.

Navarro S., Leo´N M., Roca-Pe´Rez L., Boluda R., Garcı´A-Ferriz L., Pe´Rez-Bermu´Dez P., Gavidia I., 2008. Characterisation Of Bobal And Crujidera Grape Cultivars, In Comparison With Tempranillo And Cabernet Sauvignon: Evolution Of Leaf Macronutrients And Berry Composition During Grape Ripening Food Chemistry 108, 182–190.

Orak H.H., 2007. Total Antioxidant Activities, Phenolics, Anthocyanins, Polyphenoloxidase Activities of Selected Red Grape Cultivars and Their Correlations Scientia Hort. Vol. 111, Issue 3, 5 February 2007, Pages 235-241.

Ozdemir G., Beren Sogut A., Pirinccioglu M., Kizil G., Kizil M., 2016. Changes in The Phytochemical Components in Wine Grape Varieties During The Ripening Period. Scientific Papers. Series B, Horticulture. Vol. LX, 85-93.

Ozden M., Vardin H., 2009. Quality and Phytochemical Properties of some Grapevine Cultivars Grown in Sanliurfa Conditions. Harran Journal of Agriculture and Food Science 13(2): 21-27.

Palomino O., Gomez-Serranillos M.P., Slowing K., Carretero E., Villar A., 2000. Studyof polyphenols in grape berries by reversed-phase highperfor mance liquid chromatography. Journal of Chromatography, A. 870: 449-451.

Pehlivan E.C., Uzun H.İ., 2015. Effects of Cluster Thinning on Yield and Quality Characteristics in Shiraz Grape Cultivar. Yuzuncu Yil University Journal of Agricultural Sciences 25 (2):119-126.

Pirinccioglu M., Kızıl G., M Kızıl., Ozdemir G., Kanay Z., Ketani M.A., 2012. Protective effect of öküzgözü grape (Vitis vinifera L.) juice against carbon tetrachloride induced oxidative stress in rats. Journal of Food Funct., 3, 668-673.

Rice-Evans C.A., Miller N.J., Paganda G., 1996. Structure antioxidant activity relationship of flavonoids and phenolic acids, Free Radical Biology & Medicine, 20: 933-956.

Saidani Tounsia, M., Ouerghemmi, I., Wannes, W.A., 2009. Ksouri, H.Z.; Marzouk, B.; Kchouk, M.E. Industrial Crops and Products, 30, 292–296.

Singleton V.L., 1982. Grape and wine phenolics: background and prospects. In: Webb, A.D. (ed.), Proceedings of Grape Wine Centennial Symposium University of California, Davis 215-222.

Soylemezoglu G., Atak A., Boz Y., Unal A., Saglam M., 2016. Viticulture in Turkey. Chronica Horticulturae 56 (2): 27-31.

Spayd S.E., Tarara J.M., Mee D.L., 2002. Ferguson, J.C., Separation of sunlight and temperature effects on the composition of Vitis vinifera cv. Merlot berries. Am. J. Enol. Vitic., 53(3): pp.171-182.

Toprak F.E., 2011. Phytochemical Characteristics in Kalecik Karasi Grape Cultivar (Vitis vinifera L.) Grown in Ankara and Nevsehir. Ankara University, Graduate School of Natural and Applied Sciences M.Sc. Thesis, 64p.

Uluocak E., 2010. The physical and chemical changes during ripening period of some wine grape varieties grown in Kazova (Tokat) Ecology. Gaziosmanpasa University, Graduate School of Natural and Applied Sciences M.Sc. Thesis, 89p.

22

THE EFFECTS OF DROUGHT ON THE LEVEL OF ISOFORMS OF AQUAPORIN IN CV. 'HOROZKARASI' GRAPEVINE

Mehmet KOÇ[1], İbrahim Samet GÖKÇEN[1],
Mehmet İlhan ODABAŞIOĞLU[2], Kenan YILDIZ[3]

[1]Aralık University, Faculty of Agriculture, Department of Horticulture,
Kilis 7, Turkey
[2]Harran University, Faculty of Agriculture, Department of Horticulture, Şanlıurfa, Turkey
[3]Gaziosmanpaşa University, Faculty of Agriculture, Department
of Horticulture, Tokat, Turkey
Corresponding author email: mk_mehmetkoc@outlook.com

Abstract

This study aimed to investigate the aquaporin expression of 'Horozkarası' grapevine. Therefore, own-rooted cv. 'Horozkarası' grapevines were exposed to two different irrigation treatments; well-irrigated and water-stress treatments under controlled environmental conditions in pots for six days. And then in the stressed plants lipid peroxidation (MDA), relative water content (RWC) and membrane permeability (MP) were measured in leaves. The expression patterns of different PIPs group of aquaporins (PIP2-1, PIP2-2) also were performed with root and leaf tissues. While significant decreases were observed in relative water content (RWC) and membrane permeability (MP) in stress treatment, increase was observed in lipid peroxidation (MDA) in leaves. Also, while significant decreases were observed in PIP2-1 and PIP2-2 in stress treatment in leaves, increases were observed in expressions of PIP2-1 and PIP2-2 genes in drought treatment in root.

Key words: Vitis, Aquaporins, water stress, MDA.

INTRODUCTION

'Horozkarası' grapevine has significant economic importance, and is grown in Southeastern Anatolia, especially in Gaziantep and Kilis, for using in vine and dried grape production. Because of the increase in demand on fruits having high antioxidant content that is important for human health, the demand on 'Horozkarası' grapevine that is widely consumed in dried form started to increase day by day.

Its long, elliptic, and significantly large grapes make this variety to be consumed in dried form, as well as it is also consumed freshly. 'Horozkarası' grapevine grape, which is very common especially in Gaziantep and Kilis region, is grown in almost all of the cities in GAP region (Gürsöz, 1993).

This variety is also known as Kilis Karası and Antep Karası among the producers and, among 35-40 sorts of grapes grown in Gaziantep province, it comes to the forefront with its area and amount of production.

Grapevine is a cultivation plant that has high level of adaptability to inappropriate soil conditions. Except for a narrow line in eastern Black Sea region, all of the regions of Turkey are within the arid and semi-arid climate belt. Moreover, as in entire world, the reflections of global warming also increase in Turkey. Besides the global warming, especially the decrease in usable water sources, the increase in arid and semi-arid agricultural lands, and the significant increases in duration and severity of drought have significantly stressful effect on the cultivated plants being grown. As well as it decreases the yield and quality, it might also result in increase in product losses and even in death of plants. On the other hand, it is another important research topic if the actual agricultural lands, on which the agricultural activities are performed, will be suitable for agricultural production in future. Some of the studies on this subject indicate that the patterns in agricultural production might change in future due to the climate change. It is inevitable that such a climate change will directly affect the viniculture in future. Today's vineyards will be unsuitable for vinicultural activities under the effects of increasing temperature and limited water sources. Scientists make effort for developing solution suggestions for this problem; there are 2 solution suggestions. First of them is to obtain new hybrids by crossing actual varieties (*V. vinifera* L.) with those having high drought tolerance to be used in future. The second proposal is to evaluate the new microclimate regions to occur in the regions close to the poles due to the increasing temperatures. In order both to protect the actual limited water sources we have and to obtain high crop yield and quality by using relatively lower level of water use, it is necessary to determine the drought tolerance of existing grape varieties. Thus, in near future, it would be relatively easier to obtain varieties to be grown using less water and to have quality suitable for consumption and market requirements by utilizing varieties with high tolerance to drought. Even if it is attempted to increase the efficiency of water use through the technological advancements regarding the irrigation systems and to protect the water sources, these precautions are limited, short-dated, and very expensive. For this reason, it is very important to select the grapevine varieties having high drought-tolerance and to examine the effect of their interaction with rootstock on the drought tolerance.

The leading one among the most important abiotic factors affecting the quality and yield in agricultural production is the drought. The grape gives physiological, biochemical, and molecular responses to the drought. The chemical signals coming from the roots play important role in the adaptation especially in first phase of water stress (Schachman, 2008). The signals are conveyed through the xylem to the leaves, and play role in arranging the water losses. Under the conditions of drought, many chemical signals are transmitted from the roots to the leaves. Some of these chemical signals, abscisic acid (ABA), pH, cytokine, malate, and precursor compounds of ethylene play role in the use of water during the first phase of drought stress (Schachtman, 2008). ABA was reported to act as stress hormone under the environmental conditions such as drought and salt stress (Peleg, 2011; Fukaki 2009). But,

regarding the importance of ABA for the root signals, there are debates among the studies due to the methodological differences (Schachman, 2008). It was reported that, during the drought stress, the pH changes in xylem core play role as chemical signal (Wilkinson, 1999). This change affects ABA metabolism or directly the water status of leaves. pH change in xylem causes the close of stoma by activating the ABA that is the cell protector (Wilkinson, 1999). In studies on the chlorophyll content of drought stress in grapes, when compared to irrigated plants, the increases were reported in the total chlorophyll content of unirrigated grapes by Maroco et al. (2000), while Flexas et al. (2000) reported remarkable decrease and Chaumont et al. (1994) found no change in total chlorophyll content.

In drought stress tolerance of various varieties, a close relationship was found between the antioxidant system and the decrease in oxidative damage (Zhang and Kirkham, 1996; Jiang and Zhang, 2002; Lima et al., 2002; Ramachandra et al., 2004; Sofo et al., 2005; Sanchez-Diaz et al., 2007; Aganchich et al., 2009; Ozkur et al., 2009; Wang et al., 2009). In their study on the antioxidative mechanism in drought adaptation of two varieties grown on their roots (Sabatiano and Mavrodafni), Alexandros and Angelos (2012) reported a rapid increase in hydrogen peroxide concentration of Mavrodafni variety under droughty conditions. The researchers emphasized that, while there was no significant change in CAT activity under drought stress, there was difference in APX and SOD activities.

The transportation of water in xylem in angiosperms is known to be affected from the anatomical characteristics such as vein size, distribution and intensity, core structure, vein permeability, and topology of xylem network. Besides them, the chemical signals (ABA) significantly affect the water transportation in plants in expression of aquaporin. Lovisolo et al. (2008) emphasized that ABA has important role in transpirational control in rehydrated grapes, and reported that the xylem embolism and abscisic acid hormone increased in stressed plants. In addition, they showed that aquaporin has important role in regulation of the leaf and root hydraulics. Aquaporin is a water channel protein that exists in various physiological processes among the organisms. In grapes, aquaporins play important role in drought adaptation of plants by maintaining their ion and water balance under varying environmental conditions. In studies on different plant varieties, among the plant that have not been exposed to drought stress, PIP1;1 aquaporin gene was expressed in roots at higher levels when compared to the expression in leaves (Galmes et al., 2007; Weig et al., 1997; Jang et al., 2004). It is known that, aquaporin plays important role in adaptation of grapes to drought due to varying environmental conditions. Aquaporins play important role in arranging the hydraulics in roots (Vandeleur et al., 2009) and leaves (Pou et al., 2013) of grapevines. Aquaporins play role in continuous root-to-leaf water transportation, and they might lead to rapid and inverse changes in cell's hydraulic conductivity by arranging the water permeability of membrane (Hayes et al., 2007; Surbanovski and Grand, 2014). Aquaporins play role in arranging the water movement throughout the plasma membranes in the metabolic path between the cells and in correcting the xylem embolization (Lovisolo and Schubert, 2006). Zarrouk et al. (2015) reported that, in their study on drought in cv. Toriga Nacional grape variety, the root-shoot signals responded by increasing the root hydraulic conductivity in mid-level water stress by being encouraged by the chemicals. In addition, they emphasized that the aquaporin isoforms played role as major-sub organizer and in sensing the water stress since the first phase of water stress.

MATERIALS AND METHODS

Growing the Plants, Implementation, and Taking the Samples

This study was carried out in greenhouses within the body of Kilis 7 Aralık University's Agricultural Implementation and Research Center (TUAM) and in Agricultural Engineering Faculty Laboratory of the university. The slips used in this study were collected from the vineyards in this region. The materials collected were cut into the suitable size for rooting, and then planted into rooting cases by sinking into 2% fungicide solution. The rooted healthy plants were planted into 5 L (1:1 v:v) peat/perlite mixture. In order to endure the homogeneity, they were grown on a single body with cut branches. As of the month of August, the plants were irrigated on regular basis, and then they were divided into control and drought groups. This study was carried out in experimental pattern of fully randomized coincidence parcels, in triplicated in accordance with factorial order (10 healthy saplings planted in pots in each repeat). Drought regime was implemented for 6 days. At the end of 6th day, the root and leaf samples were immediately treated with liquid nitrogen for molecular analyses, and kept at 80°C until the analyses.

Relative Moisture Content (RMC)

The leaf samples collected a little while before the harvest were immediately weighed and the wet weight (WW) was determined. By keeping the samples in pure water for 4 hours, they were transformed into turgor, and then weighed again (TW). And, finally, the leaf samples were dried in drying cabinet at 60°C for 24 hours, and dry weight (DW) was determined (Dhanda and Sethi (1998). Using the formula below, the relative moisture content was calculated.

RMC (%) = [(WW-DW)/(TW-DW)]x100

Membrane Permeability (MP)

The leaf samples taken before the harvest (0.1 g) were rinsed firstly with tap water and then with pure water. The plant samples were kept in 10 ml pure water at 40°C for 30 minutes, and EC (C1) of solution was measured (C1). EC was measured again for the sample that was kept in water bath at 100°C for 10 minutes (C2), and MSI was calculated using the formula below (Premchandra et al., 1990 and Sairam, 1994).

MP= [1-(C1/C2)] x 100

Determining the Lipid Peroxidation (MDA)

In order to show the stress effect on plant and to compare with the levels of gene expression, malondialdehyde (MDA) analysis was employed in determining the lipide peroxidation (Hodges et al., 1999). In calculating the values read on spectrophotometer and percentage MDA levels, the formulas below were used.

ABS=Absorbance

MDA=Malondialdehyde

1- ((ABS 532+TBA)-(ABS 600+TBA)-(ABS 532-TBA)-(ABS 600-TBA)= A

2- ((ABS 440+TBA)-(ABS 600+TBA)x0,0971=B

3- nmolMDA/ml= (A-B/157000)x106

RNA Isolation and qRT-PCR Analysis

All of the plant samples taken after the stress treatments were kept in refrigerator at -80°C until the RNA isolation procedure. For each of stress conditions, 3 biological replications were employed. For RNA isolation from the plant samples, the protocol of manufacturer company was followed (Total RNA Extraction Kit; Vivantis Malaysia). RNA concentration and amount were determined using the Spectrophotometer (Thermo Multiskan™ GO Microplate) with nano-drop feature. Moreover, RNA samples were swiped into 1% agarose gel. cDNA synthesis was determined using M-MuLV Reverse Transcriptase RNase H- (Vivantis, Malaysia) in accordance with the guidelines provided by the manufacturer. For real-time implementations, Thermo SYBR Green Master Mix and Real-Time PCR (LightCycler® Nano Roche; Mannheim, Germany) were used, and the reaction was performed in following order; at 95°C for 10 min. and then at 95°C for 15 s, 55°C (binding temperature varies depending on the primer) for 30 s and 72°C for 30 s for 40 cycles. Melting curve was obtained by heating the amplicon from 55°C to 95°C. Transcript abundance of PIP2.1 and PIP2.2 was analyzed using specific primers (Baiges et al. 2001). In order to determine the change in expression level, *VvActin* reference gene was used.

Statistical Analysis

The significance of differences was analyzed sing "Independent t-Test" by using JMP 13 software.

RESULTS AND DISCUSSIONS

Relative Moisture Content (RMC)

The effects on relative moisture content are seen in Figure 1 for 'Horozkarası' variety. Accordingly, it was determined that the treatments have statistically significant effect. When compared to control group, the relative moisture content of plants grown in drought treatment was found to significantly decrease ($p: < 0.05$).

In Figure 1, the change in RMC by drought is presented.

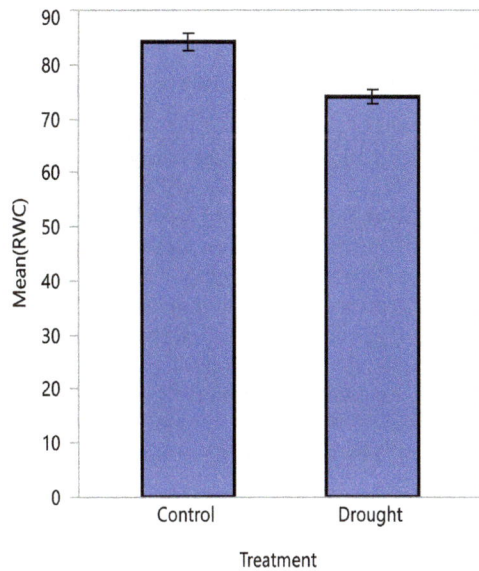

Figure 1. The effects on relative moisture content of 'Horozkarası' variety

Membrane Permeability (MP)

In Figure 2, the effects on membrane permeability for 'Horozkarası' grape variety can be observed.

Accordingly, the treatments were found to have significant effect.

When compared to control, 1% change in membrane permeability of plant grown in drought treatment was found to be statistically significant ($p: < 0.05$).

In figure 2, the change in membrane permeability (%) by drought is presented.

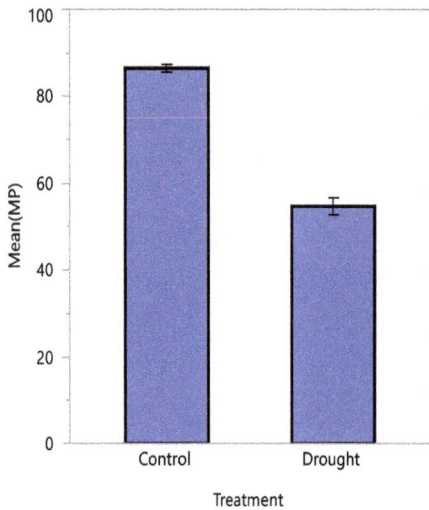

Figure 2. The effects on membrane permeability for 'Horozkarası' grape variety

Lipid Peroxidation (MDA) Level

In Figure 3, the effects on lipid peroxidation level for 'Horozkarası' grape variety can be seen. Accordingly, the treatments were found to have significant effect. It was observed that the increase in MDA level of plant grown in drought treatment was found to be statistically significant ($p:<0.05$). The change in MDA level (nmol ml^{-1}) by drought is presented.

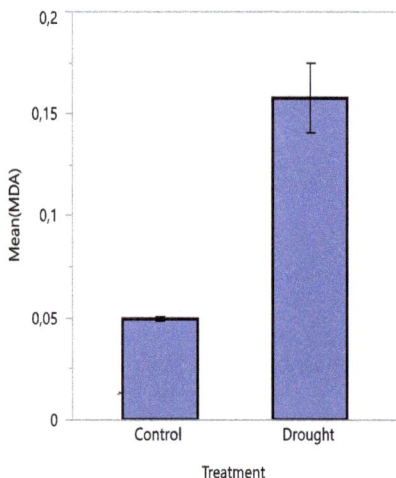

Figure 3. The effects on lipid peroxidation level for 'Horozkarası' grape variety

Aquaporin gene expression

PIP2-1 and PIP2-2 expression levels were examined in root and leaf samples taken from 'Horozkarası' grape variety exposed to stress. While significant decreases were observed in PIP2-1 and PIP2-2 in stress treatment in leaves, increases were observed in expressions of PIP2-1 and PIP2-2 genes in drought treatment in root ($p:<0.05$) (Figure 4). The role of AQPs in regulating plant's water status under water stress is a complex issue, because the expression of different AQP genes may be stimulated, decreased, or unchanged under abiotic stress (Yamaguchi et al., 1992; Maurel 1997; Kirch et al., 2000; Kawasaki et al., 2001). In grapevine, highly vigorous rootstocks have higher fine-root hydraulic conductivity partly due to the higher aquaporin expression and activity (Gambetta et al., 2012). AQPs in plants often show a tissue/organ-specific expression (Tyerman, Niemietz & Bramley, 2002). Chemical signaling such as ABA and hydraulic signaling via aquaporins regulate the stomatal conductance. Relative abundance of transcripts in roots and leaves strongly depended on which AQP and the treatment given (Galmés et al., 2007). In non-stressed plants, PIP1.1 was more abundantly expressed in the roots compared to leaves, consistent with the observations described for other PIPs in the roots of other varieties (Weig et al., 1997; Jang et al., 2004). Plants exposed to short-term water stress showed an enhanced ratio of root-to-leaf AQP expression, particularly for the moderate stress treatment. In study of Aroca et al. (2012), it was emphasized that plasma membrane intrinsic protein (PIP) sub-group played important role in absorbing water from the soil. It was found that own-rooted grapevine cultivars that differ in their response to soil water deficits via differences in the regulation of the leaf water potential also vary in their root response to water soil deficits in terms of aquaporin expression (Vandeleur et al., 2009). In parallel with the results obtained in present study, it was observed that the expression of aquaporin genes in the leaves decreased to limit water loss via transpiration, whereas the expression of the same aquaporin genes increased in the roots to enhance water uptake to avoid plant water constraints when water deficits occurred.

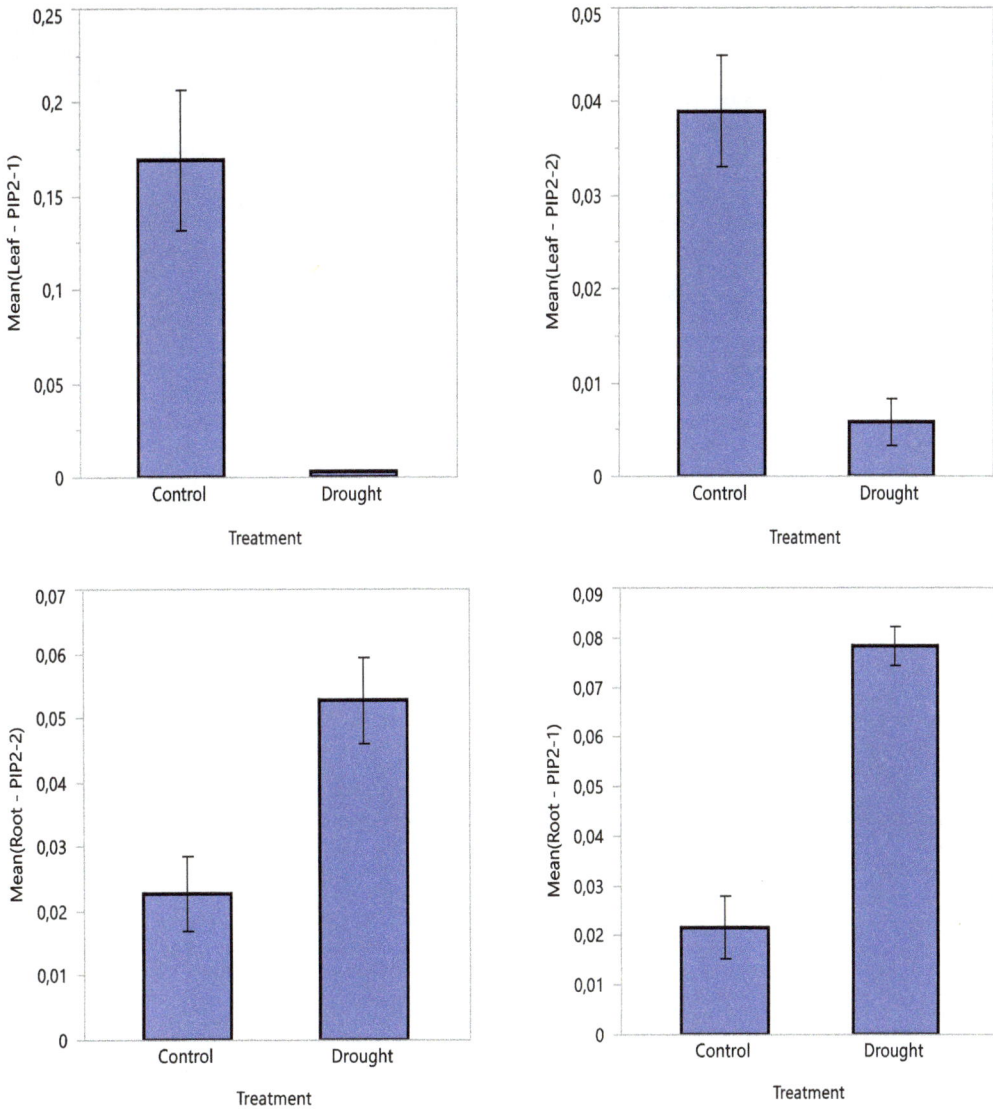

Figure 4. PIP2-1 and PIP2-2 expression levels in root and leaf samples taken from 'Horozkarası' grape variety

CONCLUSIONS

In this study, it was determined that the 'Horozkarası' grape variety, which has successfully adapted to low annual precipitation levels and arid summer conditions of Southeastern region of Turkey, gave physiological, biochemical and gene-level responses to drought and well-adapted to semi-arid regions. For this reason, it is believed to have potential to contribute to the breeding studies on developing varieties that have high tolerance to the drought.

REFERENCES

Aganchich B., Wahbi S., Loreto F., Centrito M., 2009. Partial root zine drying: regulation of phoyosynthetic limitations and antioxidant enzymatic activities in young olive (*Olea europaea*) saplings. Tree Physiol. 29, 685–696.

Alexandros Beis, Angelos Patakas, 2012. Relative contribution of photoprotection and anti oxidative mechanisms to differential droguth adaptaion ability in grapevines; Environmental and Experimenatal Botany 78, 173 – 183.

Chaumont M, Morot-Gaudry J-F, Foyer CH., 1994. Seasonal and diurnal changes in photosynthesis and carbon partitioning in *Vitis vinifera* leaves in vines with and without fruit. Journal of Experimental Botany 45, 1235–1243. doi:10.1093/jxb/45.9.1235.

Flexas J, Briantais J-M, Cerovic Z, Medrano H, Moya I., 2000. Steady-state and maximum chlorophyll fluorescence responses to water stress in grapevine leaves: a new remote sensing system. Remote Sensing of Environment 73, 283–297. doi:10.1016/S0034-4257(00)00104-8

Fukaki H, Tasaka M., 2009. Hormone interactions during lateral root formation. J. Plant Mol. Biol. 69: 437-449

Galmés J., Pou A., Alsina M. M., Tomàs M., Medrano H., Flexas J., 2007. Aquaporin expression in response to different water stress intensities and recovery in Richter-110 (*Vitis* sp.): relationship with ecophysiological status. Planta, 226(3); 671–681.

Gürsöz S., 1993. GAP Alanına Giren Güneydoğu Anadolu Bölgesi Bağcılığı Ve Özellikle Şanlıurfa İlinde Yetiştirilen Üzüm Çeşitlerinin Ampelografik Nitelikleri İle Verim Ve Kalite Unsurlarının Belirlenmesi Üzerinde Bir Araştırma. Türkiye 2. Ulusal Bahçe Bitkileri Kongresi, Cilt 2., ADANA.

Hayes M.A., Davies C., Dry I.B., 2007. Isolation, functional characterization andexpression analysis of grapevine (*Vitis vinifera* L.) hexose transporters:differential roles in sink and source tissues. J. Exp. Bot. 58, 1985–1997

Jang JK, Kim DG, Kim YO, Kim JS, Kang H., 2004. An expression analysis of a gene family encoding plasma membrane aquaporins in response to abiotic stresses in Arabidopsis thaliana. Plant Mol Biol 54:713–725.

Jiang. M., Zhang J., 2002. Water stress-induced abscisic acid accumulation triggers the increased generation of reactive oxygen species and up-regulates the activities of antioxidant enzymes in maize leaves. J. Exp. Bot. 53, 2401–2410.

Lima A.L.S., DaMatta F.M., Pinheiro H.A., Totola M.R., Loureiro M.E., 2002. Photochemical responses and oxidative stress in two clones of Coffea canephora under water deficit conditions. Environ. Exp. Bot. 47, 239–247.

Lovisolo C, Schubert A., 2006. Mercury hinders recovery of shoot hydraulic conductivity during grapevine rehydration: evidence from a whole-plant approach. New Phytol 172:469-478.

Lovisolo C, Tramontini S, Flexas J, Schubert A., 2008. Mercurial inhibition of root hydraulic conductance in *Vitis* spp. rootstocks under water stress. Environ Exp Bot 63: 178–182.

Maroco JP, Rodríguez ML, Lopes C, Chaves MM., 2002. Limitations to leaf photosynthesis in field-grown grapevine under drought – metabolic and modelling approaches. Functional Plant Biology 29, 451–459. doi:10.1071/PP01040

Ozkur O., Ozdemir F., Bor M., Turkan I., 2009. Physiochemical and antioxidant responses of the perennial xerophyte Capparis ovata Desf. to drought. Environ. Exp. Bot. 66, 487–492.

Peleg Z, Blumwald E., 2011. Hormone balance and abiotic stress tolerance in crop plants. J. Current Opinion Plant Biol. 14:290-295.

Pongrácz DP., 1983. Rootstocks for grapevines. David Philip Publisher, Cape Town, South Africa.

Pou A., Medrano H., Flexas J., Tyerman S.D., 2013. A putative role for TIP and PIP aquaporins in dynamics of leaf hydraulic and stomatal conductance ingrapevine under water stress and re-watering. Plant Cell Environ. 36, 828–843.

Ramachandra R.A., Chaitanya K.V., Jutur P.P., Sumithra K., 2004. Differential antioxidative responses to water stress among five mulberry (*Morus alba* L.) cultivars. Environ. Exp. Bot. 52, 33–42.

Sanchez-Diaz M., Tapia C., Antolin M.C., 2007. Drought-induced oxidative stress in different Canarian laurel forest tree species growing under controlled conditions. Tree Physiol. 27, 1415–1422.

Schachtman DP, Goodger JQ., 2008. Chemical root to shoot signalling under drought. Trends Plant Sci. 13: 281-287.

Schachtman DP, Goodger JQ., 2008. Chemical root to shoot signalling under drought. Trends Plant Sci. 13: 281-287.

Sofo A., Dichio B., Xiloyannis C., Masia A., 2005. Antioxidant defences in olive trees during drought stress: changes in activity of some antioxidant enzymes. Funct. Plant Biol. 32, 45–53.

Surbanovski N., Grant O.M., 2014. The emerging role of aquaporins in plants'tolerance of abiotic stress. In: Paraviz Ahmad, P., Rasool, S. (Eds.), Emerging Technologies and Management of Crop Stress Tolerance, Volume 2- A Sustainable Approach. Elsevier Inc., 431–447.

Vandeleur R.K., Mayo G., Shelden M.C., Gilliham M., Kaiser B.N., Tyerman S.D., 2009. The role of plasma membrane intrinsic protein aquaporins in watertransport through roots: diurnal and drought stress responses reveal differentstrategies between isohydric and anisohydric cultivars in grapevine. PlantPhysiol. 149, 445–460.

Wang W.B., Kim Y.H., Lee H.S., Kim K.Y., Deng X.P., Kwak S.S., 2009. Analysis of antioxidant enzyme activity during germination of alfalfa under salt and drought stresses. Plant Physiol. Biochem. 47, 570–577

Weig A, Deswarte C, Chrispeels MJ., 1997. The major intrinsic protein fanily of Arabidopsis has 23 members that form three distinct groups with functional aquaporins in each group. Plant Physiol 114:550–555.

Wilkinson S., 1999). pH as a stress signal. Plant Growth Regulation. 29:87-99.

Zarrouk O., Garcia-Tejero I., Pinto C., Genebra T., Sabir F., Prista C., Chave M.M., 2015. Aquaporins isoforms in cv. Touriga Nacional grapevine under water stress and recovery—Regulation of expression in leaves and roots. Agricultural Water Management. http://doi.org/10.1016/j.agwat.2015.08.013.

Zhang J., Kirkham M.B., 1996. Antioxidant responses to drought in sunflower and sorghum seedlings. New Phytol. 132,361–373.

23

THE USE OF A GC-ELECTRONIC NOSE FOR THE SELECTION OF A WINEMAKING PROTOCOL LEADING TO AN ENHANCED VOLATILE PROFILE IN WINES FROM AROMATIC GRAPE VARIETIES

Arina Oana ANTOCE, George Adrian COJOCARU

University of Agronomic Sciences and Veterinary Medicine of Bucharest,
Faculty of Horticulture, Department of Bioengineering of Horti-Viticultural Systems,
59 Mărăşti Blvd., District 1, 011464 Bucharest, Romania
Corresponding author emails: arina.antoce@horticultura-bucuresti.ro;_aantoce@yahoo.com

Abstract

The multiple oenological materials used in winemaking protocols have various influences on the final aromatic profile of wine. Therefore, their overall effect on a certain variety is difficult to determine, even when some of the materials are tested separately. To simplify the decision, if the winemaker would simply like to enhance the number or the concentration of the compounds forming the volatile profile of a wine, we propose in this paper a method which uses an electronic nose based on flash gas chromatographic technique to test the effect of a certain winemaking protocol. The test was made on two aromatic Romanian autochthonous varieties, Busuioaca de Bohotin and Tamâioasa românească, each vinified with 3 different winemaking protocols. The volatile profiles for each winemaking protocol and variety were recorded and compared by using multivariate statistical analysis, in order to pair the variety and protocol which can generate a more intense volatile profile.

Key words: electronic nose, volatile profile, aromatic wine, Busuioaca de Bohotin, Tamâioasa românească.

INTRODUCTION

The aromatic profile of a wine is a trait on which many consumers base their choice. For this reason many researchers have taken into account the possibilities to modify the volatile profile of a wine by using various winemaking protocols.

The main influences on the aromatic profile of the wines are imposed by the grape variety (Rocha et al., 2010). However, even for the same grape variety, the styles of wines possible to obtain vary, in accordance with the substances extracted from the skins by maceration (González-Pombo et al., 2014; Lao et al., 1997) and also by the volatile compounds released by enzymes or yeasts (Palmeri and Spagna, 2007; Cabaroglu et al., 2003) from heavier molecules, called aroma precursors, which are usually glycosides of these aroma compounds. Both advanced extraction and aroma release from precursors are achieved by treatments with specific enzymes (Piñeiro et al., 2006) or presence of specific yeasts (Loscos et al., 2007; Hernandez-Orte et al., 2008).

Other influences on aroma profile of a wine are induced by fermentation, when the secondary aroma of the wine is formed (Sumby et al., 2010). Here too we have several influences. The fermentation aroma is mainly determined by the yeast used for the winemaking process (Ubeda Iranzo et al., 2000; Swiegers and Pretorius, 2005; Swiegers et al., 2009; Samoticha et al., 2017), but raw material itself (Ghaste et al., 2015) and the fermentation activator (Marks et al., 2003; Barbosa et al., 2009; Ugliano et al., 2009) may also induce perceivable differences.

When producing wines, even from the same variety, numerous combinations of enzymes, activators and yeast are possible, all being included in what we call a winemaking protocol, and all having various influences in accordance to the grape variety and performed treatments (Piñeiro et al., 2006).

When applying a winemaking protocol, it is difficult to predict how the several oenological material and treatments entailed in this protocol are going to influence the aroma profile. Even if each oenological material is tested separately, it is not certain that the winemaker is able to predict the final result when combining the materials into a certain protocol.

Thus, to simplify the decision, if the winemaker would simply like to enhance the number or concentration of the components of the volatile profile of a wine, it is possible to test the effect of a certain winemaking protocol by the use of an electronic nose based on flash gas chromatographic technique. In this way, the testing of separate oenological materials which are part of a winemaking protocol is not anymore necessary, this method of evaluating only the final profile saving time and effort.

MATERIALS AND METHODS

Wines of aromatic grape varieties Tamâioasa românească and Busuioaca de Bohotin were produced at industrial scale in volumes of 300 hl. Each variety was vinified by using 3 different protocols (technological schemes), making use of specific enzyme treatments and specific yeast and fermentation activators. As the winemaking protocols are based on commercial products, in order to avoid conflict of interests, the brand names and producers are not disclosed in this paper.

The wine samples prepared are generically described in Table 1.

The volatile profile of each wine variant was determined by a flash gas chromatograph with two short different polarity columns, working on the principle of the electronic nose.

The apparatus from the Alpha-MOS, France, is fitted with a DB-5 (non-polar) and a BD-1701 (slightly polar) 2 m columns and for the chromatographic peak recording two flame ionization detectors, one for each column, thus resulting two simultaneous chromatograms with an acquisition time of 40 s.

Table 1. Protocol description and codification of experimental wines

Wine sample	Experimental protocol					
	Variety	Protocol code	Extraction enzyme	Clarifying enzyme	Fermentation activator	Yeast
BB_AP		AP	E1	C1	A1	Y1
BB_ED	Busuioaca de Bohotin	ED	E2	no	A2	Y2
BB_LA		LA	E3	C3	A3	Y3
TR_AP		AP	E1	C1	A1	Y1
TR_ED	Tamâioasa româneasca	ED	E2	no	A2	Y2
TR_LA		LA	E3	C3	A3	Y3

Each wine sample was injected in the e-nose in triplicate, using the method developed in our laboratory for wines (Antoce and Namolosanu, 2011; Antoce et al., 2015). The main parameters are: injection volume 2500 µl, trap (40°C, pre-purging time 5 s, preheating 20 s, baking 60 s, desorption temperature 250°C), column (heating from 40°C to 200°C with 5°C/s increment, maintaining 2 s the initial and 5 s the final temperature), injection at 200°C, detector temperature 220°C.

The data recording and processing is based on the AlphaSoft version 12.42 and Arochembase library.

Several multivariate statistical analyses are used and compared for data processing, such as the Principal Component Analysis (PCA), which can show the differences in the volatile profiles in accordance to the winemaking protocol used (evaluate discrimination performance), the Discriminant Factorial Analysis (DFA), which can be used to separate in clusters the samples with similar volatile profiles and when necessary classify the unknown samples into these clusters, and finally to determine odor distances from one cluster used as reference by Statistical Quality Control (SQC) analysis.

RESULTS AND DISCUSSION

The electronic nose is able to discriminate the samples in accordance to their raw material – the grape variety – but also to the winemaking protocol used for their preparation, being capable to identify the samples with enhanced volatile profile.

a) Discrimination of grape variety and winemaking protocol by PCA

By applying a Principal Component Analysis a good discrimination (with a discrimination index of 83) can be obtained of the wine sample clusters containing the same variety and winemaking protocol (Figure 1).

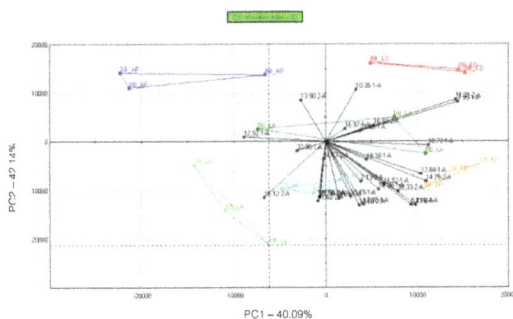

Figure 1. PCA diagram showing the discrimination by the electronic nose of the groups of samples by grape varieties (BB-dark colours and TR-light colours) and winemaking protocols (AP- light/dark blue, ED-light/dark red, LA- light/dark green)

It can be seen that the PC2 axis, which accounts for 42% of the data variability, includes the variables which are dependent on the variety; thus, the clusters of Busuioaca are placed in the upper part of the diagram, while the clusters of Tamaioasa are placed in the lower part of the diagram.

The winemaking protocols are mostly discriminated by the variables included in the PC1, which accounts for 48% of the data variability. Irrespective of the variety, the samples obtained with winemaking protocols AP and LA are placed on the left and those with winemaking protocol ED are placed on the right of the diagram.

This behaviour is most likely determined by the common volatile substances produced by the specific yeast used in each protocol, and not by the specific enzymes used, which also had some influences on the compounds extracted form the grapes or released from grape aromatic precursors.

The third PC only accounts for 4.96% of the variability (Figure 2) and is probably related to the differences induced both by the grape variety and the wine protocol - thus most likely by the type of the enzymes used in each winemaking process.

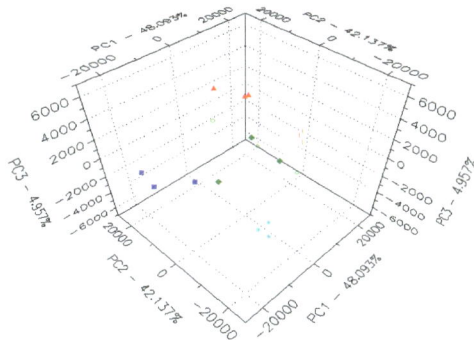

Figure 2. 3D-PCA diagram showing the discrimination by the electronic nose of the groups of samples by grape varieties and winemaking protocols

From Figure 1 we can also see that the most discriminant chromatographic peaks for Busuioaca grape variety are 13.90-2A and 10.36-1A, which were identified as ethyl 2-methyl-butanoate and buthyl acetate, respect-tively, while for Tamaioasa there were more peaks, among which we can cite 25.58-2A (linalool), 11.62-2A (ethyl butanoate), 13.49-1A (isoamyl acetate), 17.76 1A (β-pinene), 19.60-2A (ethyl hexanoate).

Even more peaks contributed to the discrimination of the winemaking protocol, their importance in discrimination being difficult de determine.

b) Discrimination of grape variety and winemaking protocol by DFA

A similar behaviour of the sample clusters as that described in the PCA diagrams can be observed when the Discriminant Factor Analysis is applied (Figure 3). Most of the variation is included in the DF 1 (79.53%), which is related to the grape variety.

The variation determined by the winemaking protocol (the yeast and to a certain degree the enzymes) is included mainly in the DF2 (16.75%) and DF3 (2.74) factors.

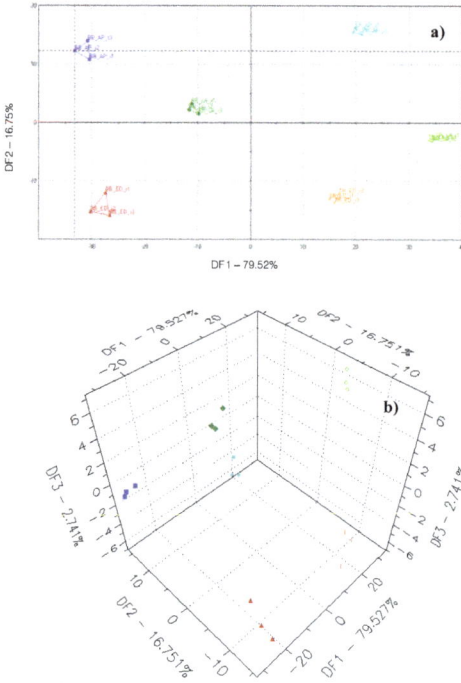

Figure 3. FDA (3a) and 3D-FDA (3b) diagram showing the discrimination by the electronic nose of the groups of samples by grape varieties and winemaking protocols

c) Discrimination of the influences induced by the winemaking protocols for Busuioaca de Bohotin

When only samples obtained based on the same variety are compared, the variability is reduced and the results show only the influence of the winemaking protocol.

The DFA analysis for the Busuioaca de Bohotin allows for discrimination among the samples clusters, but this time clusters are placed much closer to one another (Figure 4).

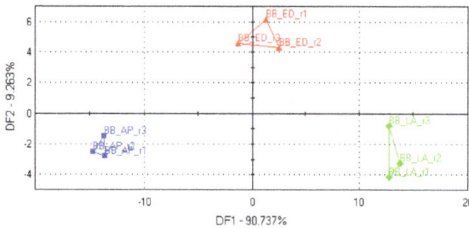

Figure 4. Discrimination by DF Analysis of the groups of Busuioaca de Bohotin samples obtained by different winemaking protocols

The influence of the winemaking protocol for Busuioaca de Bohotin is not much, most of the variability being included in the DF1=90.74%. The variability included in DF2 (9.26%) is mostly related the winemaking protocol ED, which is differentiated by the variables included in this function.

The ED protocol stands out also when it comes to the odor intensity, which was the highest among the winemaking protocols (Figure 5).

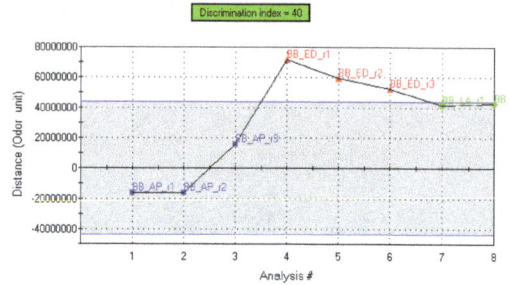

Figure 5. Odor intensity distances of Busuioaca de Bohotin samples obtained by different winemaking protocols

The clusters of BB samples produced with the ED protocol clearly differentiated from the samples produced with protocols AP and LA, when the odor intensity assessed by measuring the odor distances among the samples were determined.

This higher concentration of volatile substances in the wine profile could be a good indication of fermentation with a yeast more effective in producing secondary metabolites, but this does not necessarily indicate a better aromatic profile. On the contrary, the wines produced with the ED protocol ranked last among the three protocols on a sensory evaluation based on the OIV score sheet: BB-ED = 84, BB-LA = 88 and BB-AP = 90 points.

d) Discrimination of the influences induced by the winemaking protocols for Tamâioasa romaneasca

In the case of Tamaioasa romaneasca the winemaking protocol had a higher influence than in the case of Busuioaca de Bohotin (Figure 6). The variability induced by the winemaking protocol was 83.22 included on DF1 and 16.78% in DF2, the last one being mostly related to the LA protocol.

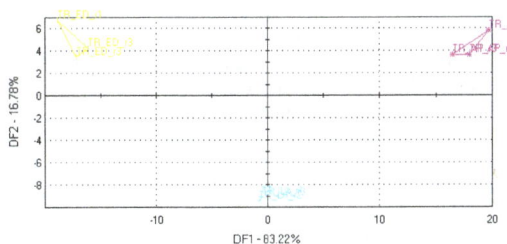

Figure 6. Discrimination by DF Analysis of the groups of Tamaioasa romaneasca samples obtained by different winemaking protocols

For this variety, the highest odor intensities were also displayed by the wines based on ED protocol (Figure 7), confirming that this protocol is clearly different and identifiable by the E-nose, irrespective of the raw material used.

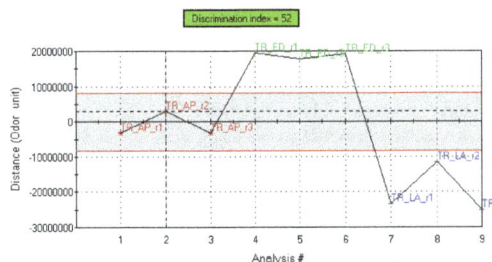

Figure 7. Odor intensity distances among the groups of Tamaioasa romaneasca samples obtained by different winemaking protocols

The sensory evaluation based on the OIV score sheet showed the following ranking: TR-ED and TR-LA = 86 points and TR-AP = 92 points. This shows that the high intensity volatile profile was not preferred by the winetasters, but neither was the lowest intensity profile (LA protocol).

CONCLUSIONS

Based on the evaluation of the volatile profiles of the wines produced from the two aromatic varieties using three different winemaking protocols, the following conclusions can be drawn:

- The variety is most likely discriminated by the electronic nose based on the terpenic volatile profile of each variety, but also by other esters and acetates either from the primary or secondary aroma.

- The winemaking protocol has two main components, with different influences: the enzyme influences the primary aroma by changing the level of extraction of several aromatic compounds and also the degree of release of the aromatic compounds from precursors, while the yeast influences the secondary/fermentation aroma, by the esters and acetates that it forms during fermentation. The yeast has the main influence in the aromatic profile of wines obtained from the same variety.

- Irrespective of the grape variety used, the ED winemaking protocol stands out from the three different protocols, with larger distances measured in odor intensity units, compared to the other two protocols. A higher concentration of the volatile substances measured for this winemaking protocol may indicate a higher aroma intensity which can be perceived sensorially, but it may not necessary mean that the consumers will prefer the resulted wines. It is however a good indication that the wines will be more intense in the nose, a trait that some consumers like.

- For the final protocol selection the results of e-nose testing should always be correlated with the sensory analysis results.

- This type of e-nose analysis may be particularly useful when more than three winemaking protocols are under evaluation and preparation of many wines on industrial scale may be not an economical option.

REFERENCES

Antoce O.A., Nămoloşanu I.C., 2011, Rapid and precise discrimination of wines by means of an electronic nose based on gas-chromatography, Revista de Chimie, Vol. 62, No. 6, 593-595.

Antoce O.A., Stroe M.V., Cojocaru G.A., 2015, Tentative application of an electronic nose to the study of the parentage of Romanian grape varieties Sarba and Alb Aromat, Agriculture and Agricultural Science Procedia, Elsevier, 6 (2015), 110 – 117.

Barbosa C., Falco V., Mendes-Faia A., Mendes-Ferreira A., 2009, Nitrogen addition influences formation of aroma compounds, volatile acidity and ethanol in nitrogen deficient media fermented by Saccharomyces cerevisiae wine strains. Journal of Bioscience and Bioengineering, 108(2), pp. 99–104.

Cabaroglu, T., Selli, S., Canbas, A., Lepoutre, J. P., & Günata, Z., 2003, Wine flavor enhancement through the use of exogenous fungal glycosidases. Enzyme and Microbial Technology, 33, 581–587.

Ghaste M., Narduzzi L., Carlin S., Vrhovsek U., Shulaev V., Mattivi F., 2015, Chemical composition of volatile aroma metabolites and their glycosylated precursors that can uniquely differentiate individual grape cultivars, Food Chemistry, Volume 188, 309-319.

González-Pombo P., Fariña L., Carrau F., Batista-Viera F., Brena B. M., 2014, Aroma enhancement in wines using co-immobilized Aspergillus niger glycosidases, Food Chemistry, Volume 143, 185-191.

Hernandez-Orte, P., Cersosimo, M., Loscos, N., Cacho, J., Garcia-Moruno, E., Ferreira, V., 2008, The development of varietal aroma from non-floral grapes by yeasts of different genera, Food Chemistry, 107, 1064–1077.

Lao, C., Lo´pez Tamanes, E., Lamuela Raventos, R. M., Buxaderas, S., Torre Boronat, M., 1997, Pectic enzyme treatment effects on quality of white grape musts and wines. Journal of Food Science, 62, 1142–1144.

Loscos, N., Hernandez-Orte, P., Cacho, J., Ferreira, V., 2007, Release and formation of varietal aroma compounds during alcoholic fermentation from non-floral grape odorless flavor precursors fractions, Journal of Agricultural and Food Chemistry, 55, 6674–6684.

Marks V., K van der Merwe G., van Vuuren H.J., 2003, Transcriptional profiling of wine yeast in fermenting grape juice: regulatory effect of diammonium, FEMS Yeast Research, Volume 3, Issue 3, 269-287.

Palmeri, R., Spagna, G., 2007, Beta-glucosidase in cellular and acellular form for winemaking application, Enzyme and Microbial Technology, 40(3), 382–389.

Piñeiro Z., Natera R., Castro R., Palma M., Puertas B., Barroso C.G., 2006, Characterisation of volatile fraction of monovarietal wines: Influence of winemaking practices, Analytica Chimica Acta, Volume 563, Issues 1–2, 165-172.

Rocha S. M., Paula Coutinho, Elisabete Coelho, António S. Barros, Ivonne Delgadillo, Manuel A. Coimbra, 2010, Relationships between the varietal volatile composition of the musts and white wine aroma quality. A four year feasibility study, LWT - Food Science and Technology, Volume 43, Issue 10,. 1508-1516.

Samoticha J., Wojdyło A., Chmielewska J., Politowicz J., Szumny A., 2017, The effects of enzymatic pre-treatment and type of yeast on chemical properties of white wine, LWT - Food Science and Technology, Volume 79, June 2017, 445-453.

Sumby, K. M., Grbin, P., & Jiranek, V., 2010. Microbial modulation of aromatic esters in wine: Current knowledge and future prospects, Food Chemistry, 121, 1–16.

Swiegers J. H., Kievit R. L., Siebert T., Lattey K. A., Bramley B. R., Francis I. L., King E. S., Pretorius I. S., 2009, The influence of yeast on the aroma of Sauvignon Blanc wine, Food Microbiology, Volume 26, Issue 2, April 2009, 204-211.

Swiegers J. H., Pretorius I. S., 2005, Yeast modulation of wine flavor, Advances in Applied Microbiology, 57(05), 131–75.

Ubeda Iranzo J.F., González Magaña F., González Viña M.A., 2000, Evaluation of the formation of volatiles and sensory characteristics in the industrial production of white wines using different commercial strains of the genus Saccharomyces, Food Control, Volume 11, Issue 2, 143-147.

Ugliano M., Fedrizzi B., Siebert T., Travis B., Magno F., Versini G., Henschke PA., 2009, Effect of nitrogen supplementation and Saccharomyces species on hydrogen sulfide and other volatile sulfur compounds in shiraz fermentation and wine, J. Agric. Food Chem. ,57(11), 4948-55.

24

RESEARCHES ON SITUATION AND TRENDS IN CLIMATE CHANGE IN SOUTH PART OF ROMANIA AND THEIR EFFECTS ON GRAPEVINE

Georgeta Mihaela BUCUR, Liviu DEJEU

University of Agronomic Sciences and Veterinary Medicine of Bucharest,
59 Marasti Blvd., District 1, Bucharest, Romania
Corresponding author email: mihaela_g_savu@yahoo.com

Abstract

In this study we analyzed data from three meteorological stations situated in Romania's south part (Bucharest, Constanta and Craiova), in the period 1977-2016. There were calculated several primary climate parameters (annual average temperature; average temperature in the growing season; average temperature in summer; average maximum temperature in the warmest month; average minimum temperature in the coldest month; annual precipitations; precipitations in summer), and bioclimatic indices (Huglin Index, Winkler Index and Cool Night Index). It has also pursued the evolution in time (19 years), of Feteasca regala, the most common grapevine cultivar in Romania, in close connection with the evolution of climatic parameters. In the period studied, it was found a highly significant trend for average temperatures (annual, during the growing season, summer, and maximum) and Huglin and Winkler indices.The results show the role of the variability, from year to year, of the warmer temperature, which affects grape production and its quality. The increase trend of the grapes sugar content was found to be highly significant (r = 0.809***). Reducing titratable acidity of the must under 4 g/L, observed during the last decade, requires acidification measures.

Key words*: grapevine, climate change, effects, trends.*

INTRODUCTION

Vitis vinifera L. is a species which is very sensitive to climate change, considered to be a bioindicator.

Studies in recent decades have highlighted, in most vineyards, significant heating influence on the development of grapevine phenophases, on the main physiological processes, vegetative growth, grape production and quality (Jones et al., 2005; Cotea et al., 2008; Ranca et al., 2008; Irimia et al., 2015).

Climate change, particularly the temperature, influences the composition of grapes and increased sugars accumulation, acidity reduction, the content of anthocyanins in grape skins and the flavor precursors (Cichi et al., 2006; Palliotti et al., 2015).

In the main vineyards of the country, there was found an increase in average air temperature in the period 2000 - 2010, with values between 0.7 and 2.1°C. The biggest differences were recorded in the vineyards Dealu Mare, Targu Bujor and Murfatlar, highlighting their tendency to aridity (Burzo, 2014).

Following global warming, there was found a change in the evolution of the annual biological cycle of grapevine, with completion faster phenophases (veraison and maturation of the grapes), which is often forced, and with significant consequences on the product quality, which are not always positive (Rotaru et al., 2013).

Grape maturation is accelerated by high temperatures. Sugar accumulation increases with temperature, but certain secondary metabolites such as anthocyanins and aroma precursors are adversely affected by high temperatures.

The annual sequence of phenological stages of grapevine is commonly observed to be accelerated with an increase in temperature (Duchêne and Schneider, 2005).

High concentrations of sugars in berries are not due to photosynthesis and their translocation from leaves and woody parts of the vine, but to water loss through evaporation (Keller, 2015).

Lowering the titratable acidity of the must under 4.0 g/L as a result of global warming requires the addition of tartaric acid to produce balanced wines and to enhance microbiological stability, causing a more expensive winemaking process.

According to the projections of the Intergovernmental Panel on Climate Change (IPCC, 2014), which considers a temperature increase of 1 - 3.7°C by the end of the century, it is necessary to elaborate strategies for the adaptation and mitigation of climate change.

The objective of the current paper is to present and to evaluate the situation and trends of climate change for the period 1977 – 2016 in three centers situated in Romania's south part (Bucharest, Constanta and Craiova) and their effects on grapevine.

MATERIALS AND METHODS

We used observation data with a complete daily series from 1977-2016 of 3 meteorological stations situated in Romania's south part (Bucharest, Constanta and Craiova).

There were calculated and analysed 10 primary climate parameters and bioclimatic indices (Table 1): annual average temperature; average temperature in the growing season (IV-X); average temperature in summer (VI-VIII); average maximum temperature in the warmest month (July): average minimum temperature in the coldest month (January); annual precipitations; precipitations in summer; Huglin Index, Winkler Index and Cool Night Index.

Table 1. List of climatic parameters and bioclimatic indices

Variable	Description (Equation)	Months
T average I-XII (°C)	Annual average temperature	I - XII
T average IV-X (°C)	Growing season temperature	IV - X
T average VI-VIII (°C)	Summer temperature	VI - VIII
T maximum VII (°C)	Average maximum temperature in July	VII
T minimum I (°C)	Average minimum temperature in January	I
P I-XII (mm)	Annual total precipitation	I - XII
P VI-VIII (mm)	Summer total precipitation	VI - VIII
Huglin Index (°C units)	$\Sigma[(Tavg-10°C) + (Tmax-10°C)] / 2 \times k$	IV- IX
Winkler Index (°C units)	$\Sigma[(Tmax+Tmin) / 2 - 10°C]$	IV - X
Cool Night Index (°C)	$Tmin_{sept}$	IX

Vine reaction to climate change was traced during the period 1998-2016 within a 1.0 ha vineyard planted in 1994 at the University of Agronomic Sciences and Veterinary Medicine of Bucharest, at 2.2/1.2 m distance, with Feteasca regala cv (clone 21 Bl) / Kober 5 BB. Vines were trained to a bilateral cordon at 0.7 m and spur pruned with a load of 10 buds / sqm.

The statistical calculations were conducted using Microsoft Excel and interpretation of the data was performed by the methodology presented by Botu and Botu, 2003.

RESULTS AND DISCUSSIONS

Climate change. Analysing primary climate parameters and bioclimatic indices, there were obtained the data presented in Table 2. During the last four decades significant warming trends were observed in the three centers studied. Similar trends were found for Huglin and Winkler indices

According Huglin Index values, the two centers from Muntenia and Oltenia (Bucharest and Craiova) are included in warm temperate climate class (HI + 1), whereas the Black Sea coast (Constanta) is at the limit between the warm temperate and the temperate climate (HI - 1) (Tonietto and Carbonneau, 2004).

The largest increase in temperature trends during 1977 - 2016 were found in Constanta (+ 2.15 ... + 3.65°C).

Following the evolution of annual precipitation reveals a non significant trend for Bucharest and a significant one for Constanta (highy significant trend) and for Craiova (significant trend). Regarding the evolution of the summer precipitations, there was observed a non-significant trend for all three centers.

Cool night Index values (represented by the average minimum temperatures in September) were not significantly affected by climate change, maintaining the areas studied in class climate with very cool nights (CI + 2, Tonietto and Carbonneau, 2004), except for the Black Sea coast (cool nights CI + 1). These conditions ensure, in addition to good grape ripening, also high aromatic and phenolic potential (varieties for red wines).

Following anomalies of average temperatures during the growing season from the multiannual average (Figure 1) there is a significant heating tendency in all 3 studied centers.

Table 2. Trend over the period 1977-2016 and trend per decade
of primary climatic parameters and bioclimatic indices

Location	Variable	Mean and standard deviation	Trend over the period 1977-2016	Trend per decade	R^2
Bucharest	T average I-XII (ºC)	11.51±0.75	+ 1.3	+ 0.35	0.223**
(44.75N; 26.11E; 91 m)	T average IV-X (ºC)	17.97±0.78	+ 1.3	+ 0.35	0.218***
	T average VI-VIII (ºC)	22.37±1.11	+ 2.1	+ 0.52	0.302***
	T maximum VII (ºC)	29.73±1.75	+ 3.1	+ 0.77	0.254***
	T minimum I (ºC)	-5.41±2.46	-	-	NS
	P I-XII (mm)	618.3±149.4	+ 72	+ 18	NS
	P VI-VIII (mm)	193.1±70.8	-	-	NS
	Huglin Index (ºC units)	2337±189	+ 375	+ 93	0.348***
	Winkler Index (ºC units)	1705±168	+ 265	+ 66	0.217***
	Cool Night Index (ºC)	10.52±1.40	-	-	NS
Constanța	T average I-XII (ºC)	11.34±0.93	+ 2.15	+ 0.54	0.521***
(44.33N; 28.43E; 25 m)	T average IV-X (ºC)	17.33±0.92	+ 2.15	+ 0.54	0.594***
	T average VI-VIII (ºC)	21.63±1.21	+ 3.25	+ 0.81	0.653***
	T maximum VII (ºC)	28.23±1.82	+ 3.65	+ 0.91	0.376***
	T minimum I (ºC)	-3.53±1.97	-	-	NS
	P I-XII (mm)	448.4±124.1	+ 200	+ 50	0.241***
	P VI-VIII (mm)	116.9±67.14	-	-	NS
	Huglin Index (ºC units)	2060±213	+ 440	+ 110	0.391***
	Winkler Index (ºC units)	1712±189	+ 445	+ 111	0.489***
	Cool Night Index (ºC)	12.30±1.53	+ 2.80	+ 0.70	0.265***
Craiova	T average I-XII (ºC)	11.51±0.77	+ 1.70	+ 0.42	0.423***
(44.31N; 23.86 E; 195m)	T average IV-X (ºC)	17.91±0.83	+ 1.50	+ 0.37	0.310***
	T average VI-VIII (ºC)	22.15±1.12	+ 2.40	+ 0.60	0.402***
	T maximum VII (ºC)	29.62±1.93	+ 3.50	+ 0.87	0.303***
	T minimum I (ºC)	-4.83±2.46	-	-	NS
	P I-XII (mm)	570.8±189.0	+ 195	+ 48	0.092*
	P VI-VIII (mm)	166.2±82.2	+ 50	+ 12.5	NS
	Huglin Index (ºC units)	2306±209	+ 435	+ 108	0.387***
	Winkler Index (ºC units)	1693±178	+ 350	+ 87	0.310***
	Cool Night Index (ºC)	11.55±1.91	-	-	NS

NS indicate trend that are not significant and *, **, and *** indicate significance at the 0.10, 0.05 and 0.01 levels, respectively.

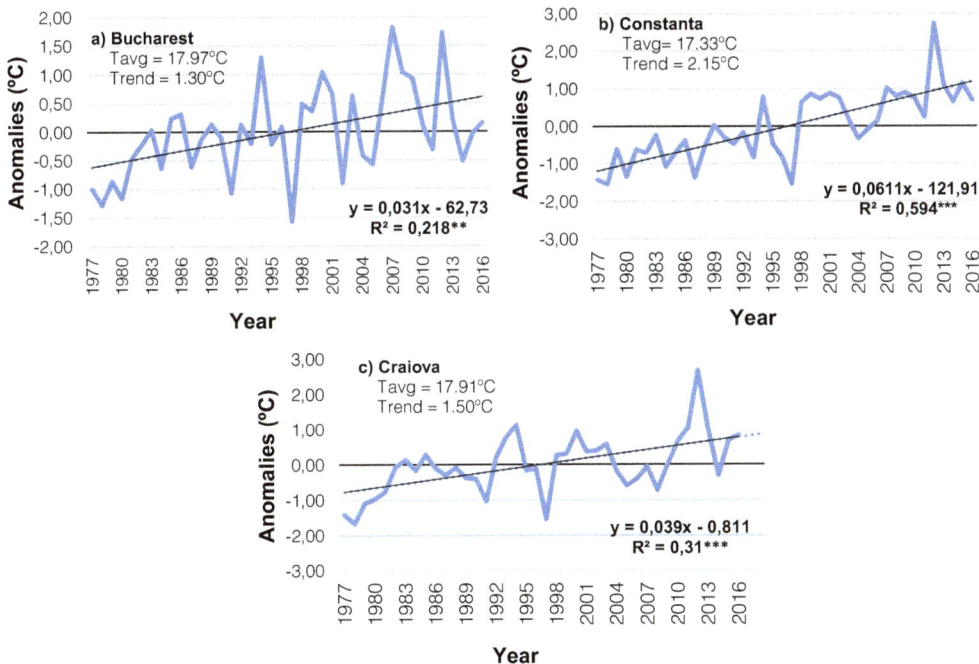

Figure 1. Registered growing season temperature anomalies in the period 1977-2016 for:
a) Bucharest; b) Constanta and c) Craiova; Tavg is the average growing temperature (April - October)

Grapevine behaviour. Climate warming has had different effects on the production of grapes (kg/vine), sugar accumulation (g/L) and titratable acidity (g/L H_2SO_4) of the grape, at Feteasca regala cultivar (Figures 2, 3, 4).

From Figure 2, it can be noted a high variability in grape yield from year to year and a slight reduction trend caused mainly by the minimum temperatures, harmful to vines during the dormant period (Tmin < -20°C) in recent years (2005; 2010; 2012; 2015 and 2016). The frequency and intensity of these

temperatures was discussed in a previous paper (Bucur and Dejeu, 2016).

From Figure 3, it is observed a highly significant trend of sugar content in berries. The statistical analysis of the data shows that sugar content depends on climatic suitability, which explains 65.5 % of the variance.

Regarding the titratable acidity of the grape must, it is found to its insignificant reduction over 19 years (Figure 4). However, in the last decade it was observed a reduction in titratable acidity of the must under 4 g/L, requiring acidification measures.

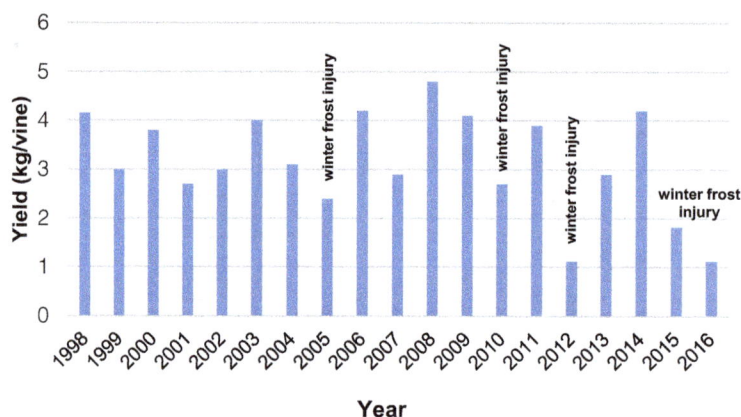

Figure 2. Evolution of grape yield (kg/vine) at Feteasca regala cultivar in the period 1998 - 2016

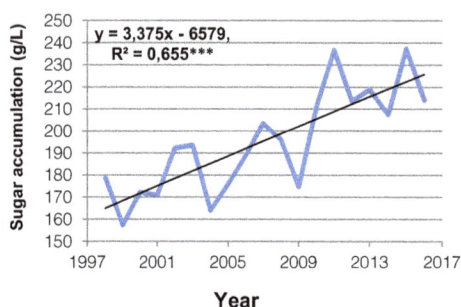

Figure 3. Evolution of sugar accumulation (g/L) at Feteasca regala cultivar in the period 1998-2016

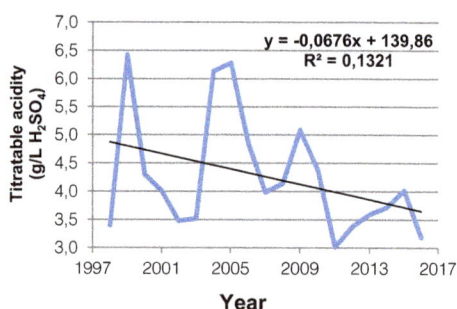

Figure 4. Evolution of titratable acidity (g/L H_2SO_4) of grape must at Feteasca regala cultivar in the period 1998-2016

CONCLUSIONS

During the last four decades, significant warming trends were observed in the three

studied centers. This trend of temperature increase is almost certainly going to continue in a future warmer climate.

The results on Feteasca regala grapevine responses to climate change (high variability and trend to reduce grape yield, a highly significant increase of sugar content, reducing must acidity) are very important for the winegrowers because this is the most widely grown cultivar in Romania.

The results of this study provide the necessary information for viticultural zoning in the new conditions. Starting from the current situation of global warming and predictions for the future, in viticulture there should be implemented mitigation and adaptation measures.

REFERENCES

Botu I., Botu M., 2003. Analiză biostatistică și design experimental în biologie și agricultură. Editura Conphys, Râmnicu Vâlcea, 338 p.

Bucur Georgeta Mihaela, Dejeu L., 2016. Research on frost injury of new romanian grapevine cultivars in the winter 2014-2015. 5th International Conference „Agriculture for Life, Life for Agriculture", Agriculture and Agricultural Science Procedia 10, 233 – 237.

Burzo I., 2014. Modificările climatice și efectele asupra plantelor horticole. Editura Sitech, 7 - 41.

Cichi Daniela Doloris, 2006. Modificările termice din ecosistemul viticol. (Cauze, efecte asupra viței de vie, studii). Editura Universitaria Craiova, 279 p.

Cotea V.V., Rotaru Liliana, Irimia L.M., Colibaba Lucia Cintia, Tudose Sandu-Ville S., 2008. The greenhouse effect on the viticultural ecoclimat in northern Moldavia, Romania. 31st World Congress of Vine and Wine, Verona, Italia.

Duchêne E., Schneider C., 2005. Grapevine and climatic changes: A glance and the situation in Alsace. Agronomy for Suistainable Developement, 25, 93, 99.

Huglin P., 1978. Nouveau mode d'évaluation des possibilités héliothermiques d'un milieu viticole. In: Symposium International sur l'Écologie de la Vigne, 1, Constanta, Roumanie, 1978. Ministère de l'Agriculture et de l'Industrie Alimentaire, 89 - 98.

IPCC, 2014. Climate Change: impacts, adaptation and vulnerability.http://www.ipcc.ch/report/ar5/wg2/. Accessed on 20 January 2017.

Irimia L.M., Patriche C.V., Quenol H., Cotea V.V., 2015. Modificarea potențialului viticol al podgoriilor, ca efect al încălzirii climatului. Studiu de caz: Podgoria Cotnari. Hortus, Nr. 14, 187 - 192.

Jones G.V., White M.A., Cooper O.R., Storchmann K., 2005. Climate change and global wine quality, Climatic Change, 73(3), 319 - 343.

Keller M., 2015. The Science of Grapevines. Anatomy and Physiology. Elsevier, Academic Press, 509 p.

Palliotti A., Poni S., , Silvestroni O., 2015. La nuova viticoltura. Innovazioni tecniche per modelli produttivi efficienti e sostenibili. Edagricole-New Business Media, - Science, 544 p.

Ranca Aurora, Boloș P., Guluță E., 2008. Climate changes of the last 10 years in the Murfatlar vineyard and their influence on the behaviour of the specific grapevine varieties of this vineyard. 31st World Congress of Vine and Wine, Verona, Italy.

Rotaru Liliana, Colibaba Lucia Cintia, Prisăcaru Anca Irina, 2013. Studies of the behavioural tendencies of some grape varieties for white wines in Moldavian vineyards, under the influence of climatic changes. Lucrări Științifice USAMV „Ion Ionescu de la Brad" Iași, Seria Horticultură, vol.56, nr. 2, 303 - 308.

Tonietto J., Carbonneau A., 2004. A multicriteria climatic classification system for grape-growing regions worldwide. Agricultural and Forest Meteorology, 124(1/2), 81 - 97.

Winkler A.J., Cook A., Kliewer W.M., Lider I.A., 1974. General Viticulture. University of California Press, Berkeley, 740 p.

25

INFLUENCE OF FERMENTOR TYPE ON POLYPHENOL EXTRACTION IN RED WINES PRODUCED FROM CABERNET SAUVIGNON

George Adrian COJOCARU, Arina Oana ANTOCE

University of Agronomic Sciences and Veterinary Medicine of Bucharest,
Faculty of Horticulture, Department of Bioengineering of Horti-Viticultural Systems,
59 Marasti Blvd., District 1, Bucharest, Romania
Corresponding author emails: arina.antoce@horticultura-bucuresti.ro; aantoce@yahoo.com

Abstract

Phenolic composition and colour of red wines produced from Cabernet Sauvignon grape variety were determined by means of specific spectrophotometric measurements in order to establish the influence of the maceration technique on the final wine quality. The results showed that the type of fermentor used during maceration-fermentation in red winemaking influence significantly the levels of anthocyanins, copigmentation colourless anthocyanins, polymeric pigments, flavones or total phenols. The age of vines had also specific influences which are thoroughly discussed in the paper. The highest values regarding total pigments (coloured, copigmented and colourless) were observed in wines from 3 years old vines, but they are less representative from the technological viewpoint. However, an increase in total phenols and total coloured pigments was observed when the 4 years old vines are compared with 5 years old vines. The type of fermentor used for the maceration-fermentation process is the most important for the total coloured pigments and total phenols extraction, these parameters ensuring structure and stable colour for wines. The highest levels of total coloured pigments were found in wines produced with horizontal fermentors with inner agitator, while the lowest values were obtained by using roto-fermentors, showing that the first are more suitable for producing well-coloured young wines. The results regarding colour intensity and hue revealed an increased colour intensity and lower hue when horizontal fermentors with inner agitator are used and lower colour intensity and higher hue when roto-fermentors are used. Although a higher hue values obtained when roto-fermentors are used means a higher oxidation, these fermentors also lead to a higher tannin extraction, being recommended when the resulted wines are intended for ageing.

Key words: spectrophotometer, anthocyanin, pigments, maceration, red wine.

INTRODUCTION

Colour, given by the anthocyanins free and combined with other polyphenols, is one of the most important intrinsic characteristic of red wines, being the first attribute evaluated during wine-tastings. Structure too, given by other specific polyphenols, is also important in red wine sensory appreciation and for the wine capacity to age. Certain technologies and treatments applied during vinification can influence the colour and the overall concentration of certain phenolic compounds in red wines. The oenologists are able to decide the appropriate technologies and treatments, in accordance with the desired final product characteristics (Cojocaru and Antoce, 2011; Antoce and Cojocaru, 2015; González-Neves et al., 2015; Gómez-Plaza et al., 2000; Busse-Valverde et al., 2011). Among the possible interventions, the effect of maceration-fermentation technique on wine colour and

phenolic compounds is widely studied (Busse-Valverde et al., 2011; Casassa and Harbertson, 2014; González-Neves et al., 2013; Gambuti et al., 2009; Koyama et al., 2007; Gil-Munoz et al., 1999). Many important parameters of wines are influenced by maceration-fermentation process, including the polyphenol compounds extraction (Casassa and Harbertson, 2014; González-Neves et al., 2013; Gambuti et al., 2009; Koyama et al., 2007; Gil-Munoz et al., 1999; Jackson, 2008; Rakonczás et al., 2015). During maceration, along with anthocyanin and tannin extraction from grape skins potassium content is also increased, which leads to an increase of pH and changes in the total titratable acidity due to the potassium hydrogen tartrate precipitation (Drăghici and Râpeanu, 2011; Rakonczás et al., 2015; Peng et al., 1996), changes that can also affect the colour. Temperature, duration, homogenisation, aeration and enzyme addition during maceration are also important factors that have direct effect on extraction rate and final concentration in polyphenols or some other cellular constituents (Casassa et al., Harbertson, 2014; González-Neves et al., 2013; Gambuti et al., 2009; Koyama et al., 2007; Gil-Munoz et al., 1999; Jackson, 2008; Ribéreau-Gayon et al., 2006). Thus, for each style of wine, these parameters should be controlled when a certain maceration industrial technology is applied.

MATERIALS AND METHODS

The grapes for this study were harvested on October 2013 in Vrancea wine region from parcels containing Cabernet Sauvignon of 3, 4 and 5 years of age. Vinification of grapes involved minimal oenological intervention to reduce the influences induced during vinification in order to assess the effect of maceration techniques on phenolic composition and colour of wine. Grape batches were treated with 30 mg/kg with sulphur dioxide for antioxidant protection and then crushed and destemmed. The resulted grape mash batches from each parcel (with grapes from vines of 3 different ages) were transferred with progressive cavity pumps and each divided for maceration-fermentation in two type of fermentors: roto-fermentors (RF) and horizontal tank with inner agitator (HF). During the first day of maceration, all analysed tanks were treated with a preparation of pectolytic enzyme in dose of 1 g/q (Zymorouge G, AEB Spindal), for an enhanced extraction of tannins and pigments and better colour stabilisation. Cap management during maceration-fermentation was programmed in both types of tanks and achieved by 5 minutes of rotation for 3 times/day. The maceration and alcoholic fermentations were conducted at 24-28°C without inoculation (using the wild grape microflora) for 15 days, after which the marc batches were pressed with a horizontal press at 0.4 Bars. The resulted wines were then gravitationally clarified, racked and analysed to determine the parameters related to colour and phenolic composition. Six types of wines were obtained and used to compare the influence of maceration technique on phenolic composition. Wine samples were analysed in triplicate by assessing the main spectrophotometric parameters usually used to describe the colour and phenolic composition. The spectrophotometric methodology requires standardization of wine pH to 3.6 and filtration with PES membrane with 0.45 μm pore size. Each spectrophotometric measurement was performed with a UV - VIS double beam spectrophotometer Specord 250 from Analytik Jena AG using the software WinAspect version 2.2.7. In accordance to the parameter determined quartz or glass cuvettes were used. All the results were calculated to account for dilution and conventionally referred to the optical path of 10 mm and expressed in absorbance units (Antoce and Cojocaru, 2015). *Monomeric anthocyanin* concentration was determined as the difference in optical densities at 520 nm of the wine diluted to 5% and buffered to pH=3.6 in order to exclude copigmented anthocyanins (Boulton, 2001; Levengood and Boulton, 2004) and the same wine treated with sulphur dioxide to bleaching in order to exclude polymeric pigments and other resistant pigments (Jacobson, 2006; Eldridge and Liles, 1997). *Copigmented anthocyanin* concentration was determined as the difference in optical densities at 520 nm of the wine sample treated with an excess of acetaldehyde and of the same wine diluted down to 5% and buffered to pH=3.6. The excess of acetaldehyde reacts with the free SO_2 in wine, preventing any bleaching effect on the existent pigments, thus giving an estimation of total colour at 520 nm (Boulton, 2016, 2010 and 1996; Levengood and Boulton, 2004; Jacobson, 2006; Eldridge and Liles, 1997). *Polymeric pigments and bleaching resistant pigments* were determined as optical densitiy at 520 nm of the wine buffered at pH=3.6 and treated with an excess of SO_2. The effect of SO_2 is to bleach all monomeric and copigmented anthocyanins, but leave unaffected (coloured) the polymeric pigments and other non-bleachable derivatives of anthocyanins, which are formed through the addition of compounds such as pyruvate, acetaldehyde, hydroxycinnamates or vinylflavanols to the C4 and 5 hydroxyl positions of anthocyanins, generally known as pyranoanthocyanins (Somers, 1971; Boulton, 1996; Levengood and Boulton, 2004;

Harbertson and Spayd, 2006; Jacobson, 2006; Remy et al., 2000; Eldridge and Liles, 1997; Bakker et al., 1997; Bakker and Timberlake, 1997; Romero and Bakker, 1999; Mateus et al., 2001; Mateus and De Freitas, 2001; Cameira-dos-Santos et al., 1996; Fulcrand et al., 1996, 1997, 1998; Es-Safi et al., 1999; Benabdeljalil et al., 2000; Lu and Foo, 2001; Hayasaka and Asenstorfer, 2002; Schwarz et al., 2003). *Colourless anthocyanins* were determined as the difference in optical densities at 520 nm of the wine sample treated with excess of HCl and the same wine treated with excess of acetaldehyde. The addition of excess of HCl allows quantifying of total anthocyanins (coloured and colourless) by destroying the effect of concentration and the equilibrium formed at certain pH, while the acetaldehyde allows quantifying of total coloured anthocyanins (Jacobson, 2006; Eldridge and Liles, 1997). *Flavone cofactors* were determined at 365 nm, being the maximum of absorption wavelength for flavonoids involved in copigmentation phenomenon, especially quercetin and kaempferol (Eldridge and Liles, 1997; Merken and Beecher, 2000; Harbertson and Spayd, 2006). *Total phenols* were determined on a diluted wine sample (1/100) at 280 nm and corrected with a factor of 3.9 representing the non-polyphenol substances absorbing at the same 280 nm wavelength (Somers, 1998). *Colour intensity* and *hue* was calculated as the sum of absorbance determined at 420, 520, and 620 nm and, respectively, the ratio between absorbance at 420 nm and 520 nm (Ribéreau-Gayon et al., 2006; OIV, 2016) with mention that samples were buffered at pH=3.6 for standardization. The detailed spectrophotometric methodology is described in previous works (Boulton, 2016 and 1996; Levengood and Boulton, 2004; Jacobson, 2006; Eldridge and Liles, 1997). The Origin 10.0 software program was used for data processing. Analysis of variance (ANOVA) was applied to the results with posthoc Tukey Test for the comparison of means, at 0.05 significance level.

RESULTS AND DISCUSSIONS

The results presented in figure 1 show the variation of coloured, copigmented and colourless anthocyanins in Cabernet Sauvignon wines produced with grapes coming from vine plantations of 3, 4 and 5 years of age, each macerated in two types of tanks. As expected, we can observe slight differences in coloured anthocyanin between vines with different ages, the highest value being observed in wines produced from 5 year old vines using the horizontal fermentor with inner agitator and the lowest being observed in wines produced from 4 year old vines using the roto-fermentor. Generally, when the horizontal fermentor with inner agitator was used, slightly higher content in coloured monomeric anthocyanin was observed in wine samples, irrespective to vine age, compared to wines produced with roto-fermentors. However, regarding the copigmented anthocyanin, the highest content was observed in wines produced from 3 year old vines, while the lowest content was observed in wines produced from 5 year old vines. This case is similar for monomeric coloured anthocyanin regarding the type of fermentor used with the exception of the copigmented anthocyanins observed in the wines produced from 5 years old vines. The lower values of copigmented anthocyanins obtained in the horizontal fermentor with inner agitator are well explained by the larger values of polymeric pigments shown in figure 2. Generally, aside of the enhancement in absorbance it is believed that the copigmented anthocyanins are do not affect the rate of polymerization reactions during wine ageing (Bimpilas et al., 2016; Boulton, 2001). A possible pathway of polymeric pigments formation was suggested by previous works, considering that the hydrophobic stacking interaction between anthocyanin chromophores and the so-called copigments, could be the first step in the formation of a covalent bond between the anthocyanin and its copigment (Brouillard and Dangles, 1994). However, the polymerization reactions during ageing are involved in formation of a more stable coloured pigments (Jackson, 2008; Somers, 1971). Between colourless hydrated form of anthocyanins and coloured flavylium cation an equilibrium forms based mainly on the wine pH, determined not only by the thermodynamic constants of anthocyanin molecules, but also by the concentration, solvent and temperature. The colourless hydrated form of anthocyanin can act as nucleophile, being involved in condensation reactions with flavan-3-ols as electrophiles in wines, leading to larger red pigments of flavanol-anthocyanin (F - A$^+$) (Cheynier et al., 2000; Brouillard and Delaporte, 1977).

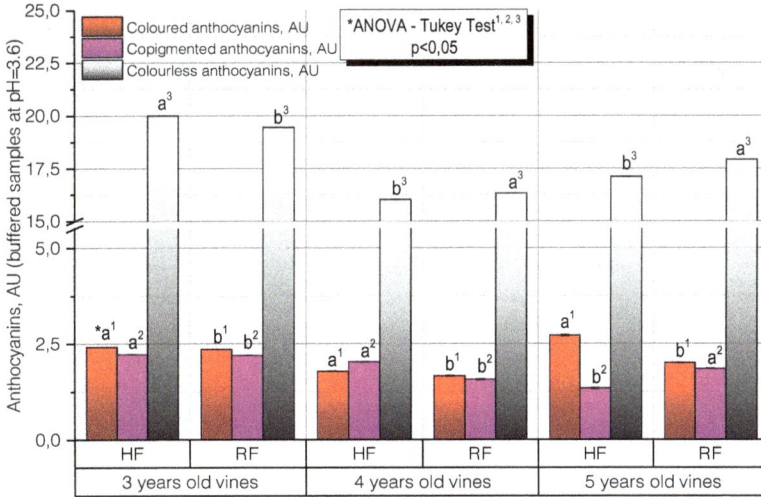

*ANOVA results: the population means for **age of vines** are significantly different; the population means for **type of fermentor** are significantly different; the interaction between **age of vines** and **type of fermentor** is significant.

**HF - Horizontal fermentor with inner agitator; RF - Roto-fermentor;

Figure 1. Variation of anthocyanin content in Cabernet Sauvignon wines produced with two types of maceration techniques from grapes with different age of vine plantations

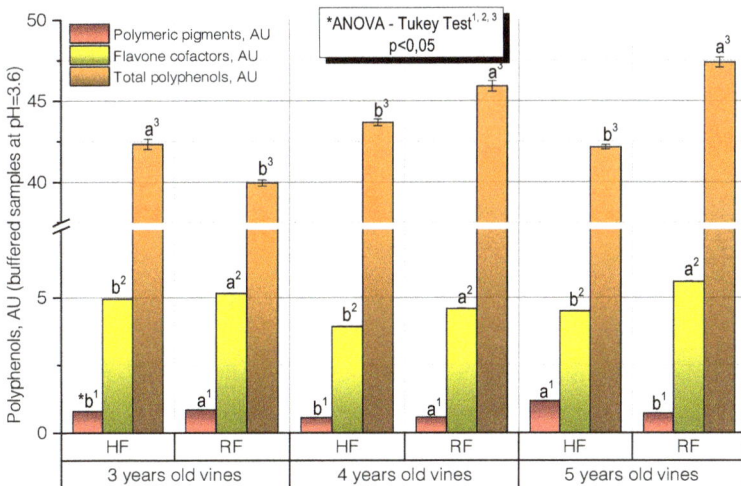

*ANOVA results: the population means for **age of vines** are significantly different; the population means for **type of fermentor** are significantly different; the interaction between **age of vines** and **type of fermentor** is significant.

**HF - Horizontal fermentor with inner agitator; RF - Roto-fermentor;

Figure 2. Variation of certain polyphenols content in Cabernet Sauvignon wines produced with two types of maceration techniques from grapes with different age of vine plantations

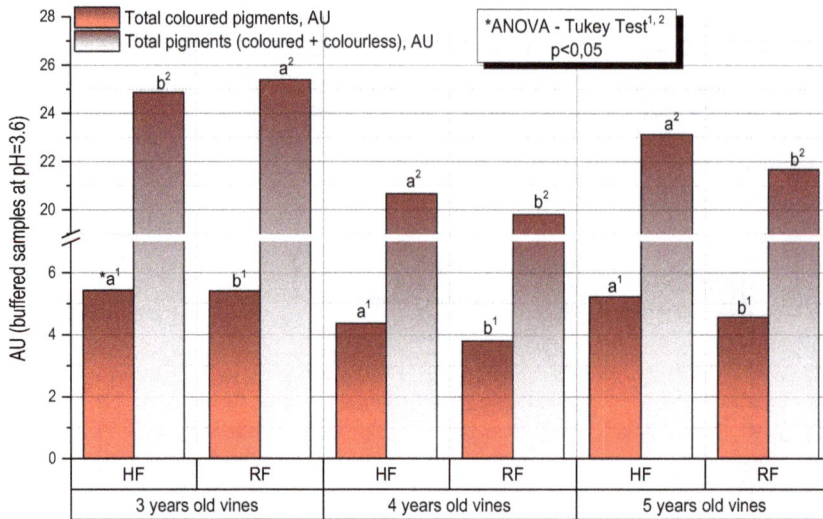

*ANOVA results: the population means for **age of vines** are significantly different; the population means for **type of fermentor** are significantly different; the interaction between **age of vines** and **type of fermentor** is significant.

**HF - Horizontal fermentor with inner agitator; RF - Roto-fermentor;

Figure 3. Variation of total coloured pigments and total pigments (coloured + colourless) in Cabernet Sauvignon wines produced with two types of maceration techniques from grapes with different age of vine plantations

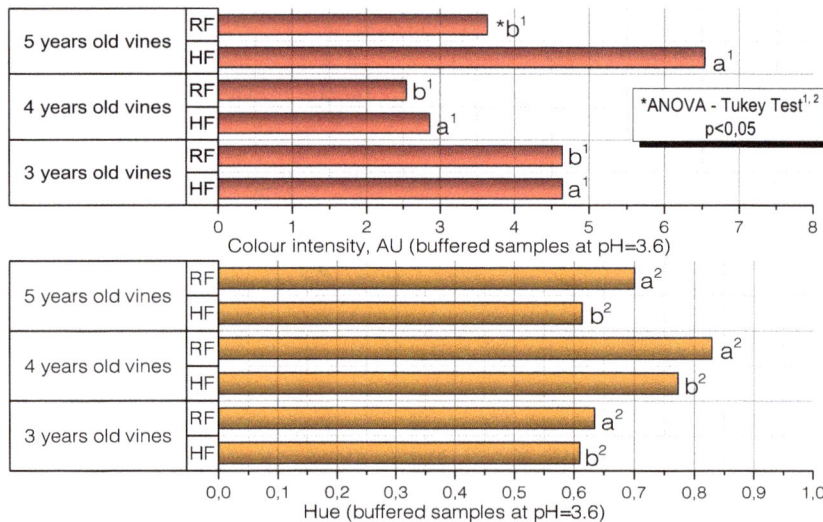

*ANOVA results: the population means for **age of vines** are significantly different; the population means for **type of fermentor** are significantly different; the interaction between **age of vines** and **type of fermentor** is significant.

**HF - Horizontal fermentor with inner agitator; RF - Roto-fermentor;

Figure 4. Variation of colour intensity and hue in Cabernet Sauvignon wines produced with two types of maceration techniques from grapes with different age vine of plantations

On the other hand, the red flavylium cation (coloured anthocyanins) can act as electrophile, being involved in condensation reactions with flavan-3-ols as nucleophile in wines, leading to larger red pigments of anthocyanin-flavanol (A^+ - F) (Cheynier et al., 2000; Remy et al., 2000).

Our results regarding colourless (hydrated) form of anthocyanins presented in figure 1 show greater values for 3 years old vines, which explains why the total content of anthocyanins (coloured, copigmented and colourless) presented in figure 3 is greater in wines produced from 3 years old vines. This phenomenon is probably due to the fact that being in the first year of production the 3 years old vines have low yields, thus accumulating more anthocyanin per grape. However, the effect of concentration, standardization of pH to 3.6 and alcoholic concentration affects this specific equilibrium of anthocyanin forms, leading to the results presented in figure 1. The polymeric pigments, flavone cofactor and total phenolic compounds are presented in figure 2.

The level of total phenols in the grapes and wines is influenced significantly by the age of vines, being greater in wines produced from 4 and 5 years old vines. Due to the high concentration of tannins (figure 2) in wines produced from 5 years old vines, in the case of a higher concentration of coloured anthocyanins (as flavylium cations) – the wines produced in horizontal fermentor (figure 1), the formation of polymeric pigments of anthocyanin-flavanol (A^+ - F) type seem to be favoured (figure 2). However, the other mechanism for the production of flavanol-anthocyanin type of polymeric pigments (F - A^+) seems to be slower in the wines from 5 years old vines. This was not the case in the wine samples produced from 3 years old vines, where we have found higher concentrations of colourless hydrated anthocyanin (figure 1), but similar values for polymeric pigments with wines produced for 5 years old vines (figure 2). Thus, based on a single study, the observed results on polymeric pigments formation are not very conclusive. Another study have already shown that certain winemaking parameters (increased temperature during storage or pH = 3.8) can lead to formation of anthocyanin-flavanol (A^+ - F) polymeric pigments (Fulcrand et al., 2006).

Generally, observations on wine samples produced from 4 and 5 years old vines, showed that the roto-fermentors tend to extract more phenols (tannins) than the horizontal fermentors with inner agitator (figure 2).

Also, the results regarding total coloured pigments presented in figure 3, are not well correlated with colour intensity of wine samples (figure 4). However, when horizontal fermentors were used, greater values were observed in both types of analyses irrespective of the age of vines and, correspondingly, lower values when roto-fermentors were used. Total pigments (coloured and colourless) are generally higher when horizontal fermentors with inner agitator were used (figure 3). Only in wines produced from 3 years old vines we have observed higher concentrations of total pigments (coloured and colourless) with the use of roto-fermentor. The results regarding hue of colour in figure 4, revealed that wines produced with roto-fermentors are more oxidized, with more yellow tones than wines produced with horizontal fermentors with inner agitator.

CONCLUSIONS

Polyphenol extraction in wines is influenced significantly by the type of fermentor used, but the age of vines, especially when they are very young, can also have a major effect on wine quality. Different equilibrium between anthocyanin forms (coloured, copigmented and colourless) could be observed in relation with age of vines and also the type of fermentor used during the process of maceration. Generally, the content of total pigment (coloured, copigmented and colourless) was higher in wine samples produced from 3 years old vines, possibly due to a low yield and higher anthocyanin accumulation/grape, but this case is less representative from the technological point of view. An increasing trend was observed in the case of total coloured pigments in the case of the 4 and 5 years old vines, with similar behaviours regarding the colour intensity, even though not correlated to the vine age. The type of fermentor used during maceration-fermentation has a significant effect on extraction of anthocyanins and tannins and implicitly big impact the structure and colour in red wines. An increased extraction of total phenols and a decreased total coloured pigments was observed in wines produced with roto-fermentors, and an opposite effect was observed in wines produced with horizontal fermentors with inner agitator. Colour intensity has a similar behaviour with total coloured pigments, more intense colour being observed when wines are produced with horizontal fermentors. The analyses regarding hue of colour showed that roto-fermentors produced wines with increased values, which suggest an advanced oxidation compared to the wines from horizontal fermentors with inner agitator. However, although the level of total coloured pigments and colour intensity are founded in greatest concentrations when wines are produced with horizontal fermentors, and also, the hue values are lower than in wines from roto-fermentors, we recommend this later type of fermentor to be used when wines with better polypehnolic structure meant for aging are desired. The horizontal fermentors, with a higher extraction of anthocyanins are more suitable for the production of intensely coloured wines designed to be commercialised young.

REFERENCES

Antoce A. O., Cojocaru G. A., 2015. Technological approaches to the vinification of Dornfelder grape variety. 38th World Congress of Vine and Wine (Part 1), BIO Web Conferences, Vol. 5, 02009.

Bakker J., Bridle P., Honda T., Kuwano H., Saito N., Terahara N., Timberlake C. F., 1997. Identification of an anthocyanin occurring in some red wines. Phytochemistry, Vol. 44, No. 7, 1375–1382.

Bakker J., Timberlake C. F., 1997. Isolation, identification, and characterization of new color-stable anthocyanins occurring in some red wines. Journal of Agricultural and Food Chemistry, Vol. 45, No. 1, 35–43.

Bakker J, Preston N.W., Timberlake C. F., 1986. The determination of anthocyanins in ageing red wine: Comparison of HPLC and Spectral methods. American Journal of Enology and Viticulture, Volume 37, No. 1, 121-126.

Benabdeljalil C., Cheynier V., Fulcrand H., Hakiki A., Mosaddak M., Moutounet, M., 2000. Evidence of new pigments resulting from reaction between anthocyanins and yeast metabolites. Sciences des Aliments, Vol. 20, No. 2, 203–220.

Bimpilas A., Panagopoulou M., Tsimogiannis D., Oreopoulou V., 2016. Anthocyanin copigmentation and color of wine: The effect of naturally obtained hydroxycinnamic acids as cofactors. Food Chemistry Vol. 197, Part A, 39-46.

Boulton R. B., 2016. The copigmentation assay for red wines. University of California, Davis, Department of Viticulture and Enology website: http://boulton.ucdavis.edu/copig.htm, accessed at 15th December 2017.

Boulton R., 2001. The copigmentation of anthocyanins and its role in the color of red wine: A critical review. American Journal of Enology and Viticulture, vol. 52, no. 2, 67-87.

Boulton R. B., 1996. Methods for the assessment of copigmentation in red wines. 47th Annual Meeting of the American Society for Enology and Viticulture, Reno, NV.

Brouillard R., Dangles O., 1994. Anthocyanin molecular interactions: the first step in the formation of new pigments during wine aging? Food Chemistry Vol. 51, Issue 4, 365-371.

Brouillard R., Delaporte B., 1977. Chemistry of anthocyanin pigments. 2. [1]Kinetic and thermodynamic study of proton transfer, hydration, and tautomeric reactions of malvidin 3-glucoside. Journal of the American Chemical Society, Vol. 99, Issue 26, 8461-8468.

Busse-Valverde N, Gómez-Plaza E., López-Roca J. M., Gil-Muñoz R., Bautista-Ortín A. B., 2011. The extraction of anthocyanins and proanthocyanidins from grapes to wine during fermentative maceration is affected by the enological technique. Journal of Agriculture and Food Chemistry Vol. 59, No. 10, 5450-5455

Cameira-dos-Santos P. J., Brillouet J. M., Cheynier V., Moutounet M., 1996. Detection and partial characterisation of new anthocyanin-derived pigments in wine. Journal of the Science of Food and Agriculture, Vol. 70, No. 2, 204–208.

Casassa L. F., Harbertson J, 2014. Extraction, Evolution, and Sensory Impact of Phenolic Compounds during Red Wine Maceration. Annual Review of Food Science and Technology, Vol. 5, No. 1, 83-109.

Cheynier V., Remy S., Fulcrand H., 2000. Mecanism of anthocyanin and tannin changes during winemaking and aging. Proceedings of the American Society for Enology and Viticulture, 50th Anniversary Annual Meeting, Seattle, Washington, USA, 337-344.

Cojocaru G. A., Antoce A. O., 2011., Color Changes Induced by Fining Treatments in Red Winemaking. Lucrări ştiinţifice U.Ş.A.M.V.B., Seria B, Vol. LV, 2011, Section Viticulture & Oenology, Congres CD, ISSN 1222-5312, 496-500.

Drăghici L., Râpeanu G., 2011. Evolution of polyphenols during the maceration of the red grapes. Journal of Agroalimentary Processes and Technologies, Vol. 17, No. 2, 169-172.

Eldridge D., Liles J., 1997 (revised in 2002 by: Bechard Andrea and Ritchie Gerry). Spectral Measures for estimating Wine Color and Phenolics. Napa Valley College Viticulture and Enology Department and Napa Valley Vintners Association Teaching Winery, Wine and Must Analyses Manual, Standard Operating Procedure.

Es-Safi N. E., Guernevé C. L., Labarbe B., Fulcrand H., Cheynier V., Moutounet M., 1999. Structure of a new xanthylium salt derivative. Tetrahedron Letters, Vol. 40, No. 32, 5869–5872.

Fulcrand H., Duenas M., Salas E., Cheynier V., 2006. Phenolic Reactions during Winemaking and Aging. American Journal of Enology and Viticulture, Vol. 57, Issue 3, 289–297.

Fulcrand H., Benabdeljalil C., Rigaud J., Cheynier V., Moutounet M., 1998. A new class of wine pigments generated by reaction between pyruvic acid and grape anthocyanins. Phytochemistry, Vol. 47, No. 7, 1401–1407.

Fulcrand H., Cheynier V., Oszmianski J., Moutounet M., 1997. An oxidized tartaric acid residue as a new bridge potentially competing with acetaldehyde in flavan-3-ol condensation. Phytochemistry, Vol. 46, No. 2, 223–227.

Fulcrand H., Cameira-dos-Santos P. J., Sarni-Manchado P., Cheynier V., Favre-Bonvin J., 1996. Structure of new anthocyanin-derived wine pigments. Journal of the Chemical Society, No. 7, 735–739.

Gambuti A, Capuano R, Lecce L., Fragasso M. G., Moio L., 2009. Extraction of phenolic compounds from 'Aglianico' and 'Uva di Troia' grape skins and seeds in model solutions: Influence of ethanol and maceration time. Vitis, Vol. 48, No. 4, 193–200.

Gil-Munoz R., Gómez-Plaza E., Martínez A., López-Roca J. M., 1999. Evolution of Phenolic Compounds during Wine Fermentation and Post-fermentation: Influence of Grape Temperature. Journal of Food Composition and Analysis, Vol. 12, 259-272.

Gómez-Plaza E., Gil-Muñoz R., López-Roca J. M., Martínez A., 2000. Color and phenolic compounds of a young red wine. Influence of wine-making techniques, storage temperature, and length of storage time. Journal of Agriculture and Food Chemistry, Vol. 48, No. 3, 736-741.

González-Neves G., Favre G., Gil G., Ferrer M., Charamelo D., 2015. Effect of cold pre-fermentative maceration on the color and composition of young red wines cv. Tannat. Journal of Food Science and Technology, Vol. 52, No. 6, 3449-3457.

González-Neves G., Gil G., Favre G., Baldi C., Hernández N., Traverso S., 2013. Influence of Winemaking Procedure and Grape Variety on the Colour and Composition of Young Red Wines. South African Journal of Enology and Viticulture, Vol. 34, No. 1, 138-146.

Harbertson J. F., Spayd S., 2006. Measuring Phenolics in the Winery. American Journal of Enology and Viticulture Vol. 57, No. 3, 280-288.

Hayasaka Y., Asenstorfer R. E., 2002. Screening for potential pigments derived from anthocyanins in red wine using nanoelectrospray tandem mass spectrometry," Journal of Agricultural and Food Chemistry, Vol. 50, No. 4, 756–761.

Jacobson J. L., 2006. Introduction to Wine Laboratory Practices and Procedures, Springer Science + Business Media, Inc.

Jackson R. S., 2008. Wine Science: Principles and Applications 3rd Edition. Academic Press is an imprint of Elsevier.

Koyama K., Goto-Yamamoto N., Hashizume K., 2007. Influence of Maceration Temperature in Red Wine Vinification on Extraction of Phenolics from Berry Skins and Seeds of Grape (Vitis vinifera). Bioscience, Biotechnology, and Biochemistry, Vol. 71, No. 4, 958-965.

Levengood J., Boulton R., 2004. The variation in the color due to copigmentation in young Cabernet Sauvignon wines. In Red Wine Color: Exploring the Mysteries. A.L. Waterhouse and J.A. Kennedy

(Eds.), 35-52. ACS Symp. Ser. 886. Am. Chemical Society, Washington, DC.

Lu Y., Foo L. Y., 2001. Unusual anthocyanin reaction with acetone leading to pyranoanthocyanin formation, Tetrahedron Letters, Vol. 42, No. 7, 1371–1373.

Mateus N, Silva A. M. S., Vercauteren J., De Freitas V., 2001. Occurrence of anthocyanin-derived pigments in red wines. Journal of Agricultural and Food Chemistry, Vol. 49, No. 10, 4836–4840.

Mateus N., De Freitas V., 2001. Evolution and stability of anthocyanin-derived pigments during port wine aging. Journal of Agricultural and Food Chemistry, Vol. 49, No. 11, 5217–5222.

Merken H. M., Beecher G. R., 2000. Measurement of food flavonoids by high-performance liquid chromatography: a review. Journal of Agriculture and Food Cemsitry, 48, 577.

OIV, 2016. Chromatic Characteristics. Section 2 – Physical analysis in Compendium of International Methods of Analysis, OIV-MA-AS2-07B.

Peng Z., Waters E. J., Pocock K. F., Williams P. J., 1996. Red wine bottle deposits, II: Cold stabilisation is an effective procedure to prevent deposit formation. Australian Journal of Grape and Wine Research, Vol. 2, Issue 1, 1-5.

Rakonczás N., Andrási D., Murányi Z., 2015. Maceration affect mineral composition and pH of wines. International Journal of Horticultural Science, Vol. 21 No. 3–4, 25–29.

Remy S., Fulcrand H., Labarbe B., Cheynier V., Moutounet M., 2000. First confirmation in red wine of products resulting from direct anthocyanin-tannin reactions. Journal of the Science of Food and Agriculture Vol. 80, No. 6, 745–751.

Ribéreau-Gayon P., Glories Y., Maujean A., Dubordieu D., 2006. Handbook of Enology (vol. 2). The Chemistry of Wine, Stabilization and Treatments 2nd Ed., John Wiley & Sons Ltd.

Romero C., Bakker J., 1999. Interactions between grape anthocyanins and pyruvic acid, with effect of pH and acid concentration on anthocyanin composition and color in model solutions. Journal of Agricultural and Food Chemistry, Vol. 47, No. 8, 3130–3139.

Schwarz M., Jerz G., Winterhalter P., 2003. Isolation and structure of Pinotin A, a new anthocyanin derivative from Pinotage wine. Vitis, Vol. 42, No. 2, 105–106.

Somers T. C., 1971. The polymeric nature of wine pigments. Phytochemistry Vol. 10, No. 9, 2175–2186.

Somers T. C., 1998. The Wine Spectrum. An Approach towards Objective Definition of Wine Quality. Hyde Park Press, Adelaide: Winetitles.

26

COMPOSITION OF PHENOLIC COMPOUNDS IN PETAL FLOWERS AND LEAF SAMPLES OF VARIOUS APPLE CULTIVARS

Fatma YILDIRIM[1], Adnan Nurhan YILDIRIM[1], Tuba DILMAÇÜNAL[1], Bekir SAN[1], Nilda ERSOY[2]

[1]Suleyman Demirel University, Agriculture Faculty, Department
of Horticulture, 32260, Isparta, Turkey
[2]Akdeniz University, Technical Sciences Vocational School,
07058, Campus/Antalya, Turkey
Corresponding author email: fatmayildirim@sdu.edu.tr

Abstract

The phenolic compounds of petals and leaf tissue samples were determined in apple cvs. 'Breaburn', 'Golden Reinders', 'Granny Smith' and 'Jonagold' grafted on M9; 'Summerred' and 'William's Pride' grafted on M.26. Petals of apple flowers were taken at the pink bloom floral stage in April. Moreover leaves were sampled from the middle part of the annual shoots in July. The phenolic compounds were analyzed by High Pressure Liquid Chromatography (HPLC) technique. The gallic acid, p-hydroxy benzoic acid, eriodictyol, quercetin, ferulic acid, chlorogenic acid, caffeic acid, syringic acid, p-coumaric acid and apigenin-7-glucoside contents were determined in petal samples. The gallic acid, p-hydroxy benzoic acid, eriodictyol, quercetin, ferulic acid and p-coumaric acid were also investigated in leaf tissues. The concentrations of the compounds were influenced by the genotypes as well as by tissue samples. While eriodictyol was the predominant phenolic compound in leaves ranging between 156.75-414.90 μg/g DW, chlorogenic acid was the predominant phenolic compound found only in petals ranging between 7784.60-19293.00 μg/g DW in all cultivars investigated in this research. It was determined that the petals of apples were quite richer than the leaf samples in terms of the phenolic contents. Among the studied cultivars the total concentrations of phenols were higher in both the petal and leaf tissues of 'Granny Smith' apple cultivar.

Key words: Malus, cultivar, leaves, petal, phenolic.

INTRODUCTION

Phenolics are aromatic compounds in which one (phenol) or more (polyphenol) hydroxyl groups are bound to a benzene ring and constitute a significant part of the biochemical components of the plant. They represent the most abundant widely spread class of plant biochemicals. They are involved in a number of physiological functions in plants. For example, flavonoids which are a large and diverse group of phenolic compounds, act as attractants for pollination and seed dispersal (Downey et al., 2006) and flavonols also help to facilitate conditional male fertility in pollen by providing pollen tube growth (Pollak et al., 1995). On the other hand phenolic compounds can protect the plant against UV-light, insects and pathogens (Vermerris and Nicholson, 2008) and have a role in mechanical support by lignification and affect the growth of neighboring plants (allelopathy) (Özeker, 1999).

Plant phenolics have positive effects on human health and nutrition (Vermerris and Nicholson. 2008), and are the basis of several plant-derived drugs and cosmetic products. Many researchers have also reported that plants are rich sources of polyphenols which have bioactive effect in human such as antioxidant (Pandey and Rizvi, 2009), antimicrobial activities (Rauha et al., 2000), anti-glycemic (Rizvi and Zaid, 2001) and anti-cancer (Yang et al., 2001), anti-inflammatory (Zhu et al., 2015) etc. Therefore, studies on the definition, extraction and purification of these compounds from different plant organs have been increasing. In this regard determination of the composition and content of phenolic compounds in different plant vegetative organs are important.

Apple (*Malus x domestica* Borkh.) is one of the most produced fruit species in the world. Its fruits are known as rich sources of phenolics, especially catechins, procyanidines, phloretin glycosides and chlorogenic acid (Matthes and Schmitz-Eiberger, 2009). Apple leaves are also rich in terms of phloridzin in which exhibit antidiabetic activity and quercetin glycosides (Liaudanskas et al., 2014). Although differences in phenolic contents of cultivars have been shown in apple leaves and fruits, there is no research on phenolic compounds of apple flowers. It is known that apple flower tea is recommended for facilitating digestion and for enhancing skin.

The aim of this study was to compare the contents of phenolic components in different apple cultivar petals and leaves to evaluate their potential as sources of bioactive compounds as well as genetic differences.

MATERIALS AND METHODS

The biological material. The trials were carried out at the experimental apple (*Malus x domestica* Borkh.) orchard of Suleyman Demirel University, Agricultural Research and Application Center located in Isparta, Turkey (37°50'23"N 30°32'02"E) in 2007. All of the trees used as plant material were planted in 2003 with 1x3 m spacing and trained as modified central leader system. The petals of the flower samples were taken at the stage of pink bloom early in the morning in April. Leaf samples were taken from the middle of one year old shoots all around the tree in July. The collected samples were brought to the laboratory immediately, washed under tap water, rinsed with distilled water, put into paper bags, and dried in an air-blowing drying oven set at 65°C. The dried samples were ground to powder with a blender. Trial I: Determining the phenolic compounds of petals and leaves of "Breaburn", "Golden Reinders", "Granny Smith" and "Jonagold" apple cultivars grafted on M9 rootstock. Trial II: Determining the phenolic compounds of petals and leaves of "Summerred" and "William's Pride" apple cultivars grafted on M26 rootstock.

Determination of phenolic compounds. Phenolic compounds were analyzed by the modified procedure of Escarpa and González (1998). 25 ml of acetone-water solution (80 % acetone and 20 % water v/v) was added to 2.5 g of ground samples. The upper phase was taken in a centrifuge tube after the extract was incubated in a water bath at 50°C for 30 min. Then, the extraction was repeated twice using 25 ml of acetone-water solution each time and

the extract was incubated in a water bath at 50°C for 30 min and the upper phase was added to the centrifuge tube again. These combined phases were centrifuged at 10.000 rpm for 5 min. The solvent was evaporated at 40°C under vacuum and samples were re-dissolved in 2 ml of methanol. Solutions were filtered by membrane filters with a pore size of 0.45 µm and then 20 µl of the solutions was injected into HPLC (High Pressure Liquid Chromatography). HPLC analysis was performed using a Shimadzu HPLC system with a diode array detector (DAD λmax=278). The column used was an Agilent Eclipse XDB-C$_{18}$ (250x4.60 mm 5 µm) operated at 30°C. Mobile phase: Solvent A (2 % solution of acetic acid in water)—Solvent B (Methanol). Flow rate: 0.8 ml min^{-1}, Injection volume: 10 µl. Peak identification was done according to the standards (*p*-hydroxybenzoik acid, eriodictyol, ferulic acid, *p*-coumaric acid, gallic acid, quercetin, apigenin 7-glucoside, chlorogenic acid, syringic acid, caffeic acid, rosmarinic acid, epicatechin, catechin, rutin, resveratrol, hesperidin, naringenin, luteolin, apigenin and acacetin). The phenolic standards were purchased from Sigma Chemical Co. The concentration of phenolics was expressed as µg g^{-1} dry matter.

Statistical analysis. The data were subjected to the analysis of variance (ANOVA) by using the Minitab software program, and the means were separated by Duncan's Multiple Range Test (5%).

RESULTS AND DISCUSSIONS

Phenolic compounds of petals. The obtained results are presented in Table 1 and Table 2. These data indicated that the petals had the higher phenolic compounds than the leaves of all apple cultivars investigated in this research. Zou et al. (2011) also reported that the petals of loquat were rich in terms of total phenolics and total flavonoids. In our study, a total of 10 kinds of phenolic compounds found in petals of apple flower samples were identified and quantified. The highest amount of phenolic compound determined in petals was the chlorogenic acid followed by apigenin-7-glucoside, quercetin, caffeic acid, eriodictyol, *p*-coumaric acid, syringic acid, gallic acid, *p*-hydroxybenzoic acid and ferulic acid.

Table 1. Average concentration of phenolic compounds in petal samples
of four apple cultivars grafted on M9 in Trail 1 (µg/g DW).

Phenolic compound	Cultivars				
	Breaburn	Golden Reinders	Granny.Smith	Jonagold	Average
Phenolic acids					
clorogenic acid	12208.00 b[*]	9840.10 c	19293.00 a	12877.00 b	13554.53
caffeic acid	160.65 d	203.25 c	400.50 a	349.30 b	278.43
p-coumaric acid	28.26 d	99.77 b	146.60 a	95.49 c	92.53
syringic acid	40.19 c	100.02 b	111.75 a	116.45a	92.10
gallic acid	89.70 b	108.40 a	83.07 c	63.98 d	86.29
p-hydroxybenzoik acid	71.50 b	29.11 c	60.34 b	141.90 a	75.71
ferulic acid	25.27 b	nd	48.54 a	nd	36.91
Flavonoids					
apigenin-7-glucoside	1272.60 c	1728.10 b	1383.90 c	2222.40 a	1651.75
quercetin	307.50 b	894.20 a	303.35 bc	236.50 c	435.39
eriodictyol	116.65 b	103.15 bc	457.25 a	95.58 c	193.16
Total	14320.32	13106.10	22288.30	16198.60	16496.78

*Means with different superscripts in the same line are statistically significantly different (p<0.05), nd: not detected.

Table 2. Average concentration of phenolic compounds in petal samples
of two apple cultivars grafted on M26 in Trail 2 (µg/g DW).

Phenolic compound	Cultivars		
	Summerred	William's Pride	Average
Phenolic acids			
clorogenic acid (phenolic acid)	7784.60 b[*]	7997.60 a	7891.10
caffeic acid	191.50 b	197.15 a	194.325
p-coumaric acid	130.25 a	44.07 b	87.16
syringic acid	81.00 a	61.54 b	71.27
gallic acid	90.91 b	94.11 a	92.51
p-hydroxybenzoik acid	110.30 a	77.60 b	93.95
ferulic acid	276.95 b	343.30 a	310.125
Flavonoids			
apigenin-7-glucoside	1735.80 a	1132.80 b	1434.30
quercetin	214.25 a	137.50 b	175.88
eriodictyol	16.99 b	23.17 a	20.08
Total	10632.55	10108.84	10370.70

*Means with different superscripts in the same line are statistically significantly different (p<0.05), nd: not detected.

These results revealed that the petals of apple flowers have a strong antioxidant capacity. Likewise a positive relationship was detected between the phenolics and antioxidant capacities of loquat flowers (Liaudanskas et al., 2014). To our knowledge there is no literature on the phenolic compounds of petals of apple flowers. Therefore, this research is also important in terms of being the first in its field. The results indicated that the contents of each phenolic component of petals were affected by the cultivars. Similarly Zhou et al. (2011) found differences between the contents of flavonoids and phenolics of the flowers of five loquat cultivars. "Granny Smith" had the highest values for total amount of the phenolic components detected in the petals, whereas the lowest value was found in "Golden Reinders" in trail 1 (Table 1). "Granny Smith" is a self-fertile, good pollinator and high productive apple cultivar. The high phenolic contents of

this cultivar can be the reason for its high fertility. Likewise, it is reported that attraction may occur through secondary phenolic compounds (flavonoids) in the petals (Shirley, 1996; Özeker, 1999). Thus phenolics may increase the fruit set and yield by playing a role in pollination as well as pollen tube growth. The contents of clorogenic acid, caffeic acid, *p*-coumaric acid, ferulic acid and eriodictyol were the highest in "Granny Smith" (Table 1). While the contents of gallic acid and quarcetin were the highest in "Golden Reinders", the contents of *p*-hydroxybenzoik acid, syringic acid and apigenin-7-glucoside were the highest in "Jonagold". In trial 2, total amount of the phenolic components of "Summerred" and "William's Pride" cultivars were found close to each other and the lowest than the cultivars investigated in trial 1. The contents of ferulic acid of both cultivars were quite higher than the other cultivars evaluated in trail 1, while the contents of eriodictyol and quercetin were quite lower. The results of this research emphasized that ferulic acid, eriodictyol and quercetin found in the petal are affected by the rootstocks as well as cultivars.

Phenolic compounds of leaves. The obtained results are presented in Table 3 and Table 4. The phenolic contents of the leaf samples were found lower than the petals of the apple cultivars. Totally 6 kinds of phenolic compounds were evaluated in this research. The highest amount of phenolic compound obtained in leaf samples was eriodictyol followed by ferulic acid, quercetin, *p*-hydroxybenzoic acid, gallic acid and *p*-coumaric acid. Similarly gallic acid, ferulic acid and *p*-coumaric acid were found in leaf samples of apple by Tao et al. (2008) and Petkovsek et al. (2009). Eriodictyol, the highest amount of phenolic compound obtained in this research from the leaf samples of apples, is known as a flavanone and has long been considered as an antioxidant and anti-inflammatory agent. Thus, apple leaves may be considered as a potential antioxidant for human diseases such as acute lung injury (Zhu et al.,

2015). The results of this research indicated that the content of each phenolic component obtained from the leaf samples of apples were affected by cultivar. Likewise, many researchers reported that the phenolic contents of the apple vary by cultivar (García et al., 2004; Mikulič-Petkovšek et al., 2004; Usenik et al., 2004; Petkovsek et al., 2009). According to the total results, the highest amount of phenolic components were found in "Granny Smith" followed by "Golden Reinders", "Jonagold" and "Breaburn" in trail 1 (Table 3), while "William's Pride" had the highest amount of these components in trail 2. The highest amount of *p*-hydroxybenzoic acid was found in "Granny Smith" and *p*-coumaric acid was found only in this cultivar. The highest amount of eriodictyol, quercetin, and ferulic acid were found in "Golden Reinders", while the amount of gallic acid was the highest in "Jonagold". The highest amount of *p*-hydroxybenzoik acid, quercetin and ferulic acid were found in "William's Pride", while the amount of gallic acid was the highest in "Summerred" in trail 2.

Table 3. Average concentration of phenolic compounds in leaf samples of four apple cultivars grafted on M9 in Trail 1 (µg/g DW).

Phenolic compound	Cultivars				
	Breaburn	Golden Reinders	Granny Smith	Jonagold	Average
Phenolic acids					
p-coumaric acid	nd	nd	17.58	nd	17.58
gallic acid	56.82 b*	47.19 bc	38.72 c	71.38 a	53.53
p-hydroxybenzoik acid	35.88 c	nd	131.20 a	48.20 b	89.70
ferulic acid	63.08 c	96.13 a	90.05 a	87.40 ab	84.17
Flavonoids					
quercetin	60.46 c	77.97 a	70.67 ab	63.42 b	68.13
eriodictyol	197.90 b	414.90 a	338.90 a	156.75 b	277.11
Total	378.26	636.19	687.12	427.15	590.22

*Means with different superscripts in the same line are statistically significantly different (p<0.05), nd: not detected.

Table 4. Average concentration of phenolic compounds in leaf samples of two apple cultivars grafted on M26 in Trail 2 (µg/g DW).

Phenolic compound	Cultivars		
	Summerred	William's Pride	Average
Phenolic acids			
gallic acid	49.92 a*	39.33 b	44.63
p-hydroxybenzoik acid	34.26 b	43.85 a	39.06
ferulic acid	53.71 b	91.18 a	72.45
Flavonoids			
quercetin	65.51 b	97.63 a	81.57
Eriodictyol	239.90	238.25	239.08
Total	443.30	510.24	572.12

*Means with different superscripts in the same line are statistically significantly different (p<0.05).

CONCLUSIONS

As a result, according to the results obtained from this research, the phenolic compounds of the apple vary by the cultivar and the parts of the plant. It was determined that the petals of an apple cultivar were quite richer than the leaf samples in terms of the phenolic contents. While chlorogenic acid was the predominant phenolic compound and found only in petals, eriodictyol was the predominant phenolic compound of leaf samples of apple. Among the studied cultivars, the total concentration of phenols was found higher in both of the tissues of "Granny Smith" apple.

REFERENCES

Downey M.O., Dokoozlian N.K., Krstic M.P., 2006. Cultural practice and environmental impacts on the flavonoid composition of grapes and wine: A Review of Recent Research. Am. J. Enol. Vitic. 57 (3): 257-268.

Escarpa A., Gonzalez M.C., 1998. High-performance liquid chromatography with diode-array detection for the determination of phenolic compounds in peel and pulp from different apple varieties. Journal of Chromatography A. 823: 331-337.

Garcia E., Rom C.R., Murphy J.B., 2004. Comparison of phenolic content of 'Liberty' apple (*Malus × domestica*) on various rootstocks. Acta Horticulturae 658: 57-60.

Liaudanskas M., Viskelis P., Raudonis R., Kviklys D., Uselis N., Janulis W., 2014. Phenolic composition and antioxidant activity of *Malus domestica* leaves. The Scientific World Journal. Article ID 306217: 10.

Matthes A., Schmitz-Eiberger M., 2009. Polyphenol content and antioxidant capacity of apple fruit: Effect of cultivar and storage conditions. Journal of Applied Botany and Food Quality 82: 152 – 157.

Özeker E., 1999. Phenolic compounds and their importance. Anadolu. J. of AARI 9 (2): 114-124.

Pandey K. B., Rizvi S.I., 2009. Plant polyphenols as dietary antioxidants in human health and disease.

Oxidative Medicine and Cellular Longevity 2 (5): 270-278.

Petkovsek M.M., Stampar F., Veberic R., 2009. Seasonal changes in phenolic compounds in the leaves of scab-resistant and susceptible apple cultivars. Canadian Journal of Plant Science 89 (4):745-753.

Pollak R.E., Hansen K., Astwood J.D., Taylor L.R., 1995. Conditional male fertility in maize. Sex Plant Reprod 8: 231-241.

Rauha J.P., Remes S., Heinonen M., Hopia A., Kahkonen M., Kujala T., Pihlaja K., Vuorela H., Vuorela P., 2000. Antimicrobial effects of Finnish plant extracts containing flavonoidsand other phenolic compounds. International Journal of Food Microbiology 56: 3–12.

Rizvi S.I., Zaid M.A., 2001. Intracellular reduced glutathionecontent in normal and type 2 diabetic erythrocytes: Effect of insulin and (-) epicatechin. J Physiol Pharmacol. 52: 483-8.

Shirley B.W., 1996. Flavonoid biosynthesis: "new" functions for an "old" pathway. Trends Plant Science 1: 377-382.

Tao F., Xue S.C., Yan M.Z., Chun Y.Z., Xiao Y.Z., Chuan J.W., 2008. Antioxidation and phenolic constituents in Xinjiang wild apple [*Malus sieversii* (Lebed.) Roem.] leaf. Scientia Agricultura Sinica 41(8):2386-2391.

Usenik V., Mikulic-Petkovsek M., Solar A., Stampar F., 2004. Flavonols of leaves in relation to apple scab resistance. Zeitschriftfür Pflanzen Kkrankheiten und Pflanzenschutz 111(2):137-144.

Vermerris W., Nicholson R., 2008. Phenolic compound biochemistry. Springer Science+Business Media B.V. e-ISBN: 978-1-4020-5164-7.

Yang C.S., Landau J.M., Huang M.T., Newmark H.L., 2001. Inhibition of carcinogenesis by dietary polyphenolic compounds. Ann Rev Nutr. 21:381-406.

Zhou C., Sun C., Chen K., Li X., 2011. Flavonoids, phenolics and antioxidant capacity in the flower of *Eriobotrya japonica* Lindl.Int. J. Mol. Sci. 12: 2935-2945.

Zhu G.F., Guo H.J., Huang Y., Wu C.T., Zhang Z.F., 2015. Eriodictyol a plant flavonoid attenuates LPS induced acute lung injury through its antioxidative and antiinflammatory activity. Experimental and Therapeutic Medicine 10: 2259-2266.

27

COMPARATIVE STUDY OF PROCESSED PRODUCTS FROM CULTIVARS OF THE NATIVE APRICOT

Constanța ALEXE[1], Marian VINTILĂ[1], Ion CAPLAN[2], Gheorghe LĂMUREANU[2], Lenuța CHIRA[3]

[1]Research and Development Institute for Processing and Marketing of the Horticultural Products - Bucharest, No. 1A, Intrarea Binelui Street, District 4, 042159, Bucharest, Romania
[2]Research Station for Fruit Growing (RSFG) Constanta, 1 Pepinierei Street, 907300, Commune Valu lui Traian, Romania
[3]University of Agronomic Sciences and Veterinary Medicine of Bucharest, 59 Marasti Blvd., District 1, 011464, Bucharest, Romania
Corresponding author email: tantialexe@yahoo.com

Abstract

The paper aimed to present the suitability of processing the seven native apricot cultivars grown at the Research Station for Fruit Growing Constanta: 'Tudor', 'Sirena', 'Orizont', 'Olimp', 'Neptun', 'Augustin' and 'Litoral', results leading to the establishment of the fruit valorization direction. Apricots were processed at the Research and Development Institute for Processing and Marketing of the Horticultural Products - Bucharest (micro lab) as compote, jam, comfiture and nectar. The cans' quality assessment was performed using Method A - STAS 12656-8, the state standard that regulates the analysis methods with unitary score scales, used to evaluate the organoleptic characteristics of food. The results show that these cultivars have in common a sweet, pleasant, aromatic flavor (which is why they are highly appreciated for fresh consumption), characteristics that, at the same time, influence the quality of processed products. Out of the studied cultivars, 'Olimp' was particularly highlighted, which is very well suitable to all processing into four types of canned analyzed: comfiture, jam, compote and nectar, the resulting product having remarkable sensory qualities. Apricots in the 'Tudor' cultivars are less suitable for processing, preferably as being able to consume as fresh fruits or, possibly, as comfiture or nectar. In conclusion: for getting canned compote, the 'Orizont', 'Olimp', 'Neptun' cultivars are recommended; for comfiture, cultivars 'Tudor', 'Olimp', 'Augustin', 'Litoral'; 'Orizont', 'Neptun', 'Litoral' cultivars for jam; for nectar, the 'Sirena', 'Orizont', 'Olimp', 'Neptun' cultivars.

Key words: compote, comfiture, jam, nectar, quality.

INTRODUCTION

Apricots are very popular with consumers, both as a dessert fruit and as well as processed in various ways.

The high demand for fruits is determined by their qualitative and technological attributes, by the complex biochemical composition and by the very pleasant taste and specific flavor etc.

There are many apricot consumption benefits that are also supported by scientific studies.

Firstly, they are a real and rich source of vitamin A, B and C, along with beta carotene (due to which the color is yellow-orange) helps maintain eyesight and nerves and tissue regeneration (www.pro-sanatate.com/caisele-beneficii...).

But in the biochemical composition of fruits there are several other important components for the human nutrition, including: 10.6 to 21.7% dry substance, 6 to 16.6% total sugar, 0.55 to 1.1% pectin , 1.09 to 1.64% protein, 0.6 to 0.86% minerals out of which: K 75-112 mg%, P 21-32 mg% 6-14 mg% Ca, mg, S, Na , 0.41 to 3.20 mg%, vitamin P, 0.72 to 1.8 mg% vitamin E, and the energy value is 21-77 calories per 100 g etc. As shown by the above data, through the biochemical composition, apricots ensure all components the human body needs to conduct metabolism in good condition. Apricots are used for their favorable effect on digestion due to alkaline reaction, in the production of hemoglobin in anemia etc. High nutritional value of apricots and apricot-based finished products, led specialists in the scientific research domain to diversify the assortment by creating or placing cultivars in the tillage that behave well in the climatic conditions from Romania. Because apricots are not suitable for fresh storage more than a short period of time, processing them as canned represents a needed and desired solution.

But fruit conservation suitability is a cultivar characteristic; therefore, studies on the potential of different cultivars to be processed in one form or another are necessary.

Lately, in our country, there have been many concerns in this direction for the species: cherries, peaches, sweet cherries (Caplan et al., 2015; Caplan et al., 2016; Lamureanu et al., 2014; Lamureanu et al. 2015; Sarbu et al., 2010; Veringa et Dumitrescu, 2016; Vintila et al., 2015).

When manufacturing, consuming and evaluating the quality of processed products, we must consider the provisions of state or professional standards that regulate quality technical conditions of raw and auxiliary materials, the organoleptic and physicochemical properties of the finished product etc.

In the present work, which final aims to establish the destination of improvement by setting the processing suitability of cultivars of apricot, these standards were taken into consideration: STAS 3164-90. Fruit compote. Standard State; STAS 3750-90. Comfiture. Standard State; STAS 3183-90. Jams. Standard State; SP 877-96. Fruit nectar. Standard Professional.

MATERIALS AND METHODS

During 2015-2016 studies and research were carried out, aiming at determining the suitability of processing of seven local native cultivars of apricots that exist in culture at Research Station for Fruit Growing Constanta: 'Tudor', 'Sirena', 'Orizont', 'Olimp', 'Neptun', 'Augustin' and 'Litoral'.

For this purpose apricots were processed as compote with whole fruits, comfiture, jam and nectar at the Research and Development Institute for Industrialization and Marketing of Horticultural Products in Bucharest, in the micro-production lab. After the period required to stabilize the product (21 days), the cans were subjected to sensory analysis, applying the evaluation method A- STAS 12656-88, which establishes the me-thods of analysis using unitary score scales (used in evaluating the organoleptic characteristics of food).

This method is applied in order to assess a combination of organoleptic properties: appearance, color, taste, texture or, where appropriate, consistency.

Each organoleptic evaluation was made by comparing with the unitary score scales from 0-5 points and got the average score given by the group of tasters, based on individual sheets of recording the marks that were given.

The score of weighted average was calculated, adding these for obtaining the overall average score and the organoleptic qualities of the products on the basis of total average, by comparison with a scale of 0 - 20 points were settled.

Finally they awarded qualifiers for each product and cultivar. In the overall score achieved by the various analyzed products, we differentiate between 5 quality classes: very good (18.1-20.0), good (15.1-18.0), satisfactory (11.1-15.0), unsatisfactory (7.1-11.0) and incompatible (0-7.0).

Before processing, fresh fruit of every cultivar were organoleptic analyzed and characterized in terms of size, shape, color and flesh peeling, texture, taste and aroma pulp, the kernel size etc.

It has also been tested and shown resistance to keeping temporary apricots of every cultivar.

RESULTS AND DISCUSSIONS

The fruit from the **'Tudor'** cultivar has a good resistance for keeping fresh, is medium-sized, averaging 40-45 g, ovoid shaped, slightly dorsal-ventrally flattened with orange peel covered with carmine red on 2/3.

The flesh is orange, non-adherent to the kernel, juicy, fragrant, and the kernel is large with bitter core.

The sensory analysis of processed products (Table 1) highlights the fact that, according to the product, their quality differs greatly from "satisfactory" in the case of compote and jam, to "very good" in the case of comfiture.

In the Figure 1, we have the deliverables: comfiture, nectar, compote and apricot jam from the 'Tudor' cultivar.

The **'Sirena'** cultivar fruit is medium to large on average 65 g, globular to ovoid shape, slightly asymmetrical, with good resistance to fresh storage.

The peel is orange with red spots on the sunny side. The orange pulp has a sturdy structure, juicy enough, pleasant taste and fine flavor.

The stone represents 6.1% of the weight of the fruit and has a sweet core.

Table 1. Sensory analysis of the apricot processed products from the 'Tudor' cultivar

MU=points

Specification	Product			
	Compote	Comfiture	Jam	Nectar
Aspect	4.08	5.72	4.80	3.84
Color	3.04	5.32	3.68	2.72
Taste	4.08	3.72	3.84	5.76
Consistency	3.04	3.72	2.40	4.00
Overall average score	14.24	18.48	14.72	16.32
Qualificative	satisfactory	very good	satisfactory	good

Fig 1. Apricot processed products from the 'Tudor' cultivar

Out of the four types of processed products (Fig. 2), the nectar is highlighted with 18.24 points and the qualification "very good".

The compote, with only 14.25 points scored "satisfactory" (Table 2).

Table 2. Sensory analysis of the apricot processed products from the 'Sirena' cultivar

MU=points

Specification	Product			
	Compote	Comfiture	Jam	Nectar
Aspect	3.81	5.06	4.56	5.12
Color	3.77	5.06	2.56	3.68
Taste	4.00	3.55	5.04	5.76
Consistency	2.67	3.55	3.36	3.68
Overall average score	14.25	17.22	15.52	18.24
Qualificative	satisfactory	good	good	very good

For the comfiture (17.22 points) and the jam (15.52 points) the qualificative that was awarded was "good".

Fig. 2. Apricot processed products from the 'Sirena' cultivar

The **'Orizont'** variety, with good resistance to fresh storage has an oblong shaped fruit, medium to large weight (45.8 to 62.3 g).

The peel is orange, with carmine red on the sunny side. The flesh is orange; fine textured, of an average firmness, aromatic, very juicy.

The kernel is medium-sized, oblong shaped, adherent to the flesh, with sweet core.

The data in Table 3 shows a good suitability at processing to 'Orizont' cultivar, compote (18.32 points), nectar (18.16 points) getting a "very good" qualificative, comfiture (17.22 points) and jam (15.84 points) the "good" qualificative.

Table 3. Sensory analysis of the apricot processed products from the 'Orizont' cultivar

MU=points

Specification	Product			
	Compote	Comfiture	Jam	Nectar
Aspect	4.80	5.32	4,32	5,.6
Color	3.52	4.86	2.72	3.20
Taste	6.00	3.49	5.76	5.52
Consistency	4.00	3.55	3.04	3.68
Overall average score	18.32	17.22	15.84	18.16
Qualificative	very good	good	good	very good

The four types of products obtained from the 'Orizont' cultivar are presented in Figure 3.

The **'Olimp'** cultivar has good resistance to preserve the fruit in fresh state. The fruit is big (65-75g) with orange skins and the flesh is bright orange, with firm texture, good flavor, very good and balanced taste. All four types obtained by processing received very high marks from tasters, which lead to the unique qualificative of "very good" (Table 4 and Figure 4).

Fig. 3. Apricot processed products
from the 'Orizont' cultivar

Table 4. Sensory analysis of the apricot processed products from the 'Olimp' cultivar

MU=points

Specification	Product			
	Compote	Comfiture	Jam	Nectar
Aspect	5.04	6.00	6.00	5,36
Color	3.36	3.20	5.32	3.84
Taste	6.00	6.00	3.90	5.76
Consistency	4.00	3.68	4.00	3.52
Overall average score	18.40	18.88	19.22	18.48
Qualificative	very good	very good	very good	very good

The compote has obtained the maximum score for taste and consistency, the comfiture for aspect and taste and the jam for aspect and consistency.

The **'Neptun'** cultivar, with good resistance for fresh storage, has a large fruit, ovoid, sharper at the top, yellow, striped and red dotted on the sunny side. The flesh is yellow-orange, firm, sweet and slightly fizzy, appreciated for meal.

Figure 4. Apricot procesed products from the 'Olimp' cultivar

The canned version is also praised (Table 5 and Figure 5), given that outside of sweetness, which received a "good" qualificative, the other assortments were employed by tasters in the "very good" column.

Table 5. Sensory analysis of the apricot processed products from the 'Neptun' cultivar

MU=points

Specification	Product			
	Compote	Comfiture	Jam	Nectar
Aspect	5.76	5.72	6.00	6.00
Color	3.36	4.39	6,00	4.00
Taste	6.00	3.46	3.77	6.00
Consistency	3.52	3.81	4.00	3.52
Overall average score	18.64	17.38	19.77	19.52
Qualificative	very good	good	very good	very good

The jam (19.77 points) as well as the nectar (19.52 points) has received score close to the highest.

Figure 5. Apricot processed products
from the 'Neptun' cultivar

The **'Augustin'** cultivar shows cordiforme fruit shape, with average size (45 to 57.5 g), with orange peel with a lot of carmine red.
The flesh is orange, with average firmness, intermediate texture, strongly scented, very juicy.
The kernel is of average size, round shaped, adherent to the pulp, sweet core.
The results of the sensory analysis for processed products indicate weaker results for this cultivar, because only in the case of comfiture (18.16 points) tasters have given the "very good" qualificative (Table 6).

Table 6. Sensory analysis of the apricot processed products from the 'Augustin' cultivar

MU=points

Specification	Product			
	Compote	Comfiture	Jam	Nectar
Aspect	5.04	5.76	4.62	4.56
Color	3.52	3.20	3.52	3.04
Taste	5.76	5.52	2.67	5.28
Consistency	3.52	3.68	3.85	3.68
Overall average score	17.84	18.16	14.66	16.56
Qualificative	good	very good	satisfactory	good

Table 7. Sensory analysis of the apricot processed products from the 'Litoral' cultivar

MU=points

Specification	Product			
	Compote	Comfiture	Jam	Nectar
Aspect	5.52	5.76	5.32	5.04
Color	3.04	3.04	5.46	3.36
Taste	5.04	5.76	3.38	5.28
Consistency	3.20	3.68	3.90	3.36
Overall average score	16.80	18.24	18.06	17.04
Qualificative	good	very good	very good	Good

The compote and the nectar have received the "good" while the jam (14.66 points) has received the "satisfactory" qualificative.

Picture 6 presents the four types of canned apricot from the 'Augustin' cultivar.

Figure 6. Apricot processed products from the 'Augustin' cultivar

The **'Litoral'** cultivar's fruit is big, ovoid, yellow-lime with red dots and streaks on the sunny side.

The flesh is yellow, moderately consistent, has a lot of dry substance and pleasant taste.

The products obtained by processing apricots of this cultivar (Table 7 and Figure 7) were well appreciated by tasters who, by the score given, made it possible that comfiture (18.24 points) and jam (18.06 points) receive the"very good" qualificative and the compote (16.80 points) and nectar (17.04 points), the rating "good".

The data shows that the compote as well the nectar has a pleasant aspect and a taste good, but the consistency has dropped below the overall average score.

Depending on the results obtained in processing the form of canned apricots, there were established the destinations of valorization in processed form of the cultivars studied (Table 8).

Figure 7. Apricot processed products from 'Litoral' cultivar

It finds that cultivar 'Olimp' is very well suitable to the processing into all 4 types of canned analyzed: comfiture, jam, compote and nectar, the resulting product having remarkable sensorial qualities.

Apricots in variety 'Tudor' are less suitable for processing, being able to consume preferably as fresh fruits or possibly as comfiture or nectar.

Table 8. – The destinations of valorization in processed form of apricot varieties

Cultivar	Option		
	I	II	III
'Tudor'	comfiture	nectar	compote, jam
'Sirena'	nectar	comfiture, jam	compote
'Orizont'	compote, nectar	comfiture, jam	
'Olimp'	compote, comfiture, jam, nectar	-	-
'Neptun'	compote, jam, nectar	comfiture	-
'Augustin'	comfiture	compote, nectar	jam
'Litoral'	comfiture, jam	compote, nectar	-

It finds that cultivar 'Olimp' is very well suitable to all the processing into four types of canned analyzed: comfiture, jam, compote and nectar, the resulting product having remarkable sensory qualities. The apricots from the 'Tudor' cultivar are less suitable for processing, being able to be consumed fresh fruits or, possibly, as comfiture or nectar.

CONCLUSIONS

The apricots from the seven cultivars studied are different in size, shape, color and fresh storage capacity. Although, all these cultivars have in common a pleasant, sweet flavor, which is why they are highly appreciated for fresh consumption. These characteristics also affect the quality of the processed products.
The following cultivars are recommended:
- for obtaining the canned compote 'Orizont', 'Olimp', 'Neptun';
- for the comfiture, the cultivars: 'Tudor', 'Olimp', 'Augustin', 'Litoral';
- for the jam, the cultivars: 'Orizont', 'Neptun', 'Litoral';
- for the nectar: 'Sirena', 'Orizont', 'Olimp', 'Neptun'.

Cultivar 'Olimp' is very well suitable to the processing into all 4 types of canned analyzed: comfiture, jam, compote and nectar, the resulting product having remarkable sensorial qualities. The apricots from the cultivar 'Tudor' are less suitable for processing, being able to be consumed preferably as fresh fruits or possibly as comfiture or nectar.

REFERENCES

Caplan I., Lamureanu Gh., Alexe Constanta, 2015. Comparative analysis of the quality of fruit, both fresh and processed as compote, of certain cherry tree cultivars. Bulletin of University of Agricultural Sciences and Veterinary Medicine Cluj-Napoca. Food Science and Technology, 72(2): 275-276.

Caplan I., Lamureanu Gh., Alexe Constanta, 2016. Suitability for processing of some sweet cherry varieties. Bulletin of University of Agricultural Sciences and Veterinary Medicine Cluj-Napoca. Food Science and Technology, 73(1): 7-13.

Lamureanu Gh., Alexe Constanta, Popescu Simona, 2012. Quality evaluation of some clingstone cultivars processed into stewed fruit. Scientific papers. Series B. Horticulture. Vol. LVI, 2012, Bucharest, 127-132.

Lamureanu Gh., Alexe Constanta, Caplan I., 2015. Quality evaluation of processed products of some peach varieties. Bulletin of University of Agricultural Sciences and Veterinary Medicine Cluj-Napoca. Food Science and Technology, 72(2): 162-168.

Sarbu S., Beceanu D., Corneanu G., Petre L., Anghel Roxana, Iurea Elena, 2010. Quality evaluation of some sweet cherry cultivars processed into stewed fruit. Lucrari stiintifice, Vol.53, 2010.University of Agricultural Sciences and Veterinary Medicine Iasi – Horticulture Series. 479-484.

Veringă Daniela, Dumitrescu L., 2016. Comparative Methods in determining the quality of peach processed products. Journal of Horticulture, Forestry and Biotechnology, vol 20 (3) Banat University of Agricultural Sciences and Veterinary Medicine Timosoara, 137-141.

Vintila M., Lămureanu Gh., Alexe Constanta, 2015. Aspects regarding the quality of processed products of some cherry varieties – 43nd Symposium "Actual Tasks on Agricultural Engineering", Opatija, Croația, ISSN 1848-4425, Thomson Reuters: Conference Proceedings, 775-785.

STAS 12656-88. Produse alimentare. Analiza senzoriala. Metode cu scari de punctaj. Standard de Stat, Editie oficiala.

STAS 3164-90. Compot de fructe. Standard de Stat, Editie oficiala.

STAS 3750-90. Dulceata. Standard de Stat, Editie oficiala.

STAS 3183–90. Gemuri. Standard de Stat, Editie oficiala.

SP 877-96. Nectar de fructe. Standard Profesional, Editie oficiala.

www.pro-sanatate.com/caisele-beneficii-sanatate/ Caisele – beneficii pentru sanatatea ta

28

A REVIEW OF HOW TO OPTIMIZE STORAGE AND SHELF LIFE EXTENDING TECHNOLOGIE OF KIWIFRUIT (*ACTINIDIA* SP.) BY USING 1-METHYLCYCLOPROPENE TO MEASURABLY REDUCE FRUIT WASTE

Ramona COTRUȚ, Anca Amalia UDRIȘTE

Research Center for the Study of Quality Food Products,
59 Mărăşti Blvd., District 1, 011464, Bucharest, Romania

Corresponding author email: ramona_cotrut@yahoo.com

Abstract

Kiwifruit (Actinidia deliciosa) are capable of long term storage only if carefully protected against deterioration prior to and during storage. They are harvested when mature but unripe and must ripen before eating. They are extremely sensitive to ethylene gas, which causes rapid flesh softening during storage, starch depletions to reduced sugars, increased susceptibility to spread of fruit rotting organisms and physiological breakdown. Hardy kiwifruit/Baby kiwi (Actinidia arguta) have smooth, edible skins and are smaller in size than 'Hayward' kiwifruit. Unlike A. deliciosa, baby kiwi fruits are very sensitive to dryness because of their smooth peels that lack hair. This phenotype characteristic is the main reason for the short-storage time and fast loss of postharvest quality. Fruits are not picked vine ripe, as they would be too soft to package and ship, instead they are picked when physiologically mature and firm, and are stored under refrigeration (0°C, 90–95% RH). Limited information exists regarding the ripening physiology of hardy kiwifruit or the ideal packaging and storage conditions for optimum quality, storage and shelf life. The objectives of this paper is to integrate existing knowledge and findings about applying technologies developed to suppress ethylene content and its effects degrading the kiwifruit post-harvest by applying 1-methylcyclopropene and periodically reviewing its effects and changes in kiwifruit quality, thereby improving storage technologies and extend the shelf life. The paper is an overview of how to optimize storage technologies of kiwifruit, managing fruit ripening by controlling naturally occurring ethylene during storage for optimal market value and more efficient harvest management, while maintaining excellent quality fruit and reduce losses and findings reveal the importance for assessing the marketing performance of the retail wine stores and the limits that generated lack of adoption on a large scale.

Key words: kiwifruit, 1-methylcyclopropene, shelf life, short-storage.

INTRODUCTION

The inhibitor of ethylene, 1-methylcyclopropene (1-MCP), is a new technology that is increasingly being used to improve storage potential and maintain quality of fruit and vegetables. 1-MCP is a synthetic cyclic olefin capable of inhibiting ethylene action. 1-MCP is a cycloalkene with the molecular formula C4H6 (Figure 1). It is a volatile gas at standard temperature and pressure with a boiling point of ~12°C. It acts as a competitor of ethylene, blocking its access to the ethylene-binding receptors (Sisler and Serek, 1997). The affinity of 1-MCP for the receptors is approximately 10-times greater than that of ethylene and, therefore, compared with ethylene, thus it is active at much lower concentrations.

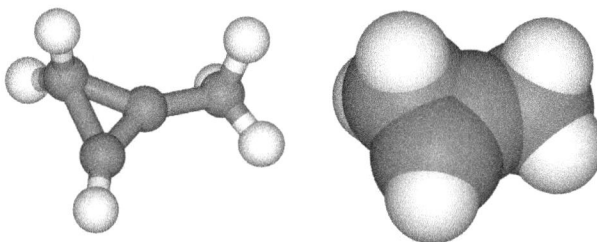

Figure 1. 1-MCP a cycloalkene with the molecular formula C4H6

1-MCP is a gaseous nontoxic product that delays softening and improves post-storage quality of several climacteric fruits (Blankenship and Dole, 2003) and it is applied to extend their postharvest life. To maintain the market position of the fruit, it is very important that techniques of storage and ripening retardant applications be used, which reduce the effect of ethylene (Watkins, 2008) and conserve a longer postharvest quality (Magaña et al., 2004; Andrade et al., 2006). Ethylene is a natural gaseous hormone (Martínez-Romero et al., 2007) that regulates the processes related to fruit ripening and senescence (Binder, 2008). The action of ethylene results from the binding of molecules to receptors located in the cell membrane of the endoplasmic reticulum (Serek et al., 2006). This binding activates the receptors, which send transduction signals, generating a gene expression and physiological response (Pereira et al., 2008). For this reason, the endogenous and/or exogenous presence of this hormone accelerates ripening and creates desirable senescence effects (fast, uniform ripening) or undesirable effects, such as reductions in the fruit's life (Giovannoni, 2004). There is a group of compounds called analogues of ethylene that competes for membrane receptors and inhibits the effect of this hormone, within which 1-methylcyclopropene (1-MCP) is notable, which has been widely used in various fruits and vegetables (Nanthachai et al., 2007; Watkins, 2008), generating changes in multiple metabolic processes, such as decreases in respiratory rates, ethylene production and volatile degradation of chlorophylls, changes of color, sugars, acidity and softening, which vary depending on the fruit, concentration and exposure time (Watkins, 2006). Sooyeon et al. (2015) used the 1-MCP treatment to inhibit hardy kiwifruits (*Actinidia arguta*) ripening by reducing respiration and ethylene production; the hardy kiwifruits could be stored for up to 5 weeks by maintaining higher fruit firmness, ascorbic acid and total phenolic contents, reducing changes in acidity, respiratory rate and color. In this sense, the employment of technologies such as 1-MCP applications has a potential use in the reduction of changes associated with quality losses during postharvest, which would extend the shelf-life of fruits and vegetables (Vilas-Boas and Kader, 2007); so the greater exposure time of the product, the lower the needed concentration will be in order to obtain the desired effect (Bassetto, 2002). Extensive literature about the effects of 1-MCP on fruit, vegetables, and ornamental products exists, and by 2007, results for over 50 fruit and vegetables, both whole and fresh-cut, as well as ornamental products, were available (Watkins and Miller, 2006). Other studies shows that the effect of 1-MCP on fruit considers the effects on factors that influence product quality using several fruit that have received the most attention in the literature, and that highlight some of the challenges that exist in commercialization of 1-MCP-based technology.

THE EFFECTS OF 1-MCP POSTHARVEST TREATMENT

For *Actinidia aruguta* (Seib. et Zucc.) Planch. Ex Miq., known as 'hardy kiwifruit' or 'baby kiwifruit' the recommended storage period of hardy kiwifruits is one to two weeks and an additional two or three days for shelf life. The main reasons for the short storage life are fruit softening, skin wrinkling due to water loss (dryness) and fruit decay. Fruit softening rapidly increases at the room-temperature ripening period, after harvest or cold storage (Krupa et al., 2011). Unlike *A. deliciosa*, the hardy kiwifruits are very sensitive to dryness because of their smooth peels that lack hair. This phenotype characteristic is the main reason for the short-storage time and fast loss of postharvest quality (Strik, 2005).

In various climacteric fruit including kiwifruit, preclimacteric application of 1-methylcyclopropene (1-MCP), a potent inhibitor of ethylene perception due to its largely irreversible binding to ethylene receptors, has been reported to delay ripening and senescence significantly, and consequently to lead to a prolonged storage life and/or shelf life (Watkins, 2006).

To extend the storage life of hardy kiwifruits different edible coating materials were used in past, consisting of mixtures of various formulas, such as calcium caseinate, chitosan, Prima Fresh 50-V and Semperfresh (Fisk et al.

(2008). Krupa et al. (2011) reported that hardy kiwifruits stored in common cold storage gradually lost physicochemical quality over 4 weeks due to decreases in ascorbic. Contrary to the fruit of *A. deliciosa*, which can be stored in cool conditions for up to 5 months, the storage time of hardy kiwifruits is usually no longer than 10–12 weeks and varies from year to year (Strik, 2005).

In Asiche et al. (2016) study, the application of 1-MCP after propylene treatment delayed the initiation and progression of ethylene biosynthesis and overall fruit ripening. 1-MCP extended the "eating window"of kiwifruit, especially in fruit treated with propylene for 48 h that had started ethylene biosynthesis. This suggests that, in kiwifruit, immediately after the commencement of propylene treatment, even before ethylene production is initiated, synthesis of cell-wall degrading enzymes isinitiated, which induces subsequent fruit softening, whereas the application of 1-MCP delayed fruit softening induced by propylene. Similar effects of 1-MCP in delaying fruit softening after ethylene or propylene treatment have also been observed in melon (Nishiyama et al., 2007) and 'La France' pear (Kubo et al., 2003).

In his study Park et. Al., (2015), provides dates where flesh firmness decreased gradually with storage time and reached lower levels at the end stage of storage in 1-MCP treatments. Fruits treated with 1-MCP enhanced firmness, about 20 % higher than control fruits. Application of 1 ppm of 1-MCP was sufficient to delay kiwi fruit softening during cold storage. In Park's experiment it has been shown that flesh firmness enhanced by 1-MCP treatment changes with increasing of starch content during storage. Several explanations are available for the tendency of firmness in correlation with starch content when the treatment was carried out. 1-MCP treated kiwi fruits showed the lesser decrease of the starch content. The relatively high content of starch is related to the enhanced kiwi fruit firmness.

In his study, Kwanhong et. al. (2017), shows that 1-MCP treatment effectively delayed the rate of fruit softening in red-fleshed kiwifruit by suppressing ethylene biosynthesis during storage at all temperatures. The application of 1-MCP was very effective in delaying fruit

ripening and senescence, especially at higher storage temperatures above 10°C, according to his study.

As shown in Hwanhong et al. study, 1-MCP treatment also influenced both SSC and TA of red-fleshed kiwifruit. The fruit reached full ripeness when the SSC level increased to 15 Brix. His study demonstrates that application of 1-MCP could delay the increase of SSC and decrease acidity. Fruit treated with 1-MCP had lower SSC and higher TA than untreated fruit during storage at all temperatures, similar to previous studies on 'Hayward'(Koukounaras and Sfakiotakis, 2007) and 'Allison' kiwifruit (Sharma et al., 2012). In which sensory evaluation regards, storage had a measurable effect on the quality of the fruit. The results show that 1-MCP-treated fruit were harder, sourer, and less juicy than untreated fruit, especially for fruit stored at 20°C. This suggests that 1-MCP treatment could delay softening and sourness in fruit stored at all temperatures. However, the 1-MCP treatment did not affect the juiciness of the fruit, except for fruit stored at 20°C which were less juicy.

Boquete et al. (2004) and Kim et al. (2001) determined that application of 1-MCP in kiwi fruits reduces ethylene production and softening during cold storage and subsequently exposed to 20 °C. Low concentrations of 1-MCP from 2.5 ppm to 1 ppm in most commodity are the most effective, but this depends also on temperature treatment. The most commonly applied is 20–25°C, but lower temperatures can be used in some commodities (Mir et al. 2001). Generally, optimal treatment durations of 12–24 h in fruits were sufficient to achieve a full response (Ku and Wills, 1999b). It was shown that 1-MCP effectiveness in fruits and vegetables were influenced by the cultivar, developmental stage, time from harvest to treatment and its concentration (Wills and Ku, 2002).

1- MCP is being used as a powerful tool to gain insights into fundamental processes that are involved in ripening and senescence, as well as to understand ethylene's action and responses (Watkins, 2006). The effects of 1-MCP in fruits are variable depending on the fruit. For example, 1-MCP induced an increase in sugars (expressed as soluble solids) in papaya (Hofman et al., 2001) and pineapple

(Selvarajah et al., 2001), but reduced sugars in kiwifruit (Boquete et al., 2004) and nectarines (Bregoli et al., 2005). Furthermore, 1-MCP had no effect on soluble solid contents of plums (Menniti et al., 2004) and mamey sapote (Ergun et al., 2005). Organic acids such as citric acid were reduced in 1-MCP-treated apple (Defilippi et al., 2004) and were increased in guava (Bassetto et al., 2005); malic acid in apple did not change due to 1-MCP treatment (Defilippi et al., 2004; Kondo et al., 2005).

Respiration rates and ethylene production are reduced in fruits treated with 1-MCP most of the time (Jiang et al., 2001; Dong et al., 2002; Mwaniki et al., 2005). Exposure of kiwifruits to exogenous ethylene (ethephon) accelerates maturation, which generates metabolic processes that reduce postharvest fruit life. The 1-MCP treatment extended the postharvest life of the kiwifruits, slowing the metabolic processes and loss of firmness and likewise decreased the respiration rate. The main method used to prolong the storage life of fruit is through reducing the fruit temperature to slow metabolism. Refrigerated storage slows the rate of ripening and senescence of the fruit, and also slows the development of any rots.

The way in which temperature management is implemented after harvest can significantly affect the quality of the fruit at the end of storage, both in the amount of ripening retardation and also the presence or absence of disorders. The basic effect of refrigerated storage on fruit can be supplemented by modification of the atmosphere in the coolstore, by reducing oxygen and increasing carbon dioxide concentrations. The application of the inhibitor of ethylene action 1-methylcyclopropene (1-MCP) has become common to slow the ripening of a range of fruit. The technologies impact on the fruit is dependent on the physiological state, or maturity, of the fruit at harvest and may differ dependent on the commercial requirements of the fruit, i.e. a short or long storage period.

Ultimately, the target for good storage is for the fruit to remain in good condition, to ripen properly, have an acceptable flavour and not to have any disorders at the end of storage and when it reaches the consumer. For commercial practice it is needed to be taken into consideration the cultivar, the maturity and ripening stage, the time between harvest and treatment, the temperature.

In **summary**, this review supports previous research on the beneficial effects of 1-MCP application in delaying ripening and postharvest quality loss of kiwifruit, and can extend its storage life.

REFERENCES

Andrade J.L., Renfigo E., Ricalde M.F., Simá J.L, Cervera J.C.,Vargas-Soto G., 2006. Microambientes de luz, crecimiento y fotosíntesis de la pitahaya (Hylocereus undatus) en un agrosistema de Yucatán, México. Agrociencia, 40: 687-697.

Asiche W. O, Mworia E. G., Oda C., Mitalo O. W., Owino W. O.,Ushijima K., Nakano R., Kubo Y, 2016. Extension of Shelf-life by Limited Duration of Propylene and 1-MCPTreatments in Three Kiwifruit Cultivars, The Horticulture Journal 85 (1): 76–85.

Bassetto E., 2002. Conservação de goiabas 'Pedro Sato' tratadas com 1-metilciclopropeno: concentrações e tempos de exposição. M.Sc. thesis. Escola Superior de Agricultura "Luiz de Queiroz", Universidade de São Paulo, Piracicaba, Brazil.

Bassetto E., Jacomino A.P., Pinheiro A.L., Kluge R.A., 2005. Delay of ripening of 'Pedro Sato' guava with 1-methylcyclopropene. Postharvest Biol. Technol., 35: 303-308. DOI: 10.1016/j.postharvbio.2004.08.003.

Binder B., 2008. The ethylene receptors: complex perception for a simple gas. Plant Sci., 175: 8-17.

Blankenship S. M., Dole J. M. 2003. 1-Methylcyclopropene: A review. Postharvest Biol. Tech. 28:1–25.

Boquete E.J., Trinchero G.D., Fraschina A.A., Vidella F., Sozzi S.G.O. 2004. Ripening of 'Hayward' kiwi fruit treated with 1- methylcyclopropene after cold storage. Postharv Biol Technol 32: 57–65.

Bregoli A.M., Ziosi V., Biondi S., Rasori A. Ciccioni M., 2005. Postharvest 1-methylcyclopropene application in ripening control of Stark Red Gold nectarines: Temperature dependent effects on ethylene production and biosynthetic gene expression, fruit quality and polyamine levels. Postharvest Biol. Technol., 37: 111-121. DOI: 10.1016/j.postharvbio.2005.04.006.

Defilippi B.G., Dandekar A.M., Kader A.A., 2004. Impact of suppression of ethylene action or biosynthesis on flavor metabolites in apple (Malus domestica Borkh) fruits. J. Agric. Food Chem., 52: 5694-5701. DOI: 0.1021/jf049504x

Dong L., Lurie S., Zhou H.W., 2002. Effect of 1-methylcyclopropene on ripening of 'Canino' apricots and 'Royal Zee' plums. Postharvest Biol. 5214(01)00130-2.

Ergun M., Sargent S.T., Fox A.J., Crane J.H., Huber D.J., 2005. Ripening and quality responses of mamey sapote fruit to postharvest wax and 1-methylcyclopropene treatments. Postharvest Biol.

Technol., 36: 127-134. DOI:10.1016/j.postharvbio.2004.12.002.

Fisk C. L., Silver A., Strik B. and Zhao Y., 2008. Postharvest quality of hardy kiwifruit (*Actinidia aruguta* 'Ananasnaya') associated with packaging and storage conditions. Postharvest Biology and Technology, 47: 338–345.

Giovannoni J.J., 2004. Genetic regulation of fruit development and ripening. Plant Cell, 16: S170-S180.

Hofman P.J., Jobin-Decor M., Meiburg G.F., Macnish A.J., Joyce D.C., 2001. Ripening and quality responses of avocado, custard apple, mango and papaya fruit to 1-methylcyclopropene. Aust. J. Exp. Agric., 41: 567-572. DOI: 10.1071/EA00152.

Jiang Y.M., Joyce D.C., Terry L.A., 2001. 1-Methylcyclopropene treatment affects strawberry fruit decay. Postharvest Biol. Technol., 23: 227-232.DOI: 10.1016/S0925-5214(01)00123-5.

Kim H. O., Hewett E. W., Lallu N. 2001. Softening and ethylene production of kiwi fruit reduced with 1-methylcyclopropene. Acta Horticult 553:167–170.

Kondo S., Setha S., Rudell D.R., Buchanan D.A., Mattheis J.P., 2005. Aroma volatile biosynthesis in apples affected by 1-MCP and methyl jasmonate. Postharvest Biol. Technol., 36: 61-68.

Koukounaras A, Sfakiotakis E., 2007. Effect of 1-MCP prestorage treatment on ethylene and CO2 production and quality of 'Hayward' kiwifruit during shelf-life after short-, medium- and long-term cold storage. Postharvest Biol Technol 46: 174-180.

Krupa T., Latocha P. and Liwin′ ska A., 2011. Changes of physiological quality, phenolics and vitamin C content in hardy kiwifruit (*Actinidia aruguta* and its hybrid) during storage. Scientia Horticulturae, 130: 410–417.

Ku V.V.V., Wills R.B.H., 1999b. Effect of 1-methylcyclopropene on the storage life of broccoli. Postharv Biol Technol 17:127–132.

Kubo Y., K. Hiwasa, Owino W. O., Nakano, Inaba A., 2003. Influence of time and concentration of 1-MCP application on the shelf life of pear 'La France' fruit. HortScience 38: 1414–1416.

Kwanhong P., Lim B. S., Lee J.S., Park H. J, Choi M. H, 2017. Effect of 1-MCP and Temperature on the Quality of Red-fleshed Kiwifruit (Actinidia chinensis), Horticultural Science and Technology 35(2):199-209.

Magaña W., Balbin M., Corrales J., Rodríguez A., Saucedo C., Cañizares E., and Sauri E., 2004. Effects of cold storage on fruit physiological behavior of pitahaya (Hylocereus undatus Haworth). Cultivos Tropicales, 25: 33-39.

Martínez-Romero D., Bailen G., Serrano M., Zapata P., Castillo S., and Valero D., 2007. Tools to maintain postharvest fruit and vegetable quality through the inhibition of ethylene action. A review. Crit. Rev. Food Sci. Nutr., 47:543-560.

Menniti A.M., Gregori R. Donati I., 2004. 1-methylcyclopropene retards postharvest softening of plums. Postharvest Biol. Technol., 31: 269-275. DOI: 10.1016/j.postharvbio.2003.09.009.

Mir N.A., Curell E., Khan N., Whitaker M., Beaudry R. M., 2001. Harvest maturity, storage temperature, and 1-MCP application frequency alter firmness retention and chlorophyll fluorescence of 'Redchief Delicious' apples. J Amer Soc Horticult Sci 126:618–624.

Mwaniki M.W., Mathooko F.M., Matsuzaki M., Hiwasa K., Tateishi A., 2005. Expression characteristics of seven members of the betagalactosidase gene family in 'La France' pear (Pyrus communis L.) fruit during growth and their regulation by 1-methylcyclopropene during postharvest ripening. Postharvest Biol. Technol., 36: 253-263. DOI:10.1016/j.postharvbio.2005.02.002.

Nanthachai N., Ratanachinakorn B., Kosittrakun M., Beaudry R.M., 2007. Absorption of 1-MCP by fresh produce. Postharv. Biol. Technol., 44:291-297.

Nishiyama I. 2007. Fruits of the Actinidia genus. Adv. Food Nutr. Res. 52: 293−324.

Park Y. S., Myeng H. I., Gorinstein S., 2015. Shelf life extension and antioxidant activity of 'Hayward' kiwi fruit as a result of prestorage conditioning and 1-methylcyclopropene treatment . J Food Sci Technol (May 2015) 52(5):2711−2720.

Pereira, G., Luiz F., Dias V., Brommonschenkel S., 2008. Influéncia do tratamento com etileno sobre o teor de sólidos solúveis e a cor de pimentas. Bragantia, 67:1031-1036.

Selvarajah S., Bauchot A.D., John P., 2001. Internal browning in cold-stored pineapples is suppressed by a postharvest application of 1- methylcyclopropene. Postharvest Biol. Technol., 23: 167-170. DOI: 10.1016/S0925-5214(01)00099-0.

Serek M., Woltering E.J., Sisler E.C., Frello S., Sriskandarajah S., 2006. Controlling ethylene responses in flowers at the receptor level. Biotechnol. Adv., 24:368-381.

Serna, L., Torres L.A, and Ayala A., 2011. Efecto del empaque y del 1-MCP sobre características físicas, químicas y fisiológicas de pitahaya amarilla. Biotecnología en el Sector Agropecuario y Agroindustrial, 9:139-149.

Sharma RR, Jhalegar MJ, Pal RK, 2012. Response of kiwifruit (Actinidia deliciosa cv. Allison) to postharvest treatment with 1-methylcyclopropene. J Hortic Sci Biotechnol 87: 278-284.

Sisler E.C., Serek M., 1997. Inhibitors of ethylene responses in plants at the receptor level: Recent developments, Physiologia Plantarum, 100, pp. 577–582.

Sooyeon S. H. H., Jeongyun K., Han J. L., Jeong G. L., Eun J. L., 2015, Inhibition of hardy kiwifruit (*Actinidia aruguta*) ripening by 1-methylcyclopropene during cold storage and anticancer properties of the fruit extract, Food Chemistry, Volume 190, 1 January 2016, Pages 150–157.

Strik B., 2005. Growing kiwifruit. A Pacific Northwest Extension Bull. 507. Corvallis, Oregon. State University, Extension Service.

Vilas-Boas E., Kader A., 2007. Effect of 1-methylcyclopropene (1-MCP) on softening of fresh-cut kiwifruit, mango and persimmon slices. Postharv. Biol. Technol., 43:238-244.

Watkins C.B., 2006. The use of 1-methylcyclopropene (1-MCP) on fruits and vegetables. Biotechnol. Adv., 24:389-409.

Watkins C.B., 2008. Postharvest effects on the quality of horticultural products: Using 1-MCP to understand the effects of ethylene on ripening and senescence processes. Acta Hort., 768:19-31.

Wang K.L.C., Li H., and Ecker J.R., 2002. Ethylene biosynthesis and signalling networks. Plant Cell, 14: S131-S151.

Wills R. B. H, Ku V. V. V., 2002. Use of 1-MCP to extend the time to ripen of green tomatoes and postharvest life of ripe tomatoes. Postharv Biol Technol 26:85–90.

29

EFFECTS OF RELATIVE HUMIDITY ON *IN VITRO* POLLEN GERMINATION AND TUBE GROWTH IN SWEET CHERRIES (*PRUNUS AVIUM* L.)

Sultan Filiz GÜÇLÜ[1], Fatma KOYUNCU[2]

[1]Süleyman Demirel University, Atabey Vocational School, Isparta, Turkey

[2] Süleyman Demirel University, Faculty of Agriculture, Department of Horticulture, Isparta, Turkey

Corresponding author email: sultanguclu@sdu.edu.tr

Abstract

Objective of this study was to examine the effects of global warming on pollen germination in fruit trees. For this purpose Bigarreu Gaucher, Celeste, Lambert, Lapins, Starks Gold, Stella and 0900 Ziraat cherry cultivar's pollens were used. Pollen germination and tube growth rates were observed at 40%, 50%, 60%, 70%, 80% and 90% relative humidity regimes. High humidity regimes were gave best results for both pollen germination and pollen tube growth. 80% was the optimum RH for in vitro germination of Starks Gold and Stella while 90% was optimum for Bigarreu Gaucher. In vitro pollen germination increased with increasing incubation relative and incubation period. 24 hours later was the optimum for all cultivars. The longest pollen tube length was obtained from 90% RH. Set fruitful could be affected by relative humidity and indirectly global warming.

Key words: *relative humidity, pollen germination, Prunus avium L.*

INTRODUCTION

Environmental conditions play a key role for the survival living organisms (Koubourıs et al., 2009). The environment can be an important determinant of pollen performance.

The effects of the environment on pollen germination may be broken down into three categories: (1) prepollination environment, (2) pollination environment, and (3) postpollination environment. The prepollination environment consists of the environmental conditions in which a pollen-donating plant grows. Pollen may also be influenced by the environment experienced during pollen transfer from donor to recipient (Travers, 1999).

Fertilization success in plants is the result of processes that take place during the progamic phase. Recently studies show that environmental conditions affect pollen devolopment and pollen germination as well as pollen tube growth.

Enviromental conditions affect steps such as pollen germination and pollen tube growth as well as development of the female structures (Stephenson et al., 1992; Delph et al., 1997; Hedly et al., 2004).

Among those environmental conditions, temperature is one of the most important factors that affect fruit and seed set. Another factor of pollen germination and tube growth is relative humidity.

Humidity and temperature has been shown to affect the chemical composition of pollen, pollen viability, and/or pollen tube growth and to stimulate the synthesis of heat and shock proteins in pollen grains (Johannson and Stephenson, 1998).

Pollen germination and pollen tube growth are important components of fertilization success in fruit trees (Janıck and Moore, 1996; Tosun (Güçlü) and Koyuncu, 2007).

Few studies, however, have examined both the environmental and the genetic effects of temperature and humidity on pollen performance (Loupassaki et al., 1997).

Temperature and humidity has a clear effect on pollen tube kinetics, expressed as the time required for pollen germination and rate of tube growth. The pollen germination and pollen tube growth are the most important properties in cherry tree fertilization. Living organisms depend on water. Water regulates their biological reactions, serves as a fluid medium and stabilizes the structure of macromolecules.

Although heavy water loss from living organisms may have deleterious effects and may lead to death (Alper and Olıver, 2002; Nepı et al., 2010).

Recent years global warming effects direct pollen germination, pollination timing, and pollen tube growth and set fruitful. We think that it is effected indirectly, reducing relative humidity. Relative humidity is very important for pollen germination stage. Partial dehydration brigs the pollen into equilibrium with the environment without fatal damage to the cytoplasm (Bassani et al., 1994) however uncontrolled water loss induced by high temperature leads to the death of the pollen (Pacini, 1996; Koubouris et al., 2009). We think that humidity rate can be affected by high temperature and indirectly fruit set can be changed.

In this study we try to explain effects of humidity on pollen germination and tube growth in sweet cherries which are grown extensively in Turkey.

MATERIALS AND METHODS

'Bigarreu Gaucher', 'Celeste', 'Lambert', 'Lapins' 'Starks Gold', 'Stella', and '0900 Ziraat' cherry cultivars were used for germination tests. Flowers at balloon stage were collected from plants in early morning. The flowers were transferred to the laboratory quickly. The anthers were removed and placed in dark colored bottles to promote dehiscence at room temperature. The 'agar in plate' method was used to assign pollen germination and pollen tube growth.

Pollen grains were sown in the medium containing 0.5 agar + 15% sucrose + 5 parts per million H_3BO_3 (Boric acid) were put consist of 40%, 50%, 60%, 70%, 80%, 90% humidity cabins at 25°C. Pollen tube long at least as its diameter was considered to be 'germinated' (Tosun (Güçlü) and Koyuncu, 2007).

The percentage of germinating pollen was determined after 2h, 6h, 12h and 24h incubation period. An ocular micrometer was used to measure pollen tube lenght, under a light microscope, at a magnification 40X. Four Petri dishes were used for germination and pollen tube growth experiments. Counts were made from 4 different microscope fields (100-

150 pollen grains per field) in each Petri dish. In all experiments, treatments were arranged according to randomized design.

Statistical analyses were performed with GLM models (General Linear Model) using SPSS (V.10; Statistical software, SPPS. Inc., USA). Percentage data were subject to arcsin root square transformation, and analysis of variance was performed. The differences among means were analysed by Duncan's multiple range test at the 0.05 level of significance.

RESULTS AND DISCUSSIONS

Pollen germination rates of seven sweet cherry cultivars are shown in Table 1 after 24 hours at different humidity regimes (40% RH, 50% RH, 60% RH, 70% RH, 80% RH, 90% RH).

Table 1. *In vitro* germination (%) of cherries pollen at different humidity regimes after 24 hours incubation, in a medium containing 0.5 agar + 15% sucrose + 5parts per million H_3BO_3

Cultivar	Humidity regimes					
	40%	50%	60%	70%	80%	90%
Bigarreu Gaucher	4.12a	11.23a	35.30a	54.09b	82.30b	91.69a
Celeste	1.33b	3.19c	14.29d	51.30c	72.46d	81.88b
Lambert	1.09b	2.00d	25.12c	46.76d	71.87d	83.45b
Lapins	1.25b	3.76c	12.30d	54.63b	76.56c	79.83b
Starks Gold	1.79b	6.90b	29.96b	55.88b	88.46a	70.05c
Stella	1.88b	3.01c	23.62c	64.03a	89.76a	71.92c
0900 Ziraat	1.03b	3.00c	8.87e	45.56d	74.36c	66.25d

Different letters in the same column indicate significant differences, according to Duncan's multiple range test (P<0.05).

All cultivars showed highest germination in 90% RH except for 'Bigarreu Gaucher'.'Noble' and '0900 Ziraat'. The lowest pollen germination rates were obtained from 40% RH. There is no difference statistically (p<0.05) except from 'Bigarreu Gaucher'.

Pollen germination rates increase with rising relative humidity. 'Bigarreu Gaucher' had the best germination rate both 50% and 60% RH. '0900 Ziraat' have had the lowest value with 8.87 germination rate at 60% RH (p<0.05).

The best results obtained from 80% and 90% RH. 80% was the optimum RH for 'Starks

Gold' and 'Stella' while 90% was optimum for 'Bigarreu Gaucher'.

It can be said that '0900 Ziraat' showed the lowest germination rates in general. Germination of cherry pollens reached own maximum percentage in 24 hours for all relative humidity regimes (Figure1).

Pollens of '0900 Ziraat' and 'Lapins' started latest germinated at 40% RH. Pollens of these cultivars germinated 24 hours later. 'Starks Gold' pollens germinated 2 hours later at 40% RH different from other cultivars (0.23). After 12 hours pollen germination rate in Celeste showed a dramatically increase from 13.96 to 69.98.

The pollen tube elongation is showed in Table 2 according to different humidity regimes.

Table 2. The effect of relative humidity on pollen tube growth (μm) cherry cultivars *in vitro* after 24 hours

Cultivar	Humidity regimes					
	40%	50%	60%	70%	80%	90%
Bigarreu Gaucher	17.96a	35.58a	52.29a	82.35a	116.33a	119.22a
Celeste	6.98c	16.23c	39.96c	69.73b	80.11c	89.74d
Lambert	5.11c	13.29d	26.34de	58.96c	71.23d	92.33d
Lapins	6.74c	15.71c	30.33d	63.72bc	88.9c	100.01c
Starks Gold	10.12b	21.79b	43.25b	66.12b	93.7b	110.23b
Stella	6.03c	14.1c	29.91d	56.73c	71.32d	90.83d
0900 Ziraat	4.79cd	10.11e	21.03e	30.54d	62.23e	80.09e

Different letters in the same column indicate significant differences, according to Duncan's multiple range test (p<0.05).

Results were found statistically significant (p<0.05). Pollen tube growth increased with increased humidity. All cultivars showed longest pollen tube growth at 90% relative humidity and 24 hours incubation period.

The longest pollen tubes were obtained from 'Bigarreu Gaucher' at all temperatures while the shortest were found for '0900 Ziraat'. 'Lambert' showed the lowest pollen tube elongation from 40% RH to 90% RH.

All cultivars showed different response for relative humidity. Oyiga et al. (2010), reported that pollen germination is species depend on environmental factors. Another study which was carried out in olive cultivars showed extreme temperature and relative humidity incidents, even for a short period, reduced pollen germination and growth capacity in 'Koroneiki', 'Kalamata', 'Mastoidis' and 'Amigdalolia' olive and may affect fruit set and yield (Koubourıs et al., 2009). *Petunia hybrida* and *Cucurbita pepo* pollens were exposed to 30 and 75% relative humidity (RH).

Water content, viability and carbohydrate content (glucose, fructose, sucrose and starch) were measured at fixed intervals over 6 h. Water content of *C. pepo* pollen decreased drastically at both RH, while *P. hybrida* pollen dehydrated slightly at RH 30% and hydrated at RH 75% (NEPI et al., 2010). 24 hours later pollen germination reached maximum.

These results of incubation duration experiments were closely parallel to findings of Koyuncu (2006), who reported that germination of strawberry pollens began within 1h at 24°C. Similarly, the pollen of 'Tsanoki' pear started to germinate after 1h incubation (Vasilikakis and Porlingis 1985). Koyuncu and Tosun (Güçlü) (2008) reported that the germinating rates increased with length of incubation period. Leech et al. (2002), investigated the responses of a period of up to 72 hours on pollen germination of different strawberry cultivars. The growth of the pollen tube in flowering plants is exceedingly rapid and its requirements, in general, seem quite unimpressive, i.e., water, oxygen and a suitable osmotic environment. Despite extensive attempts to hasten this growth process with the conventional host of growth factors, few have met with convincing success (Brewbaker and Kwack, 1963).

In vivo pollen germination and tube growth are highly sensitive to climatic factors, particularly, temperature and relative humidity (Sıngh at al., 2009). While temperature and other abiotic stresses are clearly limiting factors for crops cultivated on marginal lands, crop productivity everywhere is often at the mercy of random environmental fluctuations. Current speculation about global climate change is that most agricultural regions will experience more extreme environmental fluctuations (Solomon et al., 2007; Zinn et al., 2010). On our opinion climatic change can affect relative humidity, pollen germination, bee activity and fruitset.

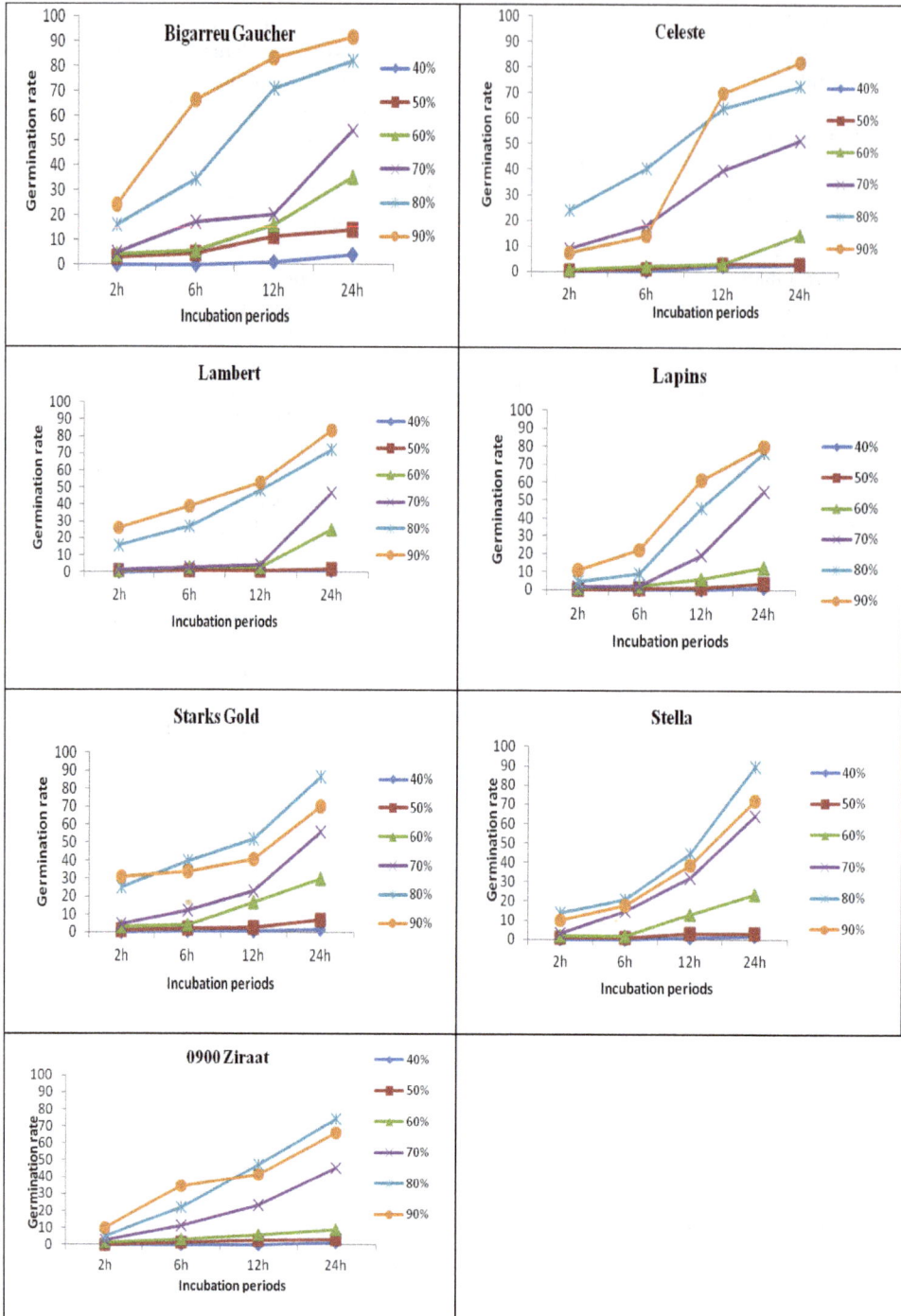

Figure 1. Germination of different cherry varieties pollens

CONCLUSIONS

Indirect effects of global warming on pollen germination and fruit set has examined in this study.

Pollen germination and pollen tube growth reached own maximum percentage at high humidity regimes.

'Starks Gold' and 'Stella' gave best resutls at 80% humidity.

90% humidity was optimum for 'Bigearreu Gaucher'.

Highest percentage was taken at 24 hours later after incubation for all cultivars.

Pollen germination and tube growth increased by incubation period.

On our opinion climatic change can affect relative humidity, pollen germination, bee activity and fruitset.

REFERENCES

Alper P., Oliver M.L(2002). Drying without dying. In: Black M, Pritchard HW (eds) Desiccation tolerance in plants. CAB International, Wallingford, pp 3–43.Assessment Report of the Intergovernmental Panel on Climate Change. New York: Cambridge University Press.

Bassanı M.,. Pacın E., Franchı G.G., 1994. Humidity stress responses in pollen of anemophilous species. Grana, 33, 146-150.

Brewbaker, J.L. and Kwack H.B 1963. The essential role of calcium ion in pollen germination and pollen tube growth. Am J Bot, 50 (9), 747-858.

Delph L., Johannson F. and Stephenson A.G., 1997. How environmental factors affect pollen performance, ecological and evolutionary perspectives. Ecology. 78, 1632-1639.

Hedly A.. Hormaza J.I and Herrero M., 2004. Effect of temperature on pollen tube kinetics and dynamics in sweet cherry, *Prunus avium* (*Rosaceae*). Am J Bot 91 (4), 558-564.

Janıck J., Moore N.J., 1996. Fruit Breeding, Tree and Tropical Fruits Vol. 1. New York, John Wiley and Sons Inc

Johannson L.F. and Stephenson A.G., 1997. How enviromental factors affect pollen performance; ecological and evolutionary perspectives. Ecology 78, 1632-1639.

Johannson H.M. and Stephenson G.A. 1998. Effect of temperature during microsporogenesis on pollen performance in *Cucurbita pepo* L. (*Cucurbitacea*). Int J Plant Scı, 159 (4), 616-626.

Koubourıs C.G., Metzıdakıs T.I. and Vasılakakıs D.M., 2009. Impact of temperature on olive (*Olea europaea* L.) pollen performance in relation to relative humidity and genotype. Envıron Exp Bot 67, 209–214.

Koyuncu F. and. Tosun (Guclu) F 2008. Evaluation of pollen viability and germinating capacity of some sweet cherry cultivars grown in Isparta, Turkey. Acta Hort.795 ISHS, (1), 71-75.

Koyuncu F., 2006. Response of *in vitro* pollen germination and pollen tube growth of strawberry cultivars to temperature. Europ J. Hor. Sci. 71(3), 125-128.

Leech L.D,, Simpson W. and Whıtehouse A.B., 2002. Effect of temperature and relative humidity on pollen germination in four strawberry cultivars. Acta Hort. 567 ISHS, 261-263

Loupassakı M., Vasılakakıs M., and Androulakıs, I. 1997. Effect of Pre-incubation Humidity and Temperature Treatment on the Germination of Avocado Pollen Grains. Euphytica, 94, 247-251.

Nepı M., Crestı. N.. Guarnıerı M. and Pacını E., 2010. Effect of relative humidity on water content, viability and carbohydrate profile of *Petunia hybrida* and *Cucurbita pepo* pollen. Plant System Evolution, 284, 57–64.

Oyiga B.C., Uguru M.I. and Aruah C.B., 2010. Pollen behaviour and fertililization impairment in Bambara groundnut (*Vigna subterrenea* [L.] Verdc.). JPBCS 2(1), 12-23.

Pacını E., 1996. Types and meaning of pollen carbohydrate reserves. Sex Plant Reprod, 9, 362-366.

Sıngh S., Rana A. and Chauan S.V.S. 2009. Impact of environmental changes on the reproductive biology in *Pyrostegia venusta* Presl. J Environ Biol, 30 (2), 271-273.

Solomon S., Qın D, Mannıng M., Marquıs M., Averyt K., Tignor M.M.B., Mıller H.L and Cheng Z. 2007. Climate change 2007: the physical science basis. Contribution of Working Group I to the Fourth

Stephenson A., Lau T.C., Quesada M. and Wınsor J.A. 1992. Factors that influence pollen performance In: Ecology and Evolution of Plant Reproduction (R. Wyatt, ed,) pp.119-134

Tosun (Güçlü), F. and Koyuncu F., 2007. Investigations of suitable pollinator for 0900 Ziraat sweet cherry cv.: pollen performance tests, germination tests, germination procedures, *in vitro* and *in vivo* pollinations. HortSci(Prague), 34, (2), 47-53

Travers S.E., 1999. Environmental effects on components of pollen performance in *Faramea occidentalis* (L.) A. Rich. (*Rubiaceae*). Biotropica, 31 (1), 159-166.

Vasılakakıs M. and. Porlıngıs I.C, 1985. Effect of temperature on pollen germination, pollen tube growth, effective pollination period, and fruit set of pear. Hortsci., 20, 733-735.

Zınn E.K., Tunç-Özdemır M., and Harper F.J., 2010. Temperature stress and plant sexual reproduction: uncovering the weakest links. J Exp Bot 61 (7), 1959–1968.

PERMISSIONS

All chapters in this book were first published in SPSBH, by University of Agronomic Sciences and Veterinary Medicine of Bucharest; hereby published with permission under the Creative Commons Attribution License or equivalent. Every chapter published in this book has been scrutinized by our experts. Their significance has been extensively debated. The topics covered herein carry significant findings which will fuel the growth of the discipline. They may even be implemented as practical applications or may be referred to as a beginning point for another development.

The contributors of this book come from diverse backgrounds, making this book a truly international effort. This book will bring forth new frontiers with its revolutionizing research information and detailed analysis of the nascent developments around the world.

We would like to thank all the contributing authors for lending their expertise to make the book truly unique. They have played a crucial role in the development of this book. Without their invaluable contributions this book wouldn't have been possible. They have made vital efforts to compile up to date information on the varied aspects of this subject to make this book a valuable addition to the collection of many professionals and students.

This book was conceptualized with the vision of imparting up-to-date information and advanced data in this field. To ensure the same, a matchless editorial board was set up. Every individual on the board went through rigorous rounds of assessment to prove their worth. After which they invested a large part of their time researching and compiling the most relevant data for our readers.

The editorial board has been involved in producing this book since its inception. They have spent rigorous hours researching and exploring the diverse topics which have resulted in the successful publishing of this book. They have passed on their knowledge of decades through this book. To expedite this challenging task, the publisher supported the team at every step. A small team of assistant editors was also appointed to further simplify the editing procedure and attain best results for the readers.

Apart from the editorial board, the designing team has also invested a significant amount of their time in understanding the subject and creating the most relevant covers. They scrutinized every image to scout for the most suitable representation of the subject and create an appropriate cover for the book.

The publishing team has been an ardent support to the editorial, designing and production team. Their endless efforts to recruit the best for this project, has resulted in the accomplishment of this book. They are a veteran in the field of academics and their pool of knowledge is as vast as their experience in printing. Their expertise and guidance has proved useful at every step. Their uncompromising quality standards have made this book an exceptional effort. Their encouragement from time to time has been an inspiration for everyone.

The publisher and the editorial board hope that this book will prove to be a valuable piece of knowledge for researchers, students, practitioners and scholars across the globe.

LIST OF CONTRIBUTORS

Volkan Okatan
Usak University, Sivasli Vocational School, Uşak, Turkey

Mehmet Polat and Mehmet Atilla Aşkin
Suleyman Demirel University, Agricultural Faculty, Department of Horticulture, Isparta, Turkey

Sezai Ercişli
Ataturk University, Agricultural Faculty, Department of Horticulture, Erzurum, Turkey

Adnan Nurhan Yildirim, Fatma Yildirim, Bekir Şan, Mehmet Polat and Tuba Dilmaçünal
Suleyman Demirel University, Faculty of Agriculture, Horticultural Science, 32260, Isparta, Turkey

GüLcan Özkan and Hatice Aşik
Suleyman Demirel University, Faculty of Engineering, Food Engineering Department, 32260, Isparta, Turkey

Mehmet Polat and Mehmet Atilla Aşkin
Süleyman Demirel University, Faculty of Agriculture, Isparta, Turkey

Abdullah Kankaya
Elma Tarım LTD Isparta, Turkey

Berna Bayar and Bekir Şan
Suleyman Demirel University, Faculty of Agriculture, Horticultural Science, 32260, Isparta, Turkey

Cristina Moale
Research Station for Fruit Growing Constanta, 25 Pepinierei Street, Valu lui Traian, Romania

Adrian Asănică
University of Agronomic Sciences and Veterinary Medicine of Bucharest, 59 Marasti Blvd., District 1, Romania

Viktorija Stamatovska, Tatjana Kalevska, Marija Menkinoska and Tatjana Blazevska
"Saint Clement of Ohrid" University of Bitola, Faculty of Technology and Technical Sciences, Dimitar Vlahov bb, 1400 Veles, Republic of Macedonia

Ljubica Karakasova
"Ss. Cyril and Methodius" University, Faculty for Agricultural Sciences and Food, Aleksandar Makedonski bb Blvd., 1000 Skopje, Republic of Macedonia

Gjore Nakov
"Angel Kanchev" University of Ruse, Department of Biotechnology and Food Technologies, Branch Razgrad, 47 Aprilsko vastanie Blvd., Razgrad 7200, Bulgaria

Sultan Filiz Guclu
Süleyman Demirel University, Atabey Vocational School, Isparta, Turkey

Fatma Koyuncu
Süleyman Demirel University, Agricultural Faculty, Isparta, Turkey

Ioana Bezdadea Cătuneanu and Liliana Bădulescu
University of Agronomic Sciences and Veterinary Medicine of Bucharest, Faculty of Horticulture, 59 Marasti Blvd., District 1, Bucharest, Romania
University of Agronomic Sciences and Veterinary Medicine of Bucharest, Research Center for Studies of Food Quality and Agricultural Products, 59 Marasti Blvd., 011464, Bucharest, Romania

Aurora Dobrin and Andreea Stan
University of Agronomic Sciences and Veterinary Medicine of Bucharest, Research Center for Studies of Food Quality and Agricultural Products, 59 Marasti Blvd., 011464, Bucharest, Romania

Dorel Hoza
University of Agronomic Sciences and Veterinary Medicine of Bucharest, Faculty of Horticulture, 59 Marasti Blvd., District 1, Bucharest, Romania

Miljan Cvetković
University of Banja Luka, Faculty of Agriculture, Bulevar Vojvode Petra Bojovića 1A, 78000 Banja Luka, Bosnia and Herzegovina

Gordana Đurić and Nikola Micic
University of Banja Luka, Faculty of Agriculture, Bulevar Vojvode Petra Bojovića 1A, 78000 Banja Luka, Bosnia and Herzegovina University of Banja Luka, Genetic Resources Institute, Bulevar Vojvode Petra Bojovića 1A, 78000 Banja Luka, Bosnia and Herzegovina

Gheorghe Petre and Daniel Nicolae Comănescu
Research and Development Station for Fruit Growing Voinesti, 387 Main Street, 137525, Dambovita, Romania

Adrian Asănică
University of Agronomic Sciences and Veterinary Medicine of Bucharest, 59 Marasti Blvd., District 1, Bucharest, Romania

Valeria Petre and Gheorghe Petre
Research and Development Station for Fruit Growing Voinesti, 387 Main Street, 137525, Dambovita, Romania

Ananie Peşteanu, Valerian Bălan and Igor Ivanov
State Agrarian University of Moldova, 44 Mircesti Street, MD-2049, Chisinau, Republic of Moldova

Roxana Ciceoi and Elena Ştefania Mardare
University of Agronomic Sciences and Veterinary Medicine of Bucharest, Laboratory of Diagnosis and Plant Protection of Research Center for Studies of Food Quality and Agricultural Products, 59 Marasti Blvd, District 1, Bucharest, Romania

Ionela Dobrin, Elena Diana Dicianu and Florin Stănică
University of Agronomic Sciences and Veterinary Medicine of Bucharest, 59 Marasti Blvd, District 1, Bucharest, Romania

Yong-Xiang Ren, Lian-Ying Shen, Xiao-Ling Wang, Yao-Wang, Chun-Mei Yan and Yong-Min Mao
Agricultural University of Hebei, Lingyusi Street, No. 289, 071001, Baoding, China

Li-Hui Mao
Renxian Agriculture Bureau, Guangming Road, NO.216, 055151, Xingtai, China

Zg Liu, J Zhao and Mj Liu
Hebei Agricultural University, 071001, Baoding, Hebei, China

Zhihui Zhao, Sujuan Gong, Lili Wang, Mengjun Liu
Jujube Research Center, Hebei Agricultural University, 289 Lingyusi Street, Baoding, China

Adrian Asănică, Valerica Tudor, Răzvan Ionuţ Teodorescu, Adrian Peticilă and Alexandru Iacob
University of Agronomic Sciences and Veterinary Medicine of Bucharest, 59 Marasti Blvd., District 1, Bucharest, Romania

Dorin Sumedrea
Research Institute for Fruit Growing Pitesti, 402 Mărului Street, Mărăcineni 117450, Romania

Reşat Esgici
Dicle University, Bismil Vacational High School, Diyarbakır, Turkey

GüLtekin Özdemir
Dicle University, Faculty of Agriculture, Department of Horticulture, Diyarbakır, Turkey

Göksel Pekitkan, Konuralp Eliçin and Abdullah Sessiz
Dicle University, Faculty of Agriculture, Department of Agricultural Machinery and Technologies Engineering, Diyarbakır, Turkey

Ferhat Öztürk
Dicle University, Faculty of Agriculture, Department of Field Crops, Diyarbakır, Turkey

Abdullah Sessiz and Fatih Goksel Pekitkan
Dicle University, Faculty of Agriculture, Department of Agricultural Machinery and Technologies Engineering, Diyarbakir, Turkey

Mihdiye Pirinççioğlu, Göksel Kizil and Murat Kizil
Dicle University, Faculty of Science, Department of Chemistry, Diyarbakir, Turkey

Mehmet Koç and İbrahim Samet Gökçen
Aralık University, Faculty of Agriculture, Department of Horticulture, Kilis 7, Turkey

Mehmet İlhan Odabaşioğlu
Harran University, Faculty of Agriculture, Department of Horticulture, Şanlıurfa, Turkey

Kenan Yildiz
Gaziosmanpaşa University, Faculty of Agriculture, Department of Horticulture, Tokat, Turkey

Arina Oana Antoce and George Adrian Cojocaru
University of Agronomic Sciences and Veterinary Medicine of Bucharest, Faculty of Horticulture, Department of Bioengineering of Horti-Viticultural Systems, 59 Mărăşti Blvd., District 1, 011464 Bucharest, Romania

Georgeta Mihaela Bucur and Liviu Dejeu
University of Agronomic Sciences and Veterinary Medicine of Bucharest, 59 Marasti Blvd., District 1, Bucharest, Romania

George Adrian Cojocaru and Arina Oana Antoce
University of Agronomic Sciences and Veterinary Medicine of Bucharest, Faculty of Horticulture, Department of Bioengineering of Horti-Viticultural Systems, 59 Marasti Blvd., District 1, Bucharest, Romania

Fatma Yildirim, Adnan Nurhan Yildirim, Tuba Dilmaçünal and Bekir San
Suleyman Demirel University, Agriculture Faculty, Department of Horticulture, 32260, Isparta, Turkey

Nilda Ersoy
Akdeniz University, Technical Sciences Vocational School, 07058, Campus/Antalya, Turkey

Constanţa Alexe and Marian Vintilă
Research and Development Institute for Processing and Marketing of the Horticultural Products - Bucharest, No. 1A, Intrarea Binelui Street, District 4, 042159, Bucharest, Romania

Ion Caplan and Gheorghe Lămureanu
Research Station for Fruit Growing (RSFG) Constanta, 1 Pepinierei Street, 907300, Commune Valu lui Traian, Romania

Lenuţa Chira
University of Agronomic Sciences and Veterinary Medicine of Bucharest, 59 Marasti Blvd., District 1, 011464, Bucharest, Romania

Ramona Cotruţ and Anca Amalia Udrişte
Research Center for the Study of Quality Food Products, 59 Mărăşti Blvd., District 1, 011464, Bucharest, Romania

Sultan Filiz Güçlü
Süleyman Demirel University, Atabey Vocational School, Isparta, Turkey

Fatma Koyuncu
Süleyman Demirel University, Faculty of Agriculture, Department of Horticulture, Isparta, Turkey

Index

www.ingramcontent.com/pod-product-compliance
Lightning Source LLC
Chambersburg PA
CBHW062002190326
41458CB00009B/2944